AF323125

Discrete Element Modelling of Particulate Media

Discrete Element Modelling of Particulate Media

Edited by

Chuan-Yu Wu
School of Chemical Engineering, University of Birmingham, Birmingham, UK
Email: C.Y.Wu@bham.ac.uk

The proceedings of the International Symposium on Discrete Element Modelling of Particulate Media held at the University of Birmingham on 29-30 March 2012.

Special Publication No. 339

ISBN: 978-1-84973-360-1

A catalogue record for this book is available from the British Library

Published by The Royal Society of Chemistry,
Thomas Graham House, Science Park, Milton Road,
Cambridge CB4 0WF, UK

Registered Charity Number 207890

For further information see our web site at www.rsc.org

Printed in the United Kingdom by CPI Group (UK) Ltd, Croydon, CR0 4YY, UK

PREFACE

The discrete element method (DEM) is a numerical technique for analysing the mechanics and physics of particulate media. It was initially developed to examine the micromechanics of granular media such as sand in the late 1970s and has been significantly advanced ever since. It is now widely employed for modelling of particulate systems across disciplines. As one of the pioneers, Colin Thornton has made enormous contributions to the application of DEM to theoretical soil mechanics and problems in particle technology over the last 25 years or so, especially in the areas of *quasi-static deformation; particle-particle interactions; agglomerate breakage; granular dynamics; liquid bridges and coupled DEM with computational fluid dynamics (CFD)*. He has published over 100 papers in this context, including a number of seminal works in the areas of

i) Quasi-static deformation:
- C. Thornton, Numerical simulations of deviatoric shear deformation of granular media, *Geotechnique*, 2000, **50**, 43-53. Times Cited: 169 *(Sources: Web of Science)*.
- C. Thornton and S.J. Antony, Quasi-static deformation of particulate media, *Philosophical Transactions of the Royal Society of London Series A-Mathematical Physical and Engineering Sciences*, 1998, **356**, 2763-2782. Times Cited: 94.
- C. Thornton and D.J. Barnes, Computer-simulated deformation of compact granular assemblies, *Acta Mechanica*, 1986, **64**, 45-61. Times Cited: 102.

ii) Particle-particle interaction::
- C. Thornton and Z. Ning, A theoretical model for the stick/bounce behaviour of adhesive, elastic-plastic spheres, *Powder Technology*, 1998, **99**, 154-162. Times Cited: 160.
- C. Thornton and K.K. Yin, Impact of elastic spheres with and without adhesion, *Powder Technology*, 1991, **65**, 153-166. Times Cited: 146.
- C. Thornton, Coefficient of restitution for collinear collisions of elastic perfectly plastic spheres, *Journal of Applied Mechanics-Transactions of The ASME*, 1997, **64**, 383-386. Times Cited: 147.
- C.-Y. Wu, L.Y. Li and C. Thornton, Rebound behaviour of spheres for plastic impacts, *International Journal of Impact Engineering*, 2003, **28**, 929-946. Times Cited: 73.

iii) Liquid bridge:
- G.P. Lian, C. Thornton and M.J. Adams, A theoretical study of the liquid bridge forces between two rigid spherical bodies. *Journal of Colloid and Interface*, 1993, **161**, 138-147. Times Cited: 212.

iv) Agglomerate breakage
- C. Thornton, Y.Y. Yin and M.J. Adams, Numerical simulation of the impact fracture and fragmentation of agglomerates, *Journal of Physics D-Applied Physics*, 1996, **29**, 424-435. Times Cited: 120.
- G.P. Lian, C. Thornton and M.J. Adams, Discrete particle simulation of agglomerate impact coalescence, *Chemical Engineering Science*, 1998, **53**, 3381-3391. Times Cited: 82.

v) Coupled DEM/CFD
- K.D. Kafui, C. Thornton and M.J. Adams, Discrete particle-continuum fluid modelling of gas-solid fluidised beds, *Chemical Engineering Science*, 2002, **57**, 2395-2410. Times Cited: 109.

Colin Thornton developed leading DEM software for the simulation of quasi-static deformation of dense-phase particulate systems, agglomerate impact coalescence, fracture and attrition, granular flow and gas-solid two phase flow, financially supported by fourteen UK Research Council Grants and eleven industrial contracts. The developed DEM code is so versatile that it not only has the facilities to simulate experiments on three-dimensional polydisperse systems of autoadhesive, elastoplastic spheres, but also incorporates viscous liquid bridges between particles and a 3D Navier-Stokes solver to model the interstitial fluid. A distinctive feature of the code is that the solid-solid particle interactions are based on theoretical contact mechanics that allow the observed phenomenological behaviour to be related to the experimentally measurable mechanical (elastic, plastic, frictional and adhesive) properties of the constituent particles.

In recognizing the outstanding scientific contributions of Colin Thornton to DEM modelling of particulate media, on the occasion of his 70th birthday, the International Symposium on Discrete Element Modelling of Particulate Media (also known as Thornton Symposium) was held in his honour at the University of Birmingham, UK on 28-30th March, 2012. A total of 92 participants attended the symposium. The symposium programme consisted of 70 contributions (10 keynote presentations, 29 oral presentation and 31 posters), with a wide range of topics including fluidisation, coupled DEM/CFD modelling, particulate flow, quasi-static deformation, cohesive particle systems and liquid-solid systems, fragmentation and electrification. The symposium offered a good opportunity for researchers and scientists to come together to discuss topics pertaining to the modelling of particulate media using DEM and to celebrate Colin Thornton's achievements on this special occasion.

This book contains a collection of papers highlighting the recent advances in discrete element modelling in four areas: i) two-phase systems; ii) cohesive systems; iii) granular flows and iv) quasi-static deformation, inspired by the pioneer work of Colin Thornton. For two-phase systems, recent developments in coupled DEM/CFD are presented, which include drag force models and techniques to enhance the capacity of DEM/CFD to simulate complex flows, including heat and mass transfer. The use of DEM-based coupling methods for modelling liquid-solid systems is discussed. For cohesive systems, the effects of liquid bridges and van der Waals forces on the mechanical behaviour of particle systems, such as lunar soils, wet granular columns and fibre filters, are discussed. For granular flows, the wide application that DEM can offer is showcased, ranging from fundamental granular physics, rock and debris avalanches, excavation of gravels, die filling, silo filling and discharge, and pebble packing in nuclear reactors. For quasi-static deformation, how DEM can be used to explore the micro-mechanics of granular materials is illustrated, especially in simulating triaxial tests and stress wave propagation.

I would like to acknowledge all authors for their efforts and contributions. I also wish to acknowledge the encouragement and guidance of Profs. Jonathan Seville, Mike Adams, Mojtaba Ghadiri and Richard Williams, Drs. Guoping Lian and Ling Zhang, and many others. I am extremely grateful to all members of the organising and scientific committees for their support and advice and the local organising committee for their hard work and dedication in successfully running the symposium.

Finally I thank Colin for his inspiration in discrete element modelling and wish him all the best!

Chuan-Yu Wu
School of Chemical Engineering, University of Birmingham, Birmingham, B15 2TT, UK

May 2012

Contents

Quasi-Static Deformation

Two-Phase Systems

FROM SINGLE PARTICLE DRAG FORCE TO SEGREGATION IN FLUIDISED BEDS

A. Di Renzo and F. P. Di Maio

Dipartimento di Ingegneria Chimica e dei Materiali, Università della Calabria
Via P. Bucci, Cubo 44A, I-87036 Rende (CS), Italy

1 INTRODUCTION

In numerical simulations of dense two-phase flow involving particulate materials the Discrete Element Method (DEM) has proved particularly effective in capturing the complex hydrodynamics of the solid phase.[1] DEM-based granular solid dynamics, including collisions and persistent contact with elaborate force-displacement laws, friction and cohesion have shown to be superior to traditional fluid-like, continuum approaches, which typically require coarse approximations and the introduction of artificial variables like solids pressure and viscosity. However, computational limitations of DEM models do not allow adding also the burden of flow simulations resolved at the level of particle-particle interstices, so that typically an averaged scale approach, with computational cell sizes of the order of a few particle diameters, is used.[2,3] As a consequence, formulations of the drag force acting on individual particles are required to close the set of equations to solve for the solid and fluid phases. While many drag force models for monodisperse systems have been proposed in the literature, as discussed below, expressions for such force on a particle in a multi-particle system is currently the subject of extensive research work.

2 DRAG FORCE AND CLOSURE IN DEM-CFD MODELS

2.1 Momentum exchange and two-way coupling

Characterisation of the relative motion between a fluid and dense particle system by a DEM-CFD approach requires the solution of the averaged equations of motion of the fluid phase and the classical Newton's second law of dynamics for each particle, where the drag force appears explicitly. The fluid flow field is obtained from the solution of the discretised locally averaged continuity and Navier-Stokes equations, which in differential terms are expressed, respectively, as:

$$\frac{\partial \varepsilon \rho_f}{\partial t} + \nabla \cdot \left(\varepsilon \rho_f \mathbf{U} \right) = 0 \tag{1}$$

$$\frac{\partial(\varepsilon\rho_f \mathbf{U})}{\partial t} + \nabla \cdot (\varepsilon\rho_f \mathbf{UU}) = -\nabla P + \nabla \cdot \boldsymbol{\tau} + \mathbf{S} + \rho_f \varepsilon \mathbf{g} \tag{2}$$

where ρ_f, $\mathbf{U}$ and P are the fluid density, fluid velocity and pressure, respectively, ε is the volumetric fraction of the fluid (or voidage), $\boldsymbol{\tau}$ is the deviatoric stress tensor, $\mathbf{S}$ is the fluid-particle inter-phase momentum exchange density and $\mathbf{g}$ the acceleration of gravity.

The corresponding equations for each particle of the solid phase follow the conventional DEM approach, i.e.:

$$m\mathbf{a} = m\mathbf{g} + \sum_i \mathbf{f}_{ci} + V \cdot \nabla P + \mathbf{f}_d \tag{3}$$

$$I\boldsymbol{\alpha} = \sum_i \left(\mathbf{f}_{ci} \times \mathbf{R}_i \right) \tag{4}$$

where m, V, I, $\mathbf{a}$ and $\boldsymbol{\alpha}$ are the particle mass, volume, moment of inertia, linear and angular acceleration, respectively. The forces considered are gravity, contact forces $\mathbf{f}_c$, pressure gradient and drag force $\mathbf{f}_d$, in the order of appearance in Equation (3). Note that the last two terms arise from the interaction with the fluid. In the rotational direction only torques arising from contact forces are considered.

Interphase coupling is achieved by connecting the momentum exchange density source term $\mathbf{S}$ in Equation (2) with the drag force acting on individual particles, i.e.:

$$\mathbf{S} = \sum_j w_j \mathbf{f}_{dj} \tag{5}$$

where the w_j coefficient plays the role of distance weighting function per unit volume.

2.2 Drag force

Expressions accounting for the influence of velocity and voidage on the drag force exerted on individual particles have been often derived based on established correlations for the pressure drop across fixed beds of a single material of diameter D. In general terms, the modulus of the dissipative pressure gradient is related to the modulus of the drag force by:

$$\nabla p = \frac{1-\varepsilon}{\varepsilon} \frac{6}{\pi} \frac{f_d}{D^3} \tag{6}$$

Extensive studies in the literature led to a number of common, relatively accurate expressions valid for monodisperse suspensions that cover many orders of magnitude of the Reynolds number and from dense packing to highly dilute systems, like the combination of Ergun[3] and Wen and Yu[5] or Di Felice's formula.[6]

In the case of disperse systems or when multiple particulate solids are present simultaneously, the drag force acting on a particle becomes much more difficult to evaluate. This is also related to the fact that experimental accessibility to such datum is very limited. The first theoretical advancements indeed appeared as a result of fully resolved simulations of fluid flow through static arrays of spheres. In particular, van der Hoef et al.,[7] based on lattice-Boltzmann simulations of the flow through random arrays of spherical particles, were able to propose the first theoretical approach in the characterisation of the phenomenon. The most significant result of their work, later used also in other papers,[8,9] is the formulation of the drag force acting on a generic particle as proportional to the average drag force in the system, and to express the coefficient as a function of a poly-dispersion index and bed voidage, as detailed in the next Section.

3 DRAG IN MULTIPARTICLE TWO-PHASE FLOW

In analogy with the relationship between individual drag force and pressure gradient across the bed, van der Hoef et al.[7] proposed the following starting point:

$$\nabla p = \frac{1-\varepsilon}{\varepsilon} \frac{6}{\pi} \sum_i \frac{x_i f_{di}}{D_i^3} \tag{7}$$

where x_i is the volumetric fraction of species i in the multi-particle mixture. To keep the derivation simple, only two solids will be considered and, without loss of generality, species 1 will be assumed to be the smaller one. The two key steps are (i) the definition of an average drag force $\overline{f_d}$ and average diameter $\overline{D}$ for the system, related to the overall pressure drop by:

$$\nabla p = \frac{1-\varepsilon}{\varepsilon} \frac{6}{\pi} \frac{\overline{f_d}}{\overline{D}^3} \tag{8}$$

and (ii) the definition of the drag force on an individual species as proportional to average drag force in the system, i.e.:

$$f_{di} = \alpha_i \overline{f_d} \tag{9}$$

Then, by introducing the average diameter, as defined by Sauter's mean, and a polydispersion index given, respectively, by:

$$\overline{D} = \left(\frac{x_1}{D_1} + \frac{x_2}{D_2} \right)^{-1} \tag{10}$$

$$y_i = \frac{D_i}{\overline{D}} \tag{11}$$

van der Hoef et al.[7] proposed the proportionality coefficient α_i to be derived based on considerations in the viscous flow regime, giving the individual drag force by:

$$f_{di} = y_i^2 \overline{f_d} \tag{12}$$

It is the case to mention that the derivation presented here is formally different, though conceptually equal, to the original treatment[7] in two aspects. Firstly, a dimensional notation is used here. Secondly, all previous considerations involve the drag force intended as the net of the pressure gradient interaction term. Additional details can be found in papers elaborating further on the presented approach, e.g. by Cello et al.[10]

4 A MODEL FOR SEGREGATION IN FLUIDISED BEDS

The capability to compute the drag force at the individual particle scale is particularly useful in dealing with fluidised beds of binary mixtures of solids and the related segregation problems. An initially mixed binary bed upon fluidisation shows a tendency for the component with the smaller size and lower density to accumulate to the bed surface, acting as *flotsam*, the reverse occurring for the other component, the *jetsam*. The matter becomes problematic when smaller and denser particles are mixed with larger but less dense particle, a case in which it is difficult even to attribute the roles of flotsam and jetsam to the mixture components.

Despite the severe consequences in process performances, the complexity of the two-phase flow and the substantial previous inability to set the force balance over individual solids species has prevented a theoretical treatment of segregation problems. We have recently proposed[11] to include the result of Equation (12) into a force balance on a particle of, by convention, species 2 immersed in a homogeneous mixture of the two solids.

Under the hypothesis of viscous flow regime, the drag force exerted by the fluid over a particle of diameter D is:

$$f_d = 30\frac{1-\varepsilon}{\varepsilon^2}\pi\mu u D \tag{13}$$

and the corresponding minimum fluidisation velocity is:

$$u_{mf} = \frac{g\varepsilon^3}{180(1-\varepsilon)\mu}\rho D^2 \tag{14}$$

Equations (13) and (14) are intended here applicable to both particle species, provided the appropriate diameter and density are used, as well as the mixture, for which the average size is defined by Equation (10) and the average density is:

$$\overline{\rho} = \rho_1 x_1 + \rho_2(1-x_1) \tag{15}$$

It is useful to recall that the pressure gradient developed as a result of fluidisation of a binary bed is:

$$\nabla p = \overline{\rho}(1-\varepsilon)g \tag{16}$$

The force balance on a particle of species 2 can be set, by comparing the ratio of the hydrodynamic action of the fluid on the particle weight to the value of one, to establish whether it will be pushed upwards or downwards, i.e. it will act as flotsam or jetsam. In formula, this reads:

$$\frac{V_2\overline{\rho}(1-\varepsilon)g + \frac{1}{6}\pi g\varepsilon\overline{\rho}\overline{D}D_2^2}{V_2\rho_2 g} = 1 \tag{17}$$

where the numerator results from the sum of the pressure gradient term (neglecting Archimedean buoyancy) and the drag force evaluated at the minimum fluidisation velocity of the mixture and the denominator is the particle weight. Introducing the dimensionless ratio of the particle-to-average density $\overline{s} = \rho_2/\overline{\rho}$ and of the average-to-particle diameter $\overline{d} = \overline{D}/D_2$, Equation (17) can be rewritten as:

$$\overline{s} = 1 - \varepsilon + \varepsilon\overline{d} \tag{18}$$

Alternatively, using the corresponding version based on the particle-to-particle density and diameter ratios $s = \rho_2/\rho_1$ and $d = D_1/D_2$, we have:

$$s = \frac{x_1}{\dfrac{1}{1-\varepsilon+\varepsilon\left(\dfrac{1}{\dfrac{x_1}{d}+1-x_1}\right)} - (1-x_1)} \tag{19}$$

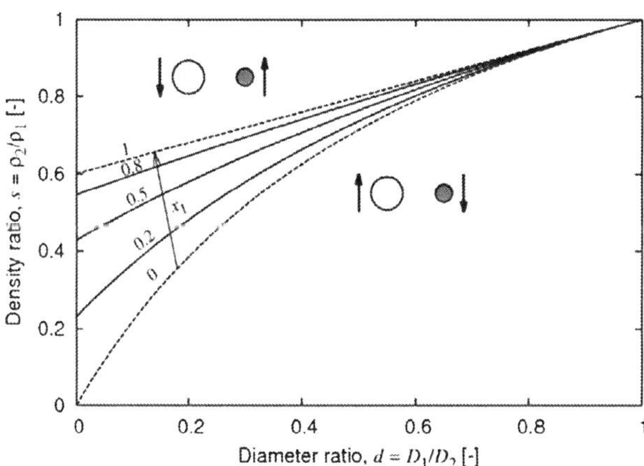

Figure 1 *Equilibrium lines as predicted by Equation (19) with sketches of the segregation direction of the two particle species. The bed voidage equals to 0.4.*

The attention will be focussed on binary mixtures composed of smaller, denser and larger, less dense materials so that values of s and d lie in the range 0 to 1. The possibility for either solid to become the flotsam component determines the occurrence of two possible *segregation directions* of an initially mixed bed. Equilibrium lines that allow discriminating between the two directions can be prescribed by Equation (19) and are shown in Figure 1 at various bed compositions. The chart with the discriminating lines can be used to predict the tendency for a given solids pair at a given composition to segregate one way or the other. This is achieved by computing s and d to locate the point on the chart and compare its position with the equilibrium line at the corresponding composition.

Similar considerations apply to the comparison of data of a given system in terms of $\bar{s}$ and $\bar{d}$ with the unique equilibrium line prescribed by Equation (18). Comparison of the predicted segregation direction with experimental data available in the literature is shown in Figure 2. Details of the examined systems and further comments are reported elsewhere.[11] However, it is evident that agreement is found for the great majority of the data, without adjustable parameters in the model, confirming the soundness of the approach and, particularly, the realistic predictions of the drag force of Equation (12). Discrete Element simulations are then expected to benefit from the adoption of Equation (12) in modelling fluid-particle flows involving multi-particle mixtures.

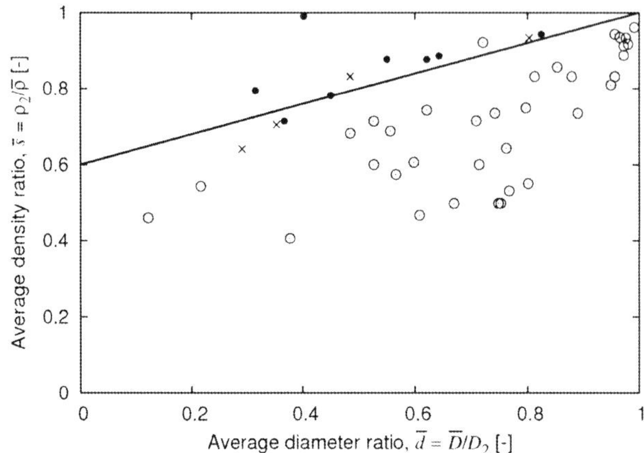

Figure 2 *Experimental data in mixing/segregating systems represented by symbols denoting the flotsam component: open circles for larger but less dense species and solid circles for the smaller and denser species. Crosses represent systems exhibiting bed mixing. The line identifies the equilibrium predicted by Equation (18).*

5 CONCLUSION

Simulations based on Discrete Element Method and averaged CFD approaches require a model for the drag force. In systems involving multi-particle mixtures such drag force is shown to require specific treatment and a recently proposed model is discussed. The same is shown to allow establishing a force balance at the particle scale that proves very useful in addressing problems related to segregation in fluidised beds. Quantitative model validation is shown for a large set of systems available in the literature.

References

1   H.P. Zhu, Z.Y. Zhou, R.Y. Yang and A.B. Yu., *Chem. Eng. Sci.*, 2008, **63**, 5728.
2   K.D. Kafui, C. Thornton and M.J. Adams, *Chem. Eng. Sci.*, 2002, **57**, 2395.

3   M.A. van der Hoef, M. van Sint Annaland, N.G. Deen and J.A.M. Kuipers, *Ann. Rev. Fluid Mech.*, 2008, **40**, 47.

4   S. Ergun, *Chem. Eng. Prog.*, 1952, **48**, 89.

5   C.Y. Wen and Y.H. Yu, *AIChE J.*, 1966, **12**, 610.

6   R. Di Felice, *Int. J. Multiphase Flow,* 1994, **20**, 153.

7   M.A. van der Hoef, R. Beetstra and J.A.M. Kuipers, *J. Fluid Mech.*, 2005, **528**, 233.

8   R. Beetstra, M.A. van Der Hoef and J.A.M. Kuipers, *AIChE J.*, 2007, **53**, 489.

9   S. Sarkar, M.A. van der Hoef and J.A.M. Kuipers, *Chem. Eng. Sci.,* 2009, **64**, 2683.

10 F. Cello, A. Di Renzo and F.P. Di Maio, *Chem. Eng. Sci.*, 2010, **65**, 3128.

11 F.P. Di Maio, A. Di Renzo and V. Vivacqua, *Powder Technol.*, 2012, in press, DOI: 10.1016/j.powtec.2012.04.040.

ENHANCING THE CAPACITY OF DEM/CFD WITH AN IMMERSED BOUNDARY
METHOD

C.-Y. Wu[1] and Y. Guo[1,2]

[1]School of Chemical Engineering, University of Birmingham, Birmingham, B15 2TT, UK
[2] Presently with Chemical Engineering Department, University of Florida, Gainesville, FL
32611, USA

1 INTRODUCTION

Discrete Element Methods (DEM) have been coupled with Computational Fluid Dynamics
(CFD) for analysing fluid-solid particle flows.[1-4] In the coupled DEM and CFD (*i.e.*
DEM/CFD), DEM is used to model the motion of particles and CFD is employed to
analyse the fluid flow, while empirical correlations for the drag forces are generally
introduced to analyse the interaction between the fluid and particles and two-way coupling
of fluid-particle interaction is considered. DEM/CFD is a computationally efficient
technique that has been widely used in modelling two-phase flows,[5-8] in which the fluid
domain is generally discretised into fluid cells using fixed and rectangular grids and all
quantities such as pressure, density and velocity are volume-averaged in the fluid cells. In
order to simulate the evolution of bubbles and the detailed fluid flow inside the bubbles,
the size of the fluid cell should be smaller than the macroscopic bubbles. On the other hand,
it has to be larger than the particle size so that the void fraction (the ratio of the volume of
void, excluding that occupied by the solid particles, to the total volume of the cell) will not
become zero. Typically, the size of the fluid cell is 3~5 times the particle diameter.[1-3,8]

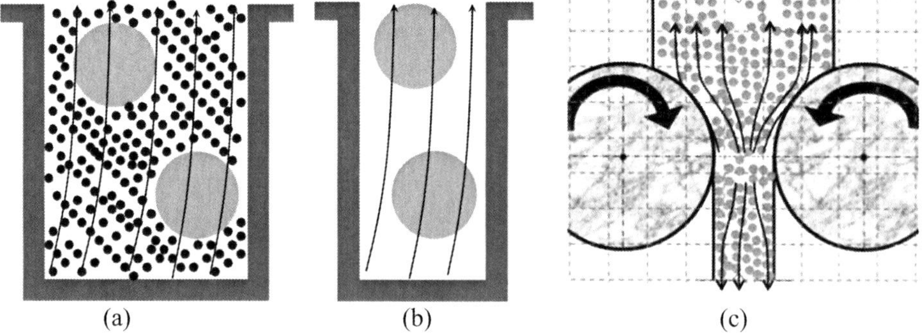

(a)   (b)   (c)

Figure 1 *Illustrations of complex two-phase flows involving (a) particles of significantly
different sizes, (b) particles whose sizes are comparable to the fluid domain and (c) moving
arbitrary shaped boundaries.*

To some extent, the treatment of fluid domain using large (compare to the size of solids) and regular grids limits the application of DEM/CFD as it is difficult to efficiently model systems with either particles of significantly different sizes (e.g. with particle size ratio higher than 10, see Figure 1a), or particles whose sizes are comparable to the fluid domain (Figure 1b), or moving or arbitrary shaped boundaries that do not conform to the regular grids (Figure 1c). In order to tackle these problems, in this study, an immersed boundary method (IBM), which is used to model the interaction between fluids and large objects or complex boundaries, is incorporated into the DEM/CFD[3] for simulating complex two-phase flows.

2   THE MODIFIED DEM/CFD WITH IBM

For the complex two-phase flow considered, it generally consists of three constituents: i) fine particles; ii) a fluid and iii) large objects and/or complex boundaries (LOCBs), such as moving or arbitrary shaped boundaries. In the modified DEM/CFD with IBM, the motion of each particle is determined using Newton's equations of motion in the DEM. For spherical elastic particles considered in this study, the particle interactions are modelled using the theory of Hertz to determine the normal force,[9] and the theory of Mindlin and Deresiewicz[10] for the tangential force. The classical JKR theory[11-14] is adopted to model the adhesive forces between particles.

The dynamics of the fluid is analysed using CFD, for which the size of the fluid cell is determined by the size of fine particles, not LOCBs (see, *e.g.*, Figure 1c). In other words, the fluid domain is divided into a number of cells of size typically in the range of 3~5 times the particle diameter, irrespective of the presence of LOCBs. The two-way coupling between the fluid and solid particles is modelled using the methods developed by Kafui *et al.*[3] An empirical model proposed by De Felice[15] is used to calculate the drag force on each particle.

To model the interaction between the fluid and LOCBs that are generally much larger than the fluid cells used, an IBM, which was initially proposed to model fluid flows with arbitrary geometrical boundaries,[16] is adopted. In the IBM, an effective void fraction ε in a fluid cell is introduced and defined as the volume fraction of the space not occupied by the fine particles, *i.e.*

$$\varepsilon = 1 - \frac{V_p}{V_c} = \frac{V_{LOCB} + V_f}{V_c} \tag{1}$$

where V_p, V_{LOCB}, V_f and V_c are the volume of the particles, LOCBs, fluid in the cell and the total volume of the cell, respectively. The interfaces between the fluid and LOCBs are assumed to be no-slip and impermeable, and the fluid and LOCBs are treated as a single continuum medium with an unified velocity field **u** that is given as

$$\mathbf{u} = \frac{V_f}{V_{LOCB} + V_f}\mathbf{u}_f + \frac{V_{LOCB}}{V_{LOCB} + V_f}\mathbf{U}_{LOCB} \tag{2}$$

where $\mathbf{u}_f$ is the local fluid velocity and $\mathbf{U}_{LOCB}$ is the local solid object velocity or the local velocity of the solid boundary.

The dynamics of the single continuum medium is governed by the following continuity and momentum equations:

$$\frac{\partial\left(\varepsilon_e\rho_f\right)}{\partial t}+\nabla\cdot\left(\varepsilon_e\rho_f\mathbf{u}\right)=0 \tag{3}$$

$$\frac{\partial\left(\varepsilon_e\rho_f\mathbf{u}\right)}{\partial t}+\nabla\cdot\left(\varepsilon_e\rho_f\mathbf{uu}\right)=-\varepsilon_e\nabla p+\varepsilon_e\nabla\cdot\boldsymbol{\tau}-\mathbf{F}_{fp}^*+\varepsilon_e\rho_f\mathbf{g}+\tilde{\mathbf{f}} \tag{4}$$

where ρ_f and p are the fluid density and the local fluid pressure, respectively, and t is the time, $\mathbf{g}$ is the gravitational acceleration. The viscous stress tensor, $\boldsymbol{\tau}$, is given as

$$\boldsymbol{\tau}=\left[\left(\mu_b-\frac{2}{3}\mu_s\right)\nabla\cdot\mathbf{u}\right]\boldsymbol{\delta}+\mu_s\left[\left(\nabla\mathbf{u}\right)+\left(\nabla\mathbf{u}\right)^T\right] \tag{5}$$

in which μ_b and μ_s are the bulk viscosity and shear viscosity, respectively, and $\boldsymbol{\delta}$ is the identity tensor. In Equation (4), the fluid-particle interaction force per unit volume is given as

$$\mathbf{F}_{fp}^*=\frac{1}{V_c}\sum_{i=1}^{n_c}\varepsilon\mathbf{f}_{di} \tag{6}$$

in which, n_c is the number of particles in that fluid cell and $\mathbf{f}_{di}$ is the drag force.

In Equation (4), $\tilde{\mathbf{f}}$ is the virtual body force field introduced in the IBM to correct the velocity field at the interfaces between the fluid and LOCBs and inside the LOCBs, so that a desired fluid velocity distribution can be imposed over the solid boundaries. Equation (4) can be discretised using a first order finite difference algorithm as follows

$$\left(\varepsilon_e\rho_f\mathbf{u}\right)^{n+1}=\left(\varepsilon_e\rho_f\mathbf{u}\right)^n+\mathbf{H}^n\cdot\Delta t-\left(\varepsilon_e\nabla p\right)^{n+1}\cdot\Delta t+\tilde{\mathbf{f}}^{n+1} \tag{7}$$

where

$$\mathbf{H}=-\nabla\cdot\left(\varepsilon_e\rho_f\mathbf{uu}\right)+\varepsilon_e\nabla\cdot\boldsymbol{\tau}-\mathbf{F}_{fp}^*+\varepsilon_e\rho_f\mathbf{g} \tag{8}$$

and Δt is the time step and the superscripts, n and $n+1$, indicates the number of time step.

If the entire fluid cell is outside the LOCB, i.e. the volume fraction of the LOCB in the cell $\alpha=V_{LOCB}/V_c=0$. The virtual body force $\tilde{\mathbf{f}}^{n+1}$ can be given as

$$\tilde{\mathbf{f}}^{n+1}=0 \tag{9a}$$

If the fluid cell is inside the LOCB, i.e. $\alpha=1$, then

$$\tilde{\mathbf{f}}^{n+1}=\frac{\left(\varepsilon_e\rho_f\mathbf{U}_{LOCB}\right)^{n+1}-\left(\varepsilon_e\rho_f\mathbf{u}\right)^n}{\Delta t}-\mathbf{H}^n+\left(\varepsilon_e\nabla p\right)^{n+1} \tag{9b}$$

If the fluid cell is partially occupied by the LOCBs, *i.e.* $0 < \alpha < 1$, we have

$$\tilde{\mathbf{f}}^{n+1} = \alpha \left[\frac{\left(\varepsilon_e \rho_f \mathbf{U}_{LOCB}\right)^{n+1} - \left(\varepsilon_e \rho_f \mathbf{u}\right)^n}{\Delta t} - \mathbf{H}^n + \left(\varepsilon_e \nabla p\right)^{n+1} \right] \tag{9c}$$

The motion of free moving LOCBs, such as the large particles shown in Figures 1a and 1b, can be obtained by solving the Newton's equations of motion given as

$$\frac{d\left(\tilde{m}\tilde{\mathbf{v}}\right)}{dt} = -\int_{V_T} \tilde{\mathbf{f}} dV + \mathbf{f}_c + \tilde{m}\mathbf{g} \tag{10a}$$

$$\frac{d\left(\tilde{I}\tilde{\omega}\right)}{dt} = -\int_{V_T} \left(\mathbf{r} \times \tilde{\mathbf{f}}\right) dV + \mathbf{T}_c \tag{10b}$$

where $\tilde{m}$, V_T and $\tilde{I}$ are the mass, total volume and the moment of inertia tensor of the LOCB, respectively, $\mathbf{f}_c$ and $\mathbf{T}_c$ are the contact force and the torque exerted by the contacts with particles, $\tilde{\mathbf{v}}$ and $\tilde{\omega}$ are the translational and angular velocities of the LOCB, respectively.

The IBM discussed above is successfully implemented in the DEM/CFD model developed by Kafui *et al.*[3] The details of the numerical implementation are given in Ref. [18]. In this paper, the model verification is presented in the next section and some case studies are given in Section 4 in order to illustrate the robustness and applicability of the modified DEM/CFD with an IBM.

3   MODEL VERIFICATION

In order to verify the modified DEM/CFD, separation of binary mixtures of equal sized bronze/glass beads in a vertically vibrating container in the presence of air, which mimics the physical experiments conducted by Burtally et al.,[19] is simulated in 2D. The model set up is shown in Figure 2a. In this simulation, the IBM is used to model the interaction between air and the moving boundaries (i.e. the walls of the vibrating container) that are assumed to be no-slip and impermeable. The binary mixture consists of 1,250 bronze particles (i.e. the dark-coloured ones) and 3,750 glass particles (the light-coloured ones), which are initially randomly generated and settled in a container of 8.25 × 13.2 mm. All particles have the same diameters of 110 μm. The interparticle and particle-wall friction coefficients are set to 0.3. The materials properties of the particles and the walls are given in Table 1. The air inside the container is assumed to have an initial pressure of 101.325 KPa, density of 1.2 Kg/m^3 and shear viscosity of 1.8X10^{-5} Kg/m.s. The temperature is kept constant at 293 K.

The container is assumed to be a closed system so that the air is enclosed during the whole process. Once all generated particles are settled in the container, it starts to vibrate in a sinusoidal motion with an amplitude of 0.9 mm and a frequency of 55 Hz. The powder patterns during the vibration are shown in Figure 2. It is clear that as the container vibrates, the heavy bronze particles start to congregate and form clusters (Figures 2b & 2c). These clusters gradually merge into a single band (Figures 2d & 2e), which is immersed in between the light glass particles, forming a sandwich structure (Figure 2f). Inside the band, some light glass particles are entrapped. These features are identical to those observed experimentally by Burtally *et al.*,[19] as shown in Figure 3. This illustrates that the modified DEM/CFD with the IBM is robust and can be used to model two-phase flows with moving boundaries accurately.

Table 1 *Material properties of the particles and the container walls*

Materials	Density (Kg/m^3)	Young's Modulus (GPa)	Poisson's ratio
Bronze particles	8900	110	0.3
Glass particles	2525	63	0.3
Container walls	2525	63	0.3

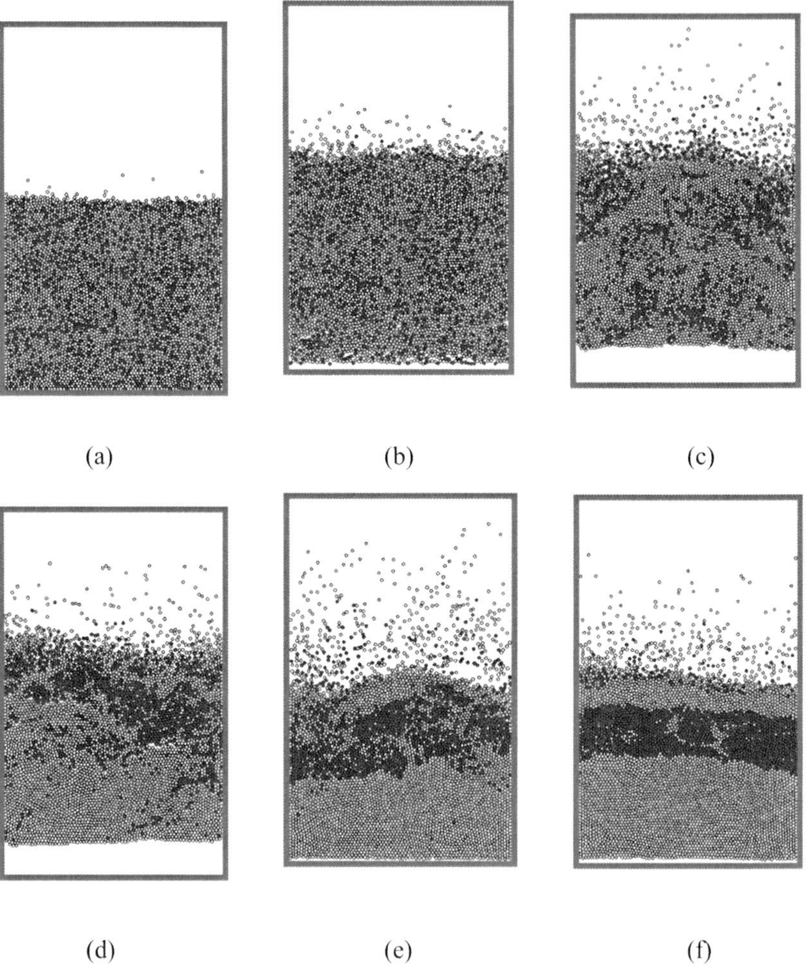

Figure 2 *Separation of a binary mixture under vertical vibration*

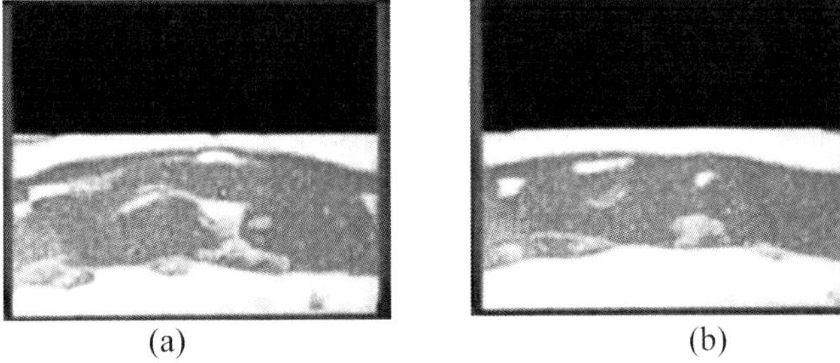
(a)   (b)

Figure 3 *Separation of a binary mixture and the formation of the sandwich structure during vertical vibration*[19]

4 NUMERICAL EXAMPLES

To further illustrate how the incorporation of the IBM enhances the capacity of the DEM/CFD, two examples of complex two-phase flows involving large objects or moving and circular shaped boundaries are presented here: i) fluidisation with a large inclusion and ii) roll compaction in the presence of air. Although both examples are simulations in 2D, the modified DEM/CFD with an IBM can be readily used to analyse problems in 3D, such as pneumatic conveying in a cylindrical pipe.[18]

4.1 Fluidisation With a Large Inclusion

For some fluidised beds used in industries, large inclusions, such as pipes, rods and disks, are used for various purposes, including process monitoring, heat and mass transfer, and structure enforcement. These inclusions are generally very large compared to the particles used in the fluidisation. It is of practical importance to understand how the presence of these inclusions affects the fluidisation behaviour. Due to the significant size difference between the inclusion and the particles, the conventional DEM/CFD struggles to handle this problem. Although attempts were made to approximate the boundaries using non-smooth stepwise boundaries, this simplification could result in significant errors in the numerical analysis. Furthermore, this approximation is unable to model the large inclusions when they are moving (i.e. moving boundaries). However, this can be easily modelled using the modified DEM/CFD with the IBM that is employed to model the interaction between the large inclusion and the fluid as illustrated in Figures 4 and 5.

Figure 4 shows the numerical model for fluidisation with a large circular inclusion that can be either stationary or moving (not shown here but details can be found in Guo *et al.*[18]). The 2D bed is 12×18 mm and the inclusion has a diameter of 5mm and located in the middle of the bed and 6.5 mm above the bottom of the bed. A monodisperse particle system consisting of 10,000 particles with a diameter of 0.1 mm, density of 2500 Kg/m^3, Young's Modulus of 8.7 GPa and Poisson's ratio of 0.3, is randomly generated in the bed and pluvially deposited to form the initial bed (Figure 4). The inclusion is assumed to be made of steel with a density of 7900 Kg/m^3, Young's Modulus of 210 GPa and Poisson's ratio of 0.29. The particles in the initial bed are color-banded according to their initial positions for visualising the fluidisation behaviour. The air with the same properties as given in Section 3 is introduced from the bottom of the bed with an initial superficial velocity of 50 mm/s. Periodic boundary conditions were assumed at the side boundaries.

The typical fluidisation behaviour is shown in Figure 5 and quantitative analysis is reported elsewhere.[18] It can be seen from Figure 5 that an air film with varying thickness is intermittently formed below the large inclusion and air bubbles form and pass through the gap between the inclusion and side boundaries. It is also clear that there constantly is a cap of defluidised particles sitting on the top of the inclusion, which is consistent with the experimental observations obtained by Glass and Harrison.[20] This cap with defluidised particles can be avoided if the large inclusion is vibrating horizontally, as demonstrated from the simulations with this modified DEM/CFD with an IBM.[18]

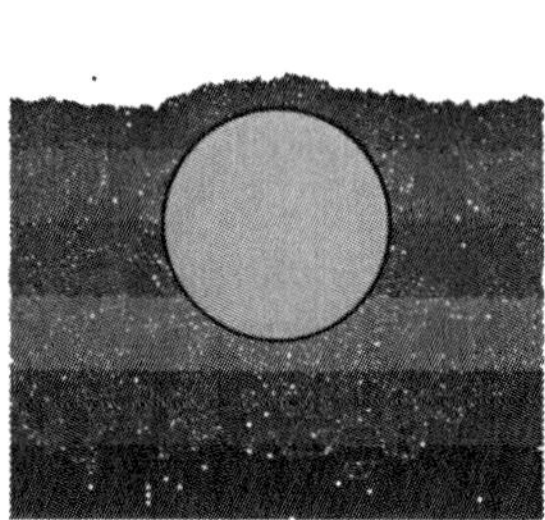

Figure 4 *Initial configuration for fluidisation with a large inclusion*

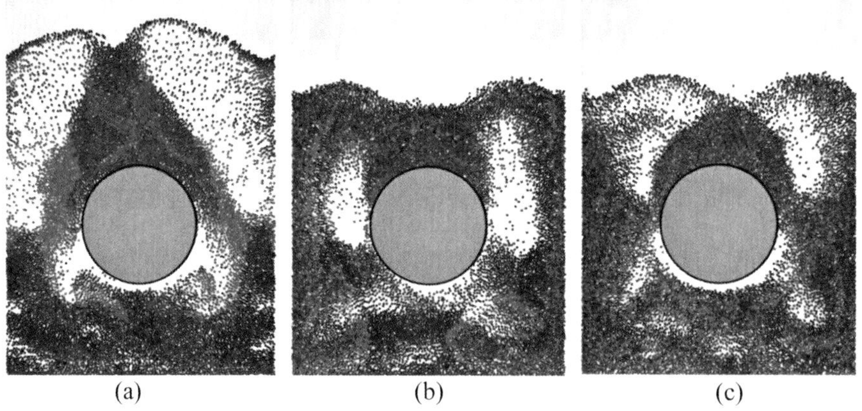

(a)   (b)   (c)

Figure 5 *Typical flow patterns for fluidisation with a stationary large inclusion*

4.2 Roll Compaction in the Presence of Air

Roll compaction is a typical agglomeration process used in pharmaceutical, fine chemicals and other industries to produce granules with increased density and flowability. It uses two counter-rotating rolls to compress dry powders through the narrow roll gap. Powders are roll compacted in the presence of air. Although roll compaction has been investigated intensively over the last two decades, how the presence of air influences roll compaction behaviour of fine powders is still unclear. As the rolls have a circular geometry and move during the roll compaction process, this cannot be analysed using the conventional DEM/CFD even with stepwise boundaries, but can be modelled using the modified DEM/CFD with IBM presented here, in such a way that the IBM is employed to model the interaction between the rolls and the air. This is illustrated in Figures 6-8.

Figure 6 shows the numerical model for roll compaction in the presence of air, in which 10,000 mono-sized particles are used. The particles have a diameter of 0.1 mm, density of 2500 Kg/m^3, Young's Modulus of 8.7 GPa and Poisson's ratio of 0.3. All friction coefficients are set at 0.3. The rolls are assumed to be steel with a density of 7900 Kg/m^3, Young's Modulus of 210 GPa and Poisson's ratio of 0.3. The particles are initially generated in the fed hopper using the similar approach as discussed in Section 4.1, then discharged from the hopper into the counter rotating rolls. The typical roll compaction behaviour for cohesionless and cohesive powders is presented in Figures 7 & 8, respectively. For the cohesive powder, a surface energy of 1.09×10^{-3} J/m^2 is used. It can be seen that for the cohesionless powder, it can flow easily through the hopper orifice and the roll gap. The powder initially accumulates in the compaction zone between two rolls as the throughput of roll compaction considered here is lower than the mass flow rate from the feed hopper (Figure 7a), while there is less material accumulation in the compaction zone when the cohesive powder is used, but arching/bridging in the hopper and between the two rolls are evident (Figure 8). It is also interesting to note that, as expected, a much higher angle of repose is obtained for the cohesive powder (Figure 8b) compared to the cohesionless powder (Figure 7b). The downflowing powder stream induces a circulation of air current, which could drag fine particles away when fine and light particles are used (Figures 7 & 8).

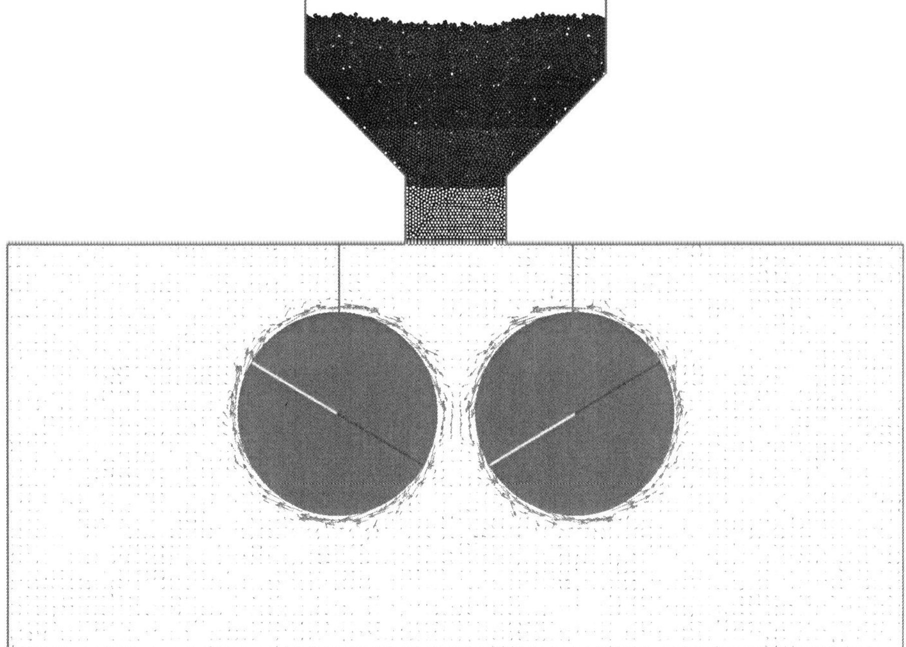

Figure 6 *Initial configuration for roll compaction in the presence of air*

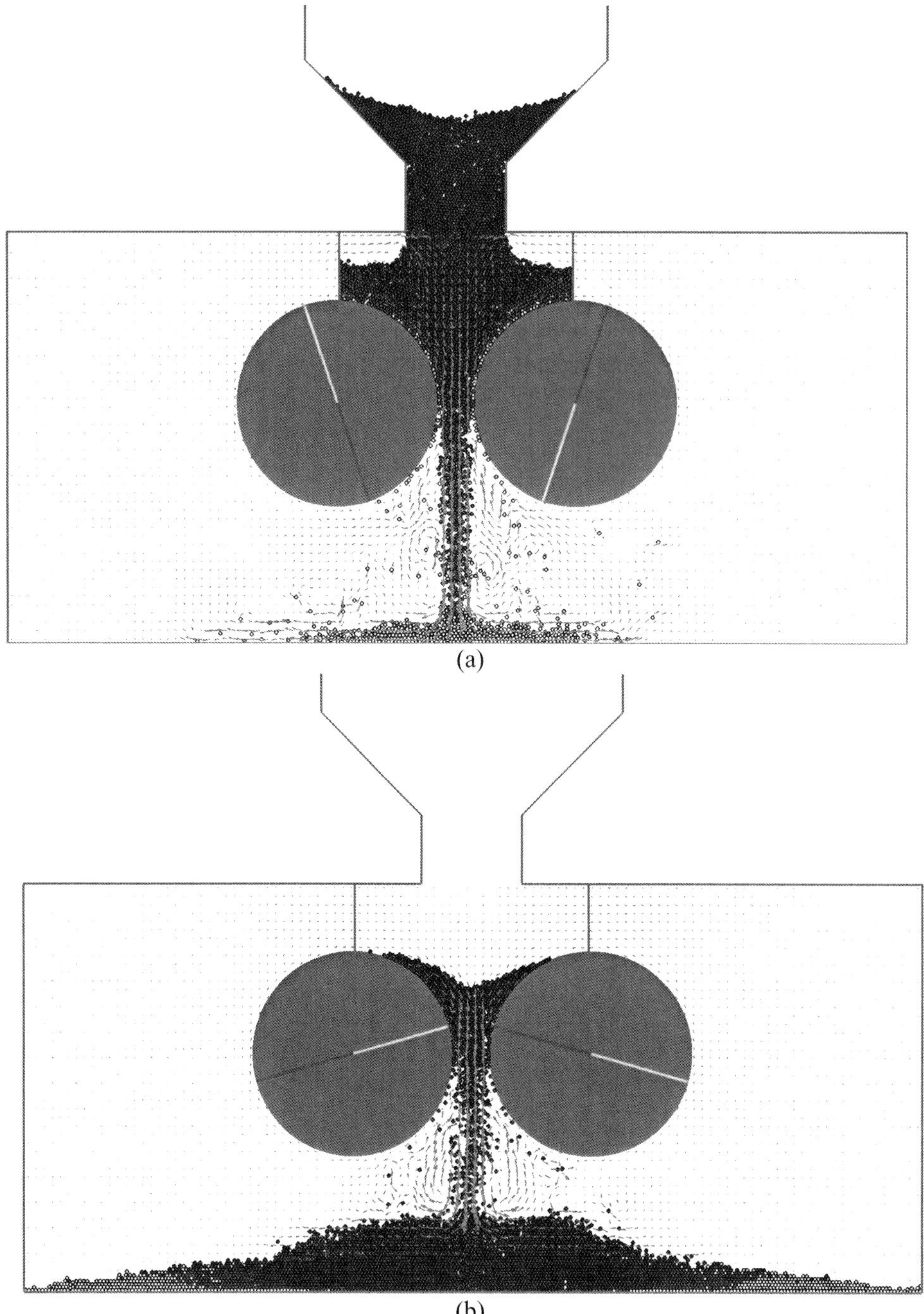

(a)

(b)

Figure 7 *Typical flow patterns of cohesionless powders during roll compaction in the presence of air*

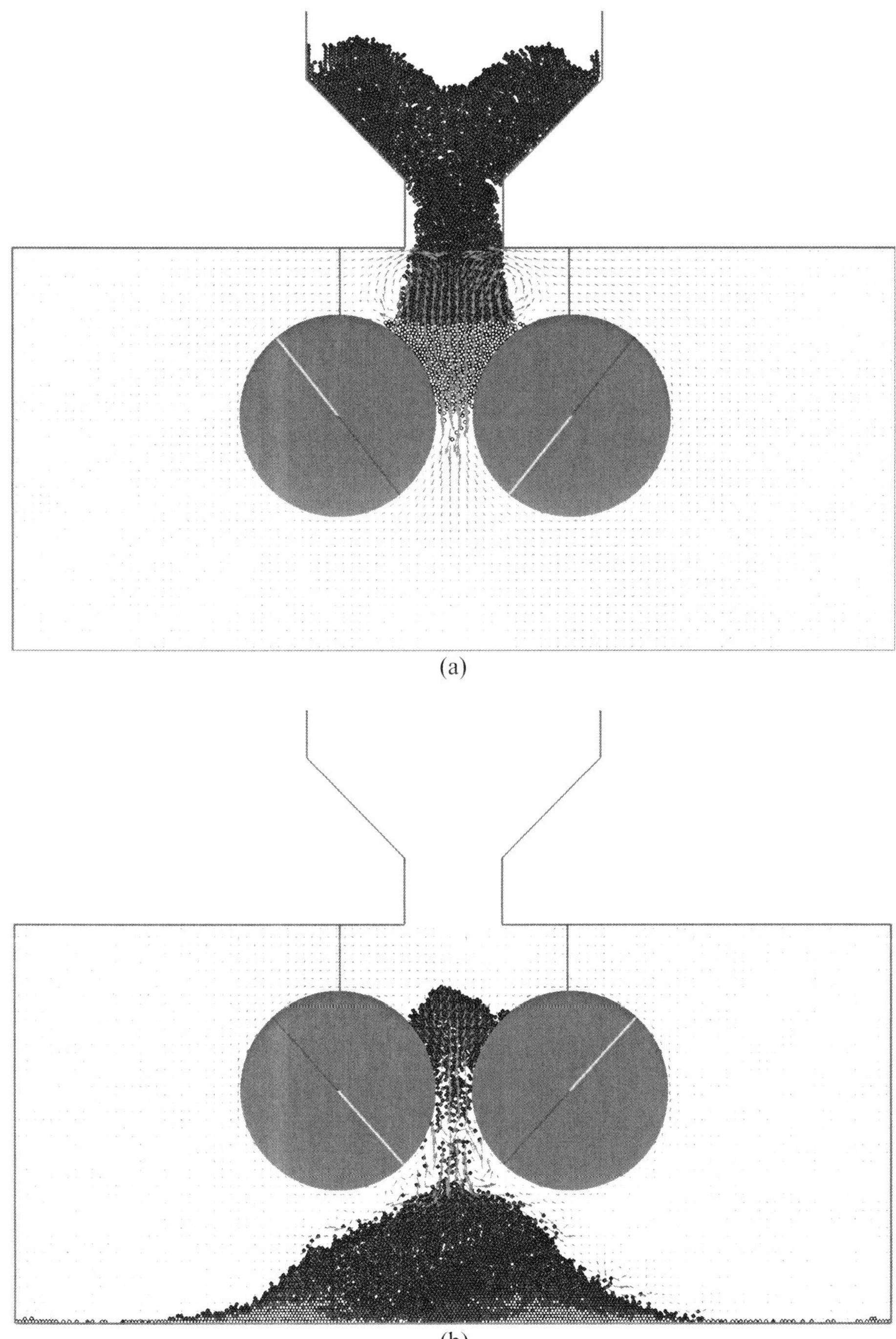

(a)

(b)

Figure 8 *Typical flow patterns of cohesive powders during roll compaction in the presence of air*

5 CONCLUSIONS

In this study, an immersed boundary method (IBM) is incorporated into an existing DEM/CFD code in order to model complex fluid/particle two-phase flows involving large objects and moving arbitrary shaped boundaries. The principle of the modified DEM/CFD with IBM is presented. The modified DEM/CFD is verified by comparing the numerical simulations with the experimental observation of the separation of binary mixtures of bronze and glass particles in a vertically vibrated container. Its capacity is further demonstrated with two case studies: i) fluidisation with a large immersed inclusion and ii) roll compaction in the presence of air. It is shown that the capacity of DEM/CFD is significantly enhanced with IBM and it can be used to simulate a range of particulate flows consisting of fluids and particles with large objects, complex and/or moving boundaries.

Acknowledgements

The authors would like to thank Colin Thornton for his constructive advices and useful discussions.

References

1   Y. Tsuji, T. Tanaka and T. Ishida, *Powder Technology*, 1992, **71**, 239.
2   Y. Tsuji, T. Kawaguchi and T. Tanaka, *Powder Technology*, 1993, **77**, 79.
3   K. D. Kafui, C. Thornton and M. J. Adams, *Chemical Engineering Science*, 2002, **57**, 2395.
4   A.B. Yu and B.H. Xu, *Journal of Chemical Technology and Biotechnology*, 2003, **78**, 111.
5   Y. Guo, C. -Y. Wu, K. D. Kafui and C. Thornton, *Powder Technol,* 2011, **206**, 177.
6   Y. Guo, C. -Y. Wu, K. D. Kafui and C. Thornton, *Powder Technol,* 2010, **197**, 111.
7   Y. Guo, C. -Y. Wu and C. Thornton, *Chemical Engineering Science,* 2011, **66**, 661.
8   Y. Guo, K.D. Kafui, C.Y. Wu, C. Thornton, J.P.K. Seville, *AIChE J.*, 2009, **55**, 49.
9   K.L. Johnson, *Contact Mechanics*, Cambridge University Press, Cambridge. 1985.
10   R.D. Mindlin and H. Deresiewicz, *ASME Journal of Applied Mechanics*, 1953, **20**, 327.
11   K.L. Johnson, K. Kendall and A.D. Roberts, *Proc. Royal Society of London A.* 1971, **324**, 301.
12   C. Thornton, *Journal of Physics D: Applied Physics*, 1991, **24**, 1942,.
13   C. Thornton and Z. Ning, *Powder Technology*, 1998, **99**, 154.
14   C. Thornton and K.K. Yin, *Powder Technology*, 1991, **65**, 153.
15   R. Di Felice, *International Journal of Multiphase Flow*, 1994, **20**, 153.
16   C. S. Peskin, *Journal of Computational Physics*, 1972,**10**, 252.
17   T. Kajishima, S. Takeuchi, H. Hamasaki and Y. Miyake, *JSME International Journal, Series B*, 2001, **44**, 526.
18   Y. Guo, C.-Y. Wu and C. Thornton, *Modelling gas-particle two-phase flows with complex and moving boundaries using DEM/CFD with an Immersed Boundary Method.* Internal report. University of Birmingham, Birmingham. 2012.
19   N. Burtally, P.J. King, M.R. Swift and M. Leaper, *Granular Matter*, 2003, **5**, 57.
20   D.H. Glass and D. Harrison, *Chemical Engineering Science*, 1964, **19**, 1001.

EFFECT OF SOLID AND LIQUID HEAT CONDUCTIVITIES ON TWO-PHASE HEAT AND FLUID FLOWS

T. Tsutsumi, S. Takeuchi and T. Kajishima

Department of Mechanical Engineering, Osaka University
2-1 Yamada-oka, Suita-city, Osaka 565-0871 Japan

1 INTRODUCTION

In solid-liquid multiphase flow, temperature distribution in the flow field often play important role, especially when the particle-fluid density ratio is very close to unity. Also, the temperature of the particles would be of great interest for improving the heat (and mass) transfer performance of the system.

There is several numerical work of heat transfer problem in (fully resolved) solid-dispersed multiphase flow using immersed boundary methods. Several approaches have been proposed to determine the temperature of the particles; Pacheco and co-workers[1, 2] and Paravento et al.[3] proposed an interpolation algorithm for temperature to match the boundary condition at the interface, and Pan[4] developed a method to approximate the interface cell temperature by volume averaging. While the above researchers simplified the problem by assuming a unique temperature for each particle, Kang et al.[5] proposed a method to consider temperature distribution within the particle by updating the boundary condition at the voxel (cuboid)-face for determining the interface temperature and heat flux.

In the present work, we show some simulation results of heat transfer problem in multiphase flow of dispersed solid particles considering the local heat conduction at liquid-solid interface (and therefore a temperature gradient within the solid object).

The interaction between fluid and particles is solved with our original immersed solid approach[6, 7] on a rectangular grid system. The method employs a simple procedure for the momentum-exchange by imposing a volume force (as interaction force) on both solid and fluid phases. The method has been applied for studying the fundamentals of clustering process with $O(10^3)$ spherical particles in a turbulent flow.[6, 8, 9] Also the usefulness of our method has been demonstrated through the analysis of sedimentation process in a 10^5-particle system by Nishiura et al.[10]

In the present study, liquid-solid two-phase flow with heat transfer is simulated. To treat the heat transfer in a liquid-solid system, heat conduction through the liquid-solid interface is carefully considered by an original heat conduction model so that the temperature distributes within the particles. A discrete element model[11] with soft-sphere collision is applied for particle-particle interaction.

The proposed method is applied to a 2-D confined flow including multiple particles under a high Rayleigh number condition. Heat transfer and particle behaviours are studied for different ratios of heat conductivity (solid to liquid) and solid volume fractions, and the highlighted effect of temperature distributions within the particles and liquid is presented.

2 GOVERNING EQUATIONS AND NUMERICAL METHODS

The fluid is assumed to be incompressible and the buoyancy force is represented by the Boussinesq approximation. The governing equations for the fluid are the equations of continuity, motion and energy:

$$\nabla \cdot \boldsymbol{u}_f = 0 \tag{1a}$$

$$\frac{\partial \boldsymbol{u}_f}{\partial t} + \boldsymbol{u}_f \cdot \nabla \boldsymbol{u}_f = -\frac{1}{\rho_f}\nabla p + v_f \nabla \cdot \left(\nabla \boldsymbol{u}_f + \left(\nabla \boldsymbol{u}_f\right)^T\right) + \boldsymbol{g}\beta(T - T_0) \tag{1b}$$

$$\rho_f c \frac{\partial T}{\partial t} + \rho_f \boldsymbol{u}_f \cdot \nabla T = \nabla \cdot \left(\lambda_f \nabla T\right) \tag{1c}$$

where $\boldsymbol{u}_f$ is the fluid velocity, p is the pressure, ρ_f the density, v_f is the kinematic viscous coefficient, $\boldsymbol{g}$ is the gravitational acceleration, β is the coefficient of thermal expansion, T is the temperature, c is the specific heat and λ_f is the heat conductivity.

2.1 Fluid-Solid Interaction Model: An Immersed Solid Approach

Momentum exchange at the fluid-solid interface, where a cell is partially occupied by a solid particle, is solved by the immersed solid approach developed by Kajishima and co-workers.[6, 8] This is briefly described below.

    A velocity field $\boldsymbol{w}$ is established through volume-averaging the local fluid velocity $\boldsymbol{u}_f$ and local particle velocity $\boldsymbol{u}_p$ in a cell, *i.e.*

$$\boldsymbol{w} = (1 - \alpha)\boldsymbol{u}_f + \alpha\boldsymbol{u}_p \tag{2}$$

where α $(0 \leq \alpha \leq 1)$ is the local solid volume fraction in the cell. The particle velocity $\boldsymbol{u}_p$ can be decomposed into translational and rotational components as $\boldsymbol{u}_p = \boldsymbol{v}_t + \boldsymbol{r} \times \boldsymbol{\omega}_p$. The velocity field $\boldsymbol{w}$ is assumed to obey the following equation:

$$\frac{\partial \boldsymbol{w}}{\partial t} = -\nabla P + \boldsymbol{H}_w + \boldsymbol{g}(T - T_0) + \boldsymbol{f}_p \tag{3}$$

where $P = p/\rho_f$ and $\boldsymbol{H}_w$ contains the convective and viscous terms. One of the simplest (explicit) time advancement schemes of $\boldsymbol{w}$ is as follows:

$$\boldsymbol{w}^{n+1} = \boldsymbol{w}_l - \Delta t \, \nabla P + \boldsymbol{H}_w + \Delta t \, \boldsymbol{f}_p \tag{4}$$

$$\boldsymbol{w}_l = \boldsymbol{w}^n + \Delta t \, \boldsymbol{H}_w + \boldsymbol{g}(T - T_0) \tag{5}$$

$$\boldsymbol{f}_p = \alpha(\boldsymbol{u}_p - \boldsymbol{w}_l)/\Delta t \tag{6}$$

where the superscripts represent the time level and Δt is the time step. The interaction term f_p works to enforce the non-slip condition at the interface.[6,8] In the practical implementation, the Crank-Nicolson method is adopted for obtaining w_l of Equation (5).

For the motion of particles, the Newton's equations of motion are solved. The same force as given by Equation (6) applies to the fraction of solid in the cell with the opposite sign. The surface integration of the hydrodynamic forces is changed to the integration of f_p over the volume of the particle V_p:

$$\frac{\mathrm{d}(m_p v_t)}{\mathrm{d}t} = -\rho_f \int_{V_p} f_p \, dV + G_p \tag{7}$$

$$\frac{\mathrm{d}(I_p \cdot \omega_p)}{\mathrm{d}t} = -\rho_f \int_{V_p} r \times f_p \, dV + N_p \tag{8}$$

This replacement from surface to volume integration considerably facilitates the computation of the solid behaviour, and also the use of the same body force f_p for both fluid and particle phases in a shared Cartesian cell ensures no leakage of momentum between the phases. Equations (7) and (8) are solved with a predictor-corrector method.[12]

2.2 Heat Flux and Interparticle Collision Models

Equation (1) is time-updated with an implicit treatment for the diffusion term, which stabilises the computation and enables simulation with particles of very high/low ratio of heat conductivity ($10^{-3} \sim 10^3$) to fluid.

Local heat flux at the fluid-solid interface is carefully discretised and incorporated into the implicit scheme of temperature. The present heat conduction model is thoroughly validated through comparison with some analytical solutions of heat conduction problems, which is the subject of future paper of the present authors.

A soft-sphere model is used to allow multiple collisions for interparticle and particle-wall collisions. A spring and dashpot model is employed to calculate the contact forces. In the present study, the same values as Tsuji *et al.*[11] are used for the spring constant (k), the restitution coefficient (e) and the friction coefficient (μ).

Tsuji *et al.*[11] suggested the following condition for determining the time step to sufficiently resolve the eigen oscillation of the mass-spring system:

$$\Delta t \leq \frac{\pi}{n}\sqrt{\frac{m_p}{k}} \quad (n \geq 10) \tag{9}$$

where m_p is the mass of the particle. In the present study, the time step is fixed to $\Delta t = 1 \times 10^{-3}$, which is smaller than the smallest value of the right hand side (1.47×10^{-3} with $n = 10$) for the smallest m_p employed.

3 RESULTS AND DISCUSSION: HEAT AND MASS TRANSFER

Heat transfer and particles behaviours are simulated with different ratios of heat conductivity and number of particles under relatively high Rayleigh number conditions. Particles are initially arranged as illustrated in Figure 1. The temperature difference between the top and bottom walls is kept constant ($\Delta T = 1$), and no heat flux is considered at the side walls. The non-slip boundary condition and the Neumann condition are applied for the velocity and the pressure, respectively, on the solid walls.

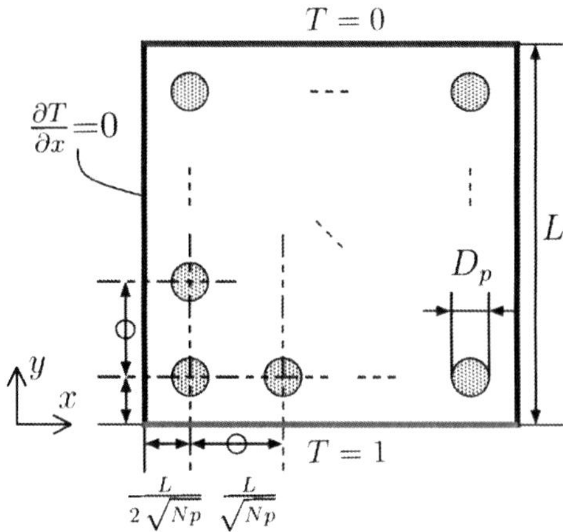

Figure 1 *Initial arrangement of particles*

Table 1 *Simulation parameters for the dilute case*

Spatial resolution	D_p/Δ	10
Num. of particles	N_p	$1, 3^2, 5^2, 6^2, 10^2$
Diameter of particles	D_p	$0.3L/\sqrt{N_p}$
Rayleigh number	Ra	1.25×10^6
Heat conductivity ratio	λ_s/λ_f	$10^{-2}, 10^0, 10^1, 10^2$

In the following analysis, the heat transfer in fixed-particle system and free (movable)-particle system are compared. In all the cases, the simulations are conducted until non-dimensional time of 1500 is reached. For evaluating heat transfer rate of each system, Nusselt number defined with the heat flux at the hot (bottom) wall is time-averaged in the range $500 \leq t \leq 1500$ (after reaching quasi-steady state). Prandtl number, density ratio and specific heat ratio are set to unity.

3.1 Dilute Case with a Solid Volume fraction of 7.1%

The particle sizes are varied by keeping the bulk solid volume fraction identical. The other simulation conditions and parameters are summarised in Table. 1.

Figure 2 shows some snapshots of the typical flow field with fixed/moving particles for different ratios of heat conductivities and numbers of particles. The cases employing particles of $\lambda_s/\lambda_f = 10^{-2}$ (near-insulated particles) exhibit large temperature gradient within the particles, whereas nearly uniform temperature distributions are observed for the particles of high conductivity.

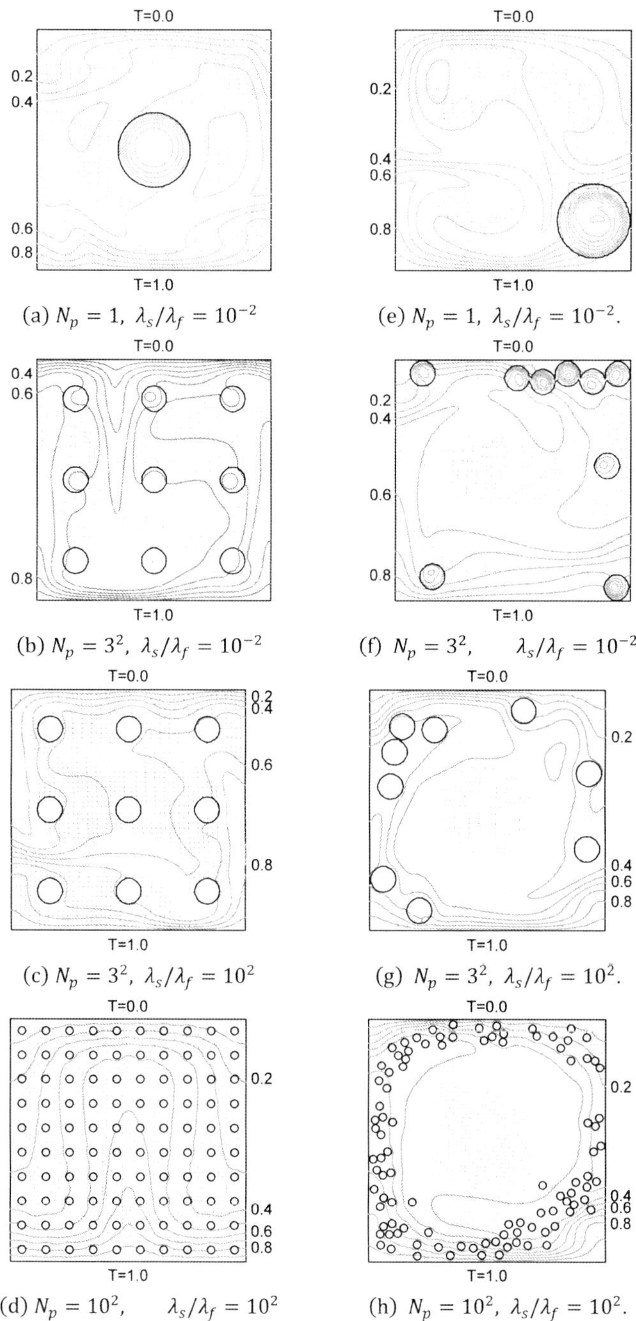

Figure 2 *Iso-contours of temperature in an instantaneous flow field (t = 1500) including the particles of $\lambda_s/\lambda_f = 10^{-2}$ (near-insulated particles) and $\lambda_s/\lambda_f = 10^2$ with fixed (a)-(d) and freely moving particles (e)-(h)*

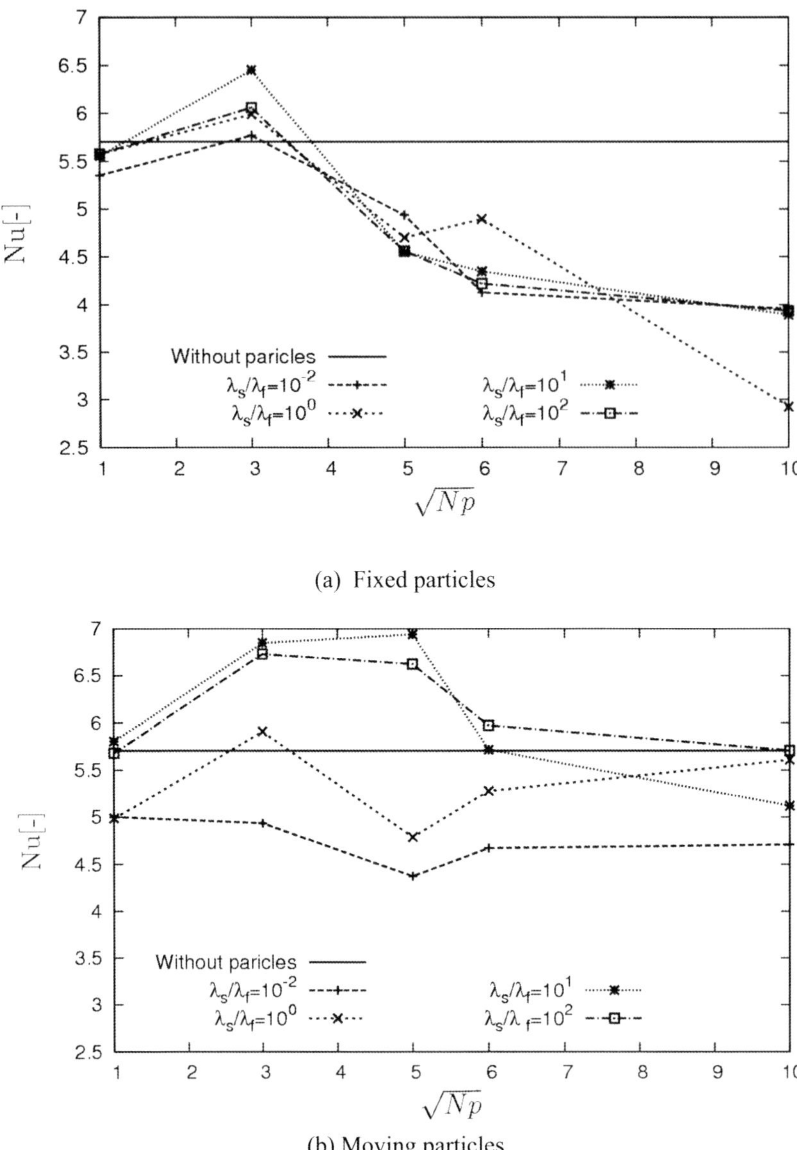

(a) Fixed particles

(b) Moving particles

Figure 3 *Relationship between average Nusselt number and the number of particles*

Figure 3 compares the average Nusselt number for fixed- and free-particle systems. In the cases with fixed particles, an overall decreasing trend of average Nusselt number is observed with increasing number of particles (regardless of heat conductivity ratio), after taking the maximum values at $N_p = 3^2$; only at this condition, heat transfer rate is larger than that of the case without particles. This trend of *Nu* may suggest that the fixed particles only interrupt the heat convection through the fluid, as shown in Figure 2d.

On the other hand, in the case with freely moving particles, a transition of profile of average Nusselt number is observed around $\lambda_s/\lambda_f = 10^0$, especially in the range of the particle number smaller than 5^2. From our observation, the interparticle collision becomes more frequent as the ratio of heat conductivity is increased, suggesting that the particle-particle interaction enhances the heat transfer through mixing of the particles. However, when the number of particles is further increased to 10^2, the Nusselt number for all cases ia lower than that without particles. Surprisingly, the particles with high heat conductivity do not necessarily enhance the heat transfer (at the present solid volume fraction). Our preliminary study suggests that bifurcation of the flow structure is responsible for this phenomenon, for which further study is ongoing.

Table 2 *Simulation conditions for the dense case*

Num. of grid points	$N_x \times N_y$	200×200
Spatial resolution	L/Δ	200
Num. of particles	N_p	14^2
Diameter of particles	D_p	$0.05L$
Rayleigh number	Ra	2.0×10^6
Heat conductivity ratio	λ_s/λ_f	$10^{-3}, 10^0, 10^1, 10^2, 10^3$

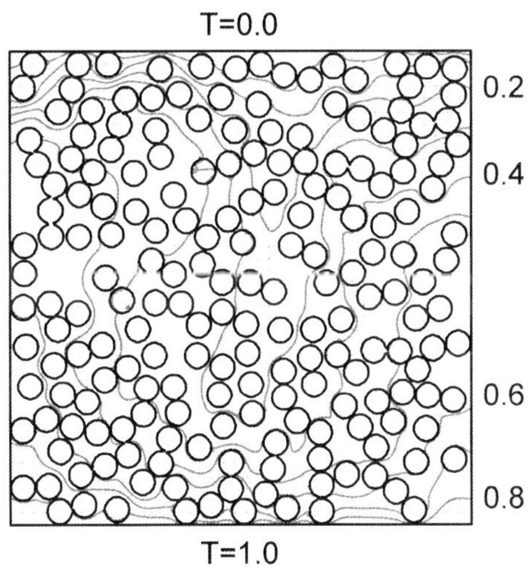

Figure 4 *Instantaneous flow field and contours of temperature* $(\lambda_s/\lambda_f = 10^3)$

3.2 Dense Case with a Solid Volume Fraction of 38.5%

A dense concentration case is also attempted under the conditions set out in Table 2. Figure 4 shows an instantaneous flow and temperature fields at $t = 1500$ by employing the particles of a very high heat conductivity. Some regions are highly concentrated with the particles, where the collision is dominant. The particles are found to form a pair of circulating flows side by side, which enhance the heat convection from bottom to top, whereas the cases of low λ_s/λ_f exhibit a single or weak double circulation(s) (and transition between them) with less mobility of particles.

Figure 5 compares the time evolution of transition of Nusselt number for three different values of λ_s/λ_f. The case with lowest conductivity ratio (10^{-3}) exhibits a very low Nusselt number over the time range tested, and the Nusselt number improves with increasing heat conductivity ratio. Note that the profiles of $\lambda_s/\lambda_f = 10^1$ and 10^2 (not shown here) fall between the curves of $\lambda_s/\lambda_f = 10^3$ and the cases without particles. Under a dense situation where the interparticle collision is dominant, the heat is transferred not only by convection but also interparticle conduction.

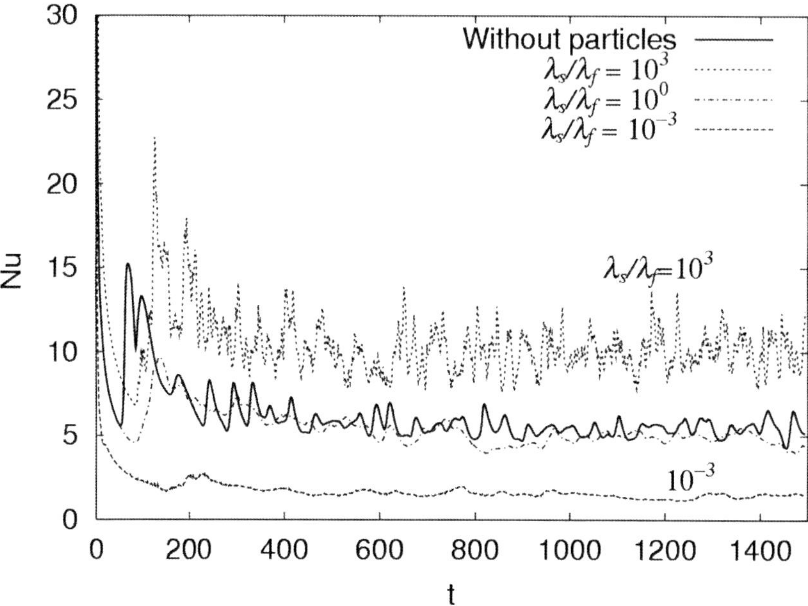

Figure 5 *Comparison of time evolutions of Nusselt number for different ratios of heat conductivity*

4 CONCLUSIONS

A method for analysing heat transfer problems through liquid-solid media was developed based on discrete element and our original immersed-solid models, and applied to liquid-solid multiphase flows from dilute to dense solid concentrations with high Rayleigh numbers. While free particles of low heat conductivity (i.e., insulated particle) only exhibit low heat transfer rate by blocking the heat convection of fluid, the free particles of high conductivity can form highly concentrated region, which may work as an independent path for heat transfer. The results highlight the effect of temperature distributions within the particles (as well as liquid) on the overall heat transfer performance in the multiphase flow.

Acknowledgments

The authors gratefully acknowledge the financial support of Grant-in-Aid for Scientific Research (B) No.23360085 and Grant-in-Aid for Young Scientist (A) No.23686030.

References

1   J.R. Pacheco, A. Pacheco-Vega, T. Rodi´c and R.E. Peck, *Numerical Heat transfer,Part B*, 2005, **48**, 1.
2   A. Pacheco-Vega, J. R. Pacheco and T. Rodi´c, *Journal of Heat Transfer*, 2007, **129**, 1506.
3   F. Paravento, M.J. Pourquie and B.J. Boersma, *Flow Turbulence Combust.*, 2008, **80**, 187.
4   D. Pan, *Numer. Heat Transfer, Part B*, 2006, **49**, 277.
5   S. Kang, G. Iaccarinoa and F. Ham, *Journal of Computational Physics*, 2009, **228**, 3189.
6   T. Kajishima, S. Takiguchi, H. Hamasaki and Y. Miyake, *JSME International Journal Series B*, 2001, **44**, 526.
7   Y. Yuki, S. Takeuchi and T. Kajishima, *Journal of Fluid Science and Technology*, 2007, **2**, 1.
8   T. Kajishima and S. Takiguchi, *International Journal of Heat and Fluid Flow*, 2002, **23**, 639.
9   T. Kajishima, *Int. Journal of Heat and Fluid Flow*, 2004, **25**, 721.
10 D. Nishiura, A. Shimosaka, Y. Shirakawa and J. Hidaka, *Kagaku Kogaku Ronbunshu*, 2006, **32**, 331 (in Japanese).
11 Y. Tsuji,T. Kawaguchi and T. Tanaka, *Powder Technology*, 1993, **77**, 79.
12 A. Ueyama, S. Moriya, M. Nakamura and T. Kajishima, *Trans. JSME Series B*, 2011, **77**, 803.

GRAVITATIONAL SEDIMENTATION AND SEPARATION OF PARTICLES IN A LIQUID: A 3D DEM/CFD STUDY

L. Qiu[1,2] and C.-Y. Wu[2]

[1] Department of Applied Mechanics, China Agricultural University, Beijing, 100083, China
[2] School of Chemical Engineering, University of Birmingham, Birmingham B25 2TT, UK

1 INTRODUCTION

Gravitational sedimentation and separation of particles in liquids are common phenomena in nature and have wide applications in many industrial processes such as slurry flows, wastewater treatment, and oil production. Due to their industrial relevance, sedimentation and separation has been a subject of many studies by engineers, chemists and physicists and there is a rich collection of literature on these subjects. In order to explore the microscopic behaviour of individual particles and their interactions, a number of numerical models have been developed, including, for example, Ladd used a discretized Boltzmann equation to simulate particulate suspensions,[1,2] Hoogerbrugge *et al.* applied a dissipative particle dynamics (DPD) for simulating microscopic hydrodynamic phenomena,[3] Yang *et al.* adopted the discrete particle method for simulation of sedimentation of microparticle,[4] and Alexander *et al.* modelled liquid-solid flows using smoothed particle hydrodynamics and the discrete element method (DEM).[5] Over the last two decades, a coupled DEM with computational fluid dynamics (CFD), *i.e.* DEM/CFD, in which two-way coupling of fluid-particle interaction was considered, has been developed and advanced for modelling fluid-solid flows,[6,7] and can be used to efficiently analyse the gas-solid flows.[7,8] However, most DEM/CFD methods were developed for compressible fluids (i.e. gases) based upon the ideal gas law, which cannot be used for sedimentation and separation in liquids that are generally incompressible. The objective of this study is to extend the DEM/CFD methodology to consider the compressibility of liquids so that a robust numerical method can be developed for modelling sedimentation and separation of solid particles in liquids.

2. THEORETICAL

In order to model incompressible fluids in DEM/CFD, instead of using the ideal gas law, an equation of state for weakly compressible fluids is adapted. This is incorporated into the DEM/CFD model originally developed by Kafui *et al.* [7] while the original algorithms are intact and its capability for gas-solid flows is retained. For completeness, the theoretical aspect of the modified DEM/CFD is presented in this section.

2.1. Governing Equation for Particles

The motion of particles is governed by the Newton's second law of motion, *i.e.*

$$m_i \frac{d\mathbf{v}_i}{dt} = \mathbf{f}_{ci} + \mathbf{f}_{fpi} + m_i \mathbf{g} \tag{1}$$

$$I_i \frac{d\boldsymbol{\omega}_i}{dt} = \mathbf{T}_i \tag{2}$$

in which m_i and I_i are the mass and moment of inertia of the particle, respectively, $\mathbf{v}_i$ and $\boldsymbol{\omega}_i$ are the linear and angular velocity of the particle, $\mathbf{T}_i$ is the torque arising from the tangential components of the contact force, and $\boldsymbol{g}$ is the acceleration due to the gravity, $\boldsymbol{f}_{ci}$ is the particle–particle contact force, $\boldsymbol{f}_{fpi}$ is the fluid-particle interaction force. Following Anderson and Jackson,[9] $\boldsymbol{f}_{fpi}$ is given as:

$$\mathbf{f}_{fpi} = -v_{pi}\nabla.p + v_{pi}\nabla.\boldsymbol{\tau}_f + \varepsilon \mathbf{f}_{di} \tag{3}$$

where v_{pi} is the volume of particle, ε is the void fraction, p is the fluid pressure, τ_f is viscous stress tensor, and $\mathbf{f}_{di}$ is the drag force in the direction of the relative velocity between fluid and particle, for which the continuous single-function correlation proposed by Di Felice is used.[10]

The interactions between particles are modelled using classical contact mechanics, in which the Hertz theory is employed to analyse the normal interaction,[11] and the theory of Mindlin and Deresiewicz is used to model the tangential interaction.[12] The evolution of contacts can thus be determined explicitly using realistic physical properties, such as Young's modulus and Poisson's ratio. This is superior to the DEM codes using spring-dashpot type contact models, in which artificial contact stiffness are used, and the link with the actual physical properties of the particles is obscure.

2.2. Governing Equations for Fluids

The equations of continuity and momentum conservation are given as follows:

$$\frac{\partial(\varepsilon \rho_f)}{\partial t} + \nabla.(\varepsilon \rho_f \mathbf{u}) = 0 \tag{4}$$

$$\frac{\partial(\varepsilon \rho_f \mathbf{u})}{\partial t} + \nabla.(\varepsilon \rho_f \mathbf{u}\mathbf{u}) = -\nabla.p + \nabla.\boldsymbol{\tau}_f - \mathbf{F}_{fp} + \varepsilon \rho_f \mathbf{g} \tag{5}$$

where $\mathbf{u}$ and ρ_f are the fluid velocity and density, respectively. The fluid-particle interaction force per unit volume $\mathbf{F}_{fp}$ is

$$\mathbf{F}_{fp} = \frac{\displaystyle\sum_{i-1}^{n_c} \mathbf{f}_{fpi}}{\Delta V_c} \tag{6}$$

where n_c is the number of particles in a fluid cell of volume ΔV_c.

For closure of system, an equation of state (EOS) relating density to pressure is required. In the original gas-solid flow model developed by Kafui *et al.*,[7] the ideal gas EOS was employed in evaluating the gas density. For the liquid-solid flow systems to be considered in this study, a pressure–density relationship of the modified Tait equation of state is incorporated,[13] *i.e.*

$$\frac{p+B}{p_0+B}=\left(\frac{\rho_f}{\rho_0}\right)^n \tag{7}$$

where B and n are constants, p_0 and ρ_0 denote the undisturbed value of pressure and density, respectively. The Tait EOS is independent of entropy and hence the energy equation is not required for solving the weakly compressible liquid flow.

2.3 Numerical Strategy

In the present numerical scheme, the computational domain is discretised into a number of fluid cells. All quantities such as fluid pressure and velocity are averaged in the fluid cells. As argued by Tsuji *et al.*,[6] the size of the fluid cells should be smaller than the macroscopic motion of bubbles, but larger than the particle size. A fluid cell size of 3~5 time particle diameter was recommended by Tsuji *et al.*[6] and was followed by many others in the simulations of fluidized bed.[7,8] The size of the fluid cells is set as 3 times the average particle diameter in all simulations reported here. The numerical scheme employed here is identical to that given by Kafui *et al.*[7] At each time step, the equations of motion for discrete particles are first solved yielding the particle positions and velocities. This enables the porosity of each computational cell to be calculated. Using the known particle positions and velocities, the fluid-phase hydrodynamic equations are then solved to give the fluid velocity and pressure fields, from which the drag forces acting on particles can then be computed so is the fluid-particle interaction force per unit volume $\mathbf{F}_{fp}$. The forces acting on particles and each fluid cell are then updated in order to determine the positions and velocities of particles at the next time step. This procedure is repeated until a specified time (or a specific number of time steps) is reached.

3. MODEL VALIDATION

In order to validate the modified DEM/CFD model, settling of a sphere in a closed container with a square cross section filled with liquids is analysed and compared with experimental results reported by Cate *et al.*[14] The sphere has a diameter of $d = 15$ mm, Young's modulus $E_p = 8.7\times10^9$ Pa, Poisson's ratio $v_p = 0.3$ and density $\rho_p = 1120$ kg/m^3. Reynolds number in the range of 1.5 and 31.9 was considered by varying the density and viscosity of the liquid. The fluid properties used and corresponding Reynolds numbers were given in Table 1. For all simulations reported in this study, $B=3.0\times10^8$ and $n=7$ are assumed in the equation of state (7) as suggested by Brujan.[12] The container has a dimension of $100\times100\times160$ mm, filled with a liquid at rest (see Figure 1). All the walls are treated as impermeable with no slip boundaries. The sphere is released without initial velocity at time $t = 0$. The non-perturbed flow is simply the hydrostatic state. The settling velocity of the sphere is monitored from which the terminal velocity is determined. The variation of terminal velocity with Reynolds number was shown in Figure 2, in which the experimental results of Cate *et al.*[14] were also superimposed. It is clear that the terminal velocity increases as the Reynolds number increases. In addition, the numerical results are in excellent agreement with the experimental data obtained by Cate *et al.*[14] This demonstrates that the modified DEM/CFD can accurately capture the settling behaviour of spheres in liquids.

Table 1 Fluid properties used in the experiment[14]

Case No.	ρ_f [kg/m^3]	μ_f [Ns/m^2]	Re
Case E 1	970	373	1.5
Case E 2	965	212	4.1
Case E 3	962	113	11.6
Case E 4	960	58	31.9

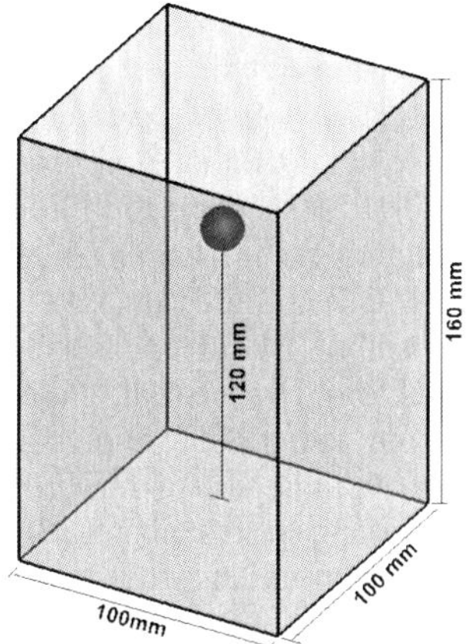

Figure 1 *Schematic diagram for the settling of a sphere in liquids.*

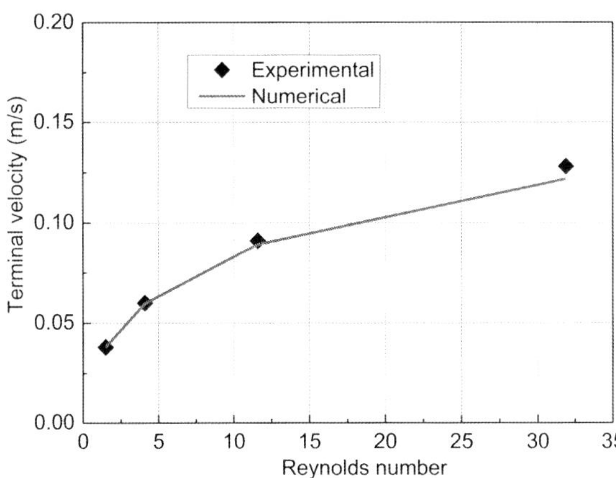

Figure 2 *Variation of terminal velocity with Reynolds number.*

4. SETTLING OF SPHERES IN LIQUIDS

In order to explore the effects of particle size, particle density, particle concentration and liquid viscosity on the settling and sedimentation behaviour of a sphere in liquids, simulations of the settling of spherical particles in a closed container (with the similar set-up to that shown in Figure 1) with dimension of 100×100×360 mm fully filled with a liquid are performed. The material properties considered are listed in Table 2. The initial pressure is set to the atmospheric pressure and particles are released without initial velocity. All the walls are treated as impermeable with no slip boundaries.

<table>
<tr><td colspan="2" align="center">Table 2 Material properties considered</td></tr>
<tr><td colspan="2">Particles and walls:</td></tr>
<tr><td>Young's modulus (GPa)</td><td>8.7</td></tr>
<tr><td>Poisson's ratio</td><td>0.3</td></tr>
<tr><td>Density (kg/m^3)</td><td>2000</td></tr>
<tr><td>Friction coefficient</td><td>0.3</td></tr>
<tr><td colspan="2">Liquid:</td></tr>
<tr><td>Density (kg/m^3)</td><td>1000</td></tr>
<tr><td>Shear viscosity (kg/ms)</td><td>0.001</td></tr>
</table>

4.1 The Effect of Particle Density

A spherical particle located at a distance of 225 mm from the bottom of the container is released to settle vertically in the water. The diameter of the particle is 10 mm but its density is varied to investigate how it affects the settling behaviour. Figure 3 shows the evolution of settling velocities for the particles with various densities. It is clear that once the particle is released, it accelerates under gravity until a certain velocity (i.e. the terminal velocity) is reached, at which the gravitational force is equal to the drag and buoyant forces. Thereafter, the particle travels at the terminal velocity. It can be seen that the terminal velocity increases as the particle density increases.

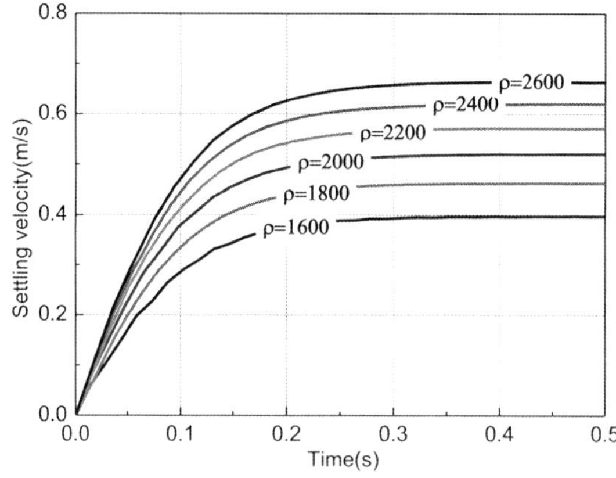

Figure 3 *Evolutions of settling velocities for particles with different densities.*

4.2 The Effect of Particle Size

In order to explore the effect of particle size on settling behaviour, the particle diameter is varied from 8 mm to 16 mm while the particle density was kept constant at 2000 kg/m^3. All other simulation parameters are identical to those used in section 4.1. Figure 4 shows the evolutions of settling velocities for particles of various sizes. It is clear that when the particles are released, initially they travel at the same velocity (say when t<0.5 s), but their velocities start to deviate as the settling proceeds. This is due to the fact that the initial motion of the particle is primarily governed by the gravity force as the drag force and the buoyant force are relatively smaller. As the drag force develops, the smaller particles start to decelerate earlier and reach a lower terminal velocity more rapidly, when compared to the larger particles. Consequently, the terminal velocities for small particles are lower than those for larger particles.

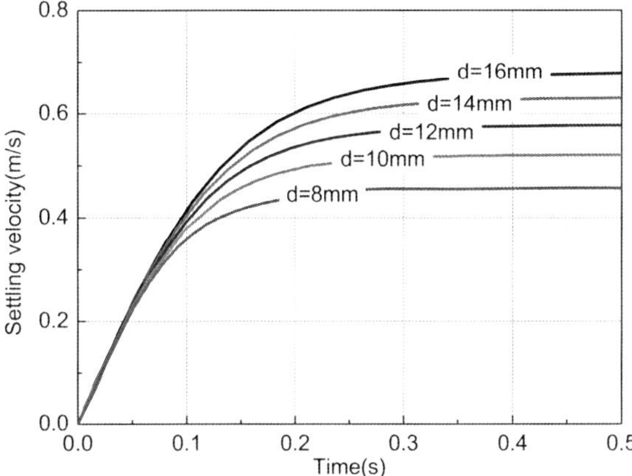

Figure 4 *Evolution of settling velocities for particles of various sizes.*

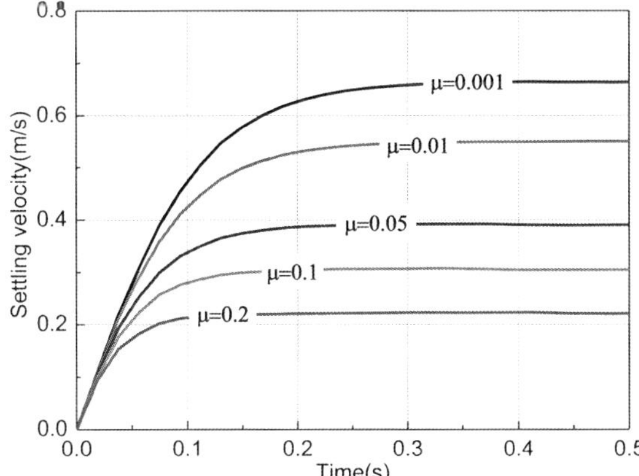

Figure 5 *Evolution of settling velocities of a particle in liquids of various viscosities.*

4.3 The Effect of Liquid Viscosity

In order to examine the influence of liquid viscosity on settling behaviour, liquids with various viscosities but of the density of water are considered. The particle has a diameter of 10 mm and a density of 2000 kg/m^3. The evolutions of settling velocity of the particle in these liquids are shown in Figure 5. It is clear that the higher the viscosity of the liquid, the lower the terminal velocity of the particle. In addition, the time taken to reach the terminal state is reduced as the liquid viscosity increases.

4.4 The Effect of Particle Concentration

In order to explore the influence of particle concentration on settling behaviour, the number of particles is systematically increased from 1 to 1000 so that the concentration of particles is changed. The particles are identical with a diameter of 10 mm and the particle density is kept constant at 2000 kg/m^3. The liquid is assumed to be water with the properties given in Table 2. Figure 6 illustrates the effect of particle concentration on the sedimentation velocity of particles, in which the evolutions of the average settling velocity for various particle concentrations are presented. It can be seen that the average terminal velocity (i.e. the apparent terminal velocity) decreases slightly with the increasing particle number. In suspensions containing many mono-disperse particles, the settling behaviour of individual particles is altered due to the presence of other particles that interact with one another indirectly through the fluid phase. This leads to the fluctuation in the average settling velocity and a decrease in the apparent terminal velocity, as shown in Figure 6.

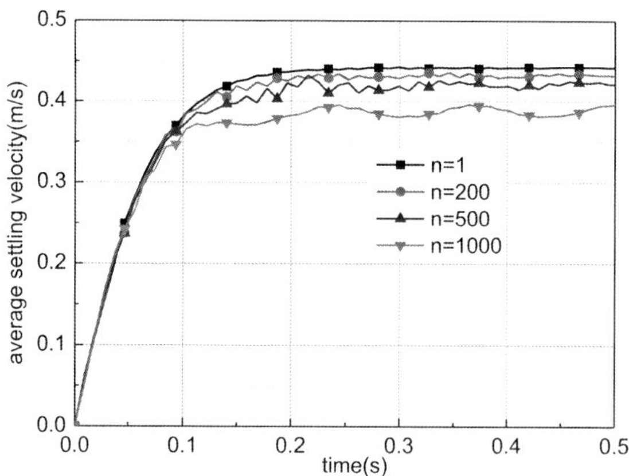

Figure 6 *Effects of particle concentration on the average settling velocity.*

5 DENSITY-DRIVEN SEPARATION

In suspensions with particles of different densities, the particles often sediment differently in the vertical direction thus the well-known density-driven separation phenomenon occurs. A 3D simulation of density-driven separation is also performed using the modified DEM/CFD. A container of a dimension of 100×100×160 mm is fully filled with water and

5000 mono-sized particles of a diameter of 5 mm are randomly distributed in the water, among which 2500 particles have a density of 1200 kg/m^3 (i.e. heavy particles) and the other 2500 particle have a density of 800 kg/m^3 (i.e. buoyant particles). All other parameters are identical to those listed in Table 2. All walls are treated as impermeable with no slip boundaries. All particles are initially stationary. Figure 7a shows the initial packing patterns of the particles in the container. Once the particles are released, the heavy particles move towards the bottom of the container while the buoyant ones move upwards (Figure 7b). Eventually these particles are completely separated, as clearly shown in Figure 7c. Figure 8 shows the evolutions of average vertical positions of two groups of particles. It is shown that the separation rates for the two groups of particles considered, especially with the chosen particle densities, are essentially identical.

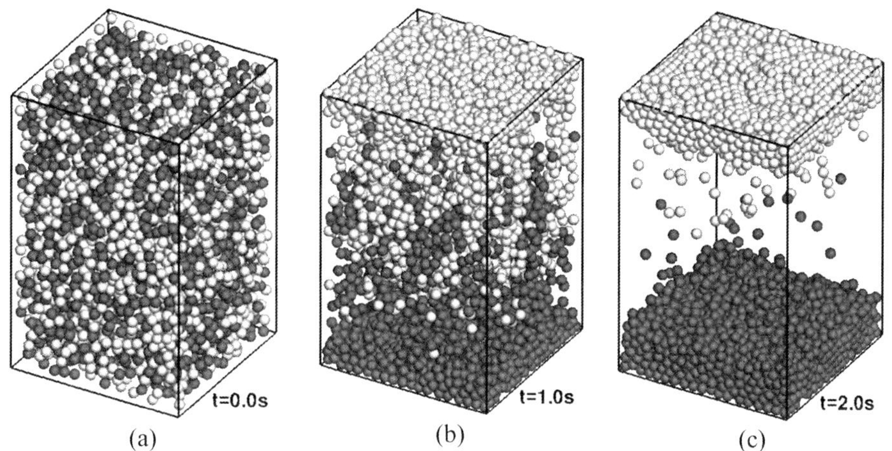

Figure 7 *Snapshots of density driven particle separation in a liquid.*

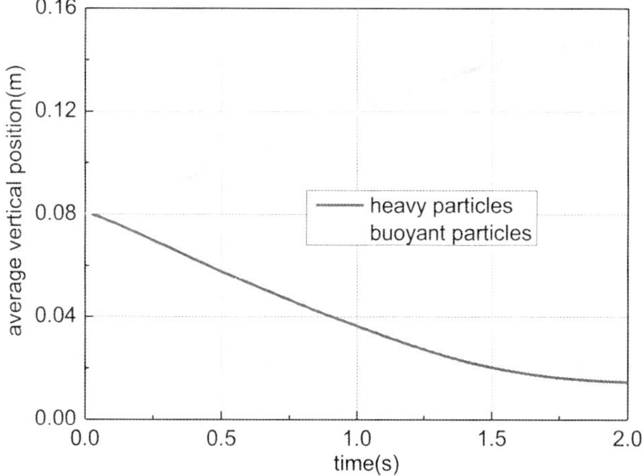

Figure 8 *Evolutions of average vertical positions of two groups of particles.*

6 CONCLUSIONS

A coupled DEM and CFD model was developed for investigating the sedimentation behaviour of particles in weakly compressible liquids. The DEM/CFD model was validated with experimental data reported in the literature. The influences of particle size, particle density, particle concentration, and fluid viscosity on the particle sedimentation behaviour were investigated. It is found that terminal velocity increases as particle size and density increase, and as particle concentration and fluid viscosity decrease. The numerical results also demonstrate the modified DEM/CFD is capable of modelling the complex sedimentation and separation behaviour in liquids in 3D.

Acknowledgements

The first author gratefully acknowledges the financial support from National Science Foundation of China (Grant No. 11172321) and from the Chinese Universities Scientific Fund (Grant No. 2011JS047).

References

1   A.J.C. Ladd, *Journal of Fluid Mechanics*, 1994, **271**, 285.
2   A.J.C. Ladd, *Journal of Fluid Mechanics*, 1994b. **271**, 311.
3   P. J. Hoogerbrugge and J. M. V. A. Koelman, *Europhysics Letter*, 1992, **19**, 155.
4   C.Y. Yang, Y.L. Ding, D. York and W. Broeckx, *Particuology*, 2008, **6**, 38.
5   A.V. Potapov, M.L. Hunt and C.S. Campbell, *Powder Technology*, 2001, **116**, 204.
6   Y. Tsuji, T. Kawaguchi and T. Tanaka, *Powder Technology*, 1993, **77**, 79.
7   K.D. Kafui, C. Thornton and M.J. Adams, *Chemical Engineering Science*, 2002, **57**, 2395.
8   Y. Guo, K.D. Kafui, C.Y. Wu, C. Thornton, J.P.K. Seville, *AIChE J.*, 2009, **55**, 49.
9   T.B. Anderson and R. Jackson, *Ind. Eng. Chem. Fundam.*, 1967, **6**, 527.
10   R. Di Felice, *International Journal on Multiphase Flow*, 1994, **20**, 153.
11   K.L. Johnson, *Contact Mechanics*. Cambridge University Press, Cambridge, 1985.
12   R.D. Mindlin and H. Deresiewicz, *ASME J Appl. Mech.*, 1953, **20**, 327.
13   E.A. Brujan, *J. Non-Newt. Fluid Mech.*, 1999, **84**, 83.
14   T.A. Cate, C.H. Nieuwstad, J.J. Derksen and H.E.A.V. den Akker, *Physics of Fluids*, 2002, **14**, 4012.

DEM SIMULATION OF MIGRATION PHENOMENA IN SLOW, DENSE SLURRY FLOW WITH BROWNIAN MOTION EFFECTS

M.A. Koenders, M. Ibrahim and S.Vahid

Division of Civil, Chemical and Environmental Engineering, University of Surrey, Guildford GU2 7XH. UK

1   INTRODUCTION

Migration phenomena in dense, neutrally buoyant, slow-flowing, non-colloidal particle-fluid mixtures have been widely reported in experiments in the fluid mechanics literature on conduit flow.[1-3] Typically, migration is observed in situations where a shear gradient is present with particles migrating from regions of high shear to regions of low shear. Continuum theories to describe this effect and to identify mechanisms that offer an explanation have also been put forward[5-10] and their implications reviewed[11].

    There is a real theoretical difficulty here. In order for migration to take place a particulate-phase pressure gradient should balance the concentration gradient. A particle pressure can only exist when particle pairs are either able to engage in solid interaction, or interact via a repulsive potential. In 'classical' fluid mechanics for non-colloidal particles neither can take place because of the properties of the lubrication limit. In this limit two smooth bodies with radii of curvature R_1 and R_2 and gap width h in a Newtonian fluid with viscosity η experience a force[12]

$$F_L = 6\pi \frac{\eta}{\left(R_1^{-1} + R_2^{-1}\right)^2} \frac{\dot{h}}{h} \tag{1}$$

For $h \to 0$ at finite interactive velocity $\dot{h}$ this evidently becomes singular. The symmetry in the interaction is broken when it is assumed that the surfaces of the bodies are rough. In that case while the gap is squeezed fluid can escape through the interstices of the roughness features[13]. The interaction (1) is replaced by

$$F_L = 6\pi \frac{\eta}{\left(R_1^{-1} + R_2^{-1}\right)^2} \frac{\dot{h}}{h + h_0} \tag{2}$$

where h_0 is a measure for the typical dimension of the asperities. Denoting averages by an overbar, then generally speaking the regime under consideration is one in which $h_0 << \overline{h} << \overline{R}$.

For colloidal particles the migration pattern is less pronounced. The Péclet number characterises the influence of Brownian motion at temperature T in a shear rate field of strength $\dot{\gamma}$ (k_B is Boltzmann's constant)

$$Pe = \frac{6\pi\eta\dot{\gamma}\overline{R}^3}{k_B T} \tag{3}$$

Figure 1 shows the concentration profile in conduit flow at various Péclet numbers as reported by Frank et al.[14] Clearly, the Brownian motion destroys the migration pattern. In this paper these results will be reproduced by means of a DEM simulation.

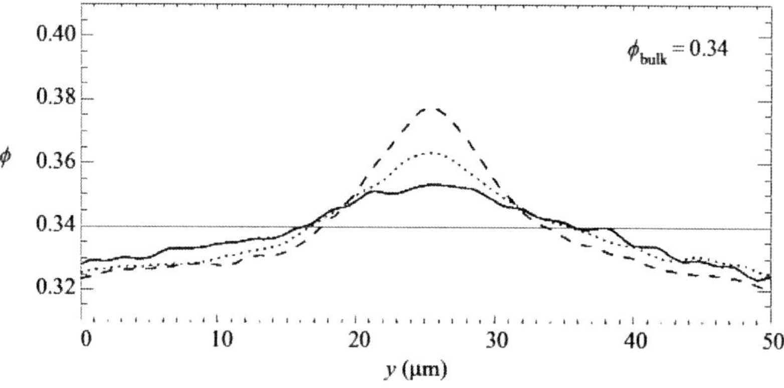

Figure 1 *Solidosity profile of a Brownian suspension as measured by Frank et al.[14] at various Péclet numbers: solid line Pe=69, dotted line Pe=550 and dashed line Pe=4400.*

2   SIMULATION DETAILS

Literature on other methods to simulate slurries include the Stokesian Dynamics method[8] and the Lattice Boltzmann method.[15] In these methods the low-Reynolds number fluid flow field in the spaces between the particles is solved in approximation. Here another approach is taken. The flow field between the particles is not explicitly solved, but a particle pair interaction (as mediated by the fluid) is used. The motivation is that for dense slurries the forces on the particles are dominated by those transmitted by other particles in close proximity, while the drag force that is associated with the bulk fluid motion is negligible. This is obviously only true for dense slurries at low Reynolds numbers.

When the only relevant force on the particles follows from a particle-pair interaction, the Discrete Element Method is a highly suitable procedure. The intention is to simulate flow in a conduit to reproduce the experimental results reported in Section 1. The dominant contribution to the interactive force is given by Equation (2). In order to make the method work, additional measures need to be taken to indicate what happens when two particles collide at finite normal velocity, as is permitted by Equation (2). A collision rheology is invoked, which obeys the rules that the tangential motion is unaffected, while the normal relative velocity is ruled by a coefficient of restitution-type collision law.

In this way a very simple DEM procedure is generated. The boundary conditions are made up by invoking periodic boundaries in the flow direction at the ends of the conduit. The walls are made of static particles. For simplicity the diameters of these are made equal to the average particle diameter in the bulk of the flow, as it is not expected that the phenomena are much affected by the details of the roughness of the boundaries.

The flow is driven by a pressure gradient in the streamwise direction. This is achieved by operating a force on the particles in the bulk. For a straight conduit the force is a constant, no matter where the particle resides.

3   MODELLING BROWNIAN MOTION

In order to introduce the effects of Brownian motion into the simulation a fluctuating force needs to be introduced. This force must be applied in such a way that the properties of Brownian motion are reproduced. For dilute media these properties – the mean quadratic displacement and mean quadratic velocity fluctuation – are well-known.[16] The properties of the fluctuating force are studied for a dilute medium first. The results are then carried over to a dense slurry.

For a dilute medium the equation of motion of a (generic) particle with mass m in a fluid damping environment with damping factor γ is

$$m\frac{d^2q}{dt^2} + \gamma\frac{dq}{dt} = f(t),\tag{4}$$

where $f(t)$ is the fluctuating force due to the collisions with fluid molecules. The well-known Langevin ensemble averaging approach leads to the classic result[17]

$$\frac{1}{2}m\left(\frac{dq}{dt}\right)^2 = \frac{1}{2}kT;\quad \overline{q^2} = \overline{q^2}(0)e^{-\gamma t/m} + \frac{2kT}{\gamma}t\tag{5}$$

The problem has a time constant of m/γ.

In order to obtain the properties of the fluctuating force, Equation (4) is Fourier transformed over a truncated time τ, then multiplied with its complex conjugate and divided by τ, so that the term on the right hand side can be identified as the spectral intensity ($\tau \gg m/\gamma$)

$$\left(\omega^4 m^2 + \omega^2 \gamma^2\right)\frac{\hat{q}_\tau(\omega)\hat{q}_\tau^*(\omega)}{\tau} = \frac{\hat{f}_\tau(\omega)\hat{f}_\tau^*(\omega)}{\tau}\tag{6}$$

The mean kinetic energy is obtained from this via integration over all frequencies ω

$$\overline{\left(\frac{dq}{dt}\right)^2} = \frac{1}{2}kT = \frac{1}{4\pi}m\int_{-\infty}^{\infty}d\omega\frac{1}{\left(\omega^2 m^2 + \gamma^2\right)}\frac{\hat{f}_\tau(\omega)\hat{f}_\tau^*(\omega)}{\tau} \tag{7}$$

The character of the fluctuating force, which is entirely thermal in nature, is assumed to be such that the spectral intensity is 'white'; the integral is then elementary and an expression for the intensity is obtained in terms of the temperature

$$\frac{1}{2}m\frac{\hat{f}_\tau(\omega)\hat{f}_\tau^*(\omega)}{\tau}\left(\frac{1}{2\gamma m}\right) \rightarrow \frac{\hat{f}_\tau(\omega)\hat{f}_\tau^*(\omega)}{\tau} = 2\gamma kT \tag{8}$$

In the (time-stepping) numerical simulation of Brownian motion the force spectrum cannot be directly employed, as there is no information on the exact pulse train that originates in the collisions between fluid molecules and slurry particles. The Brownian motion can only be simulated by applying a series of random pulses. Each pulse has a duration λ and intensity I. For pulses, the force in equation has the form

$$f(t) = \sum_j I_j H_\lambda(t,t_j), \text{ where } H_\lambda(t,t_j) = 1 \text{ for } t_j \leq t \leq t_j + \lambda \text{ and}$$

$$H_\lambda(t,t_j) = 0 \text{ otherwise} \tag{9}$$

Fourier transforming this force leads to the spectrum

$$\hat{f}_T(\omega)\hat{f}_T^*(\omega) = 2\frac{1-\cos(\omega\lambda)}{\omega^2}\sum_j I_j e^{-i\omega t_j}\sum_k I_k e^{i\omega t_k} \tag{10}$$

The double sum is approximated using the assumption that the magnitudes of the pulses are uncorrelated and therefore $\sum_j I_j e^{-i\omega t_j}\sum_k I_k e^{i\omega t_k} \simeq \sum_j I_j^2$.

Analogously to Equation (7) the kinetic energy is obtained as

$$\frac{1}{2}kT = \frac{1}{2}\frac{\sum_j I_j^2}{\tau}\frac{m}{\gamma^2}\left(\frac{m}{\gamma}e^{-\lambda\gamma/m} + \lambda - \frac{m}{\gamma}\right) \tag{11}$$

Now the average quadratic pulse intensity is known and it depends on the pulse duration.

The pulse duration may be chosen equal to the time step of the simulation. Numerical simulation of Equation (4), employing (9) with (11) for the intensities, shows that the Langevin value of the quadratic displacements is achieved for correlation times that are substantially greater than the time constant m/γ (while the time step is of course chosen to be much smaller than this value). Equation (11) may be approximated to give

$$\frac{1}{\tau}\sum_j I_j^2 \rightarrow \frac{2kT\gamma}{\lambda^2} \tag{12}$$

The velocity distribution of the velocities of the molecules is given by the Maxwell-

Boltzmann distribution, that is, the probability $p(v)$ to find a molecule with velocity between v and $v + dv$ is proportional to

$$p(v)\,dv \sim \exp\left(-\frac{mv^2}{2kT}\right)dv \tag{13}$$

The magnitude of the impulses is proportional to the velocity $I = bv$ and therefore the probability to encounter a value I between I and $I + dI$ is

$$p(I)\,dI \sim \exp\left(-\frac{mI^2}{2b^2kT}\right)dI \tag{14}$$

Equating the quadratic mean value of the pulses to Equation (12) yields a value for b

$$b = \sqrt{2m\gamma / \lambda} \tag{15}$$

The treatment for dense slurries is carried out along similar lines and the result is easily generalised to higher dimensions. In the dense case the damping factor is a priori unknown. However Equations (14) and (15) are used to generate pulses and the resulting kinetic energy (as recorded during the simulation) is equated to kT.

4   SIMULATION RESULTS

Simulations of conduit flow have been carried out for assemblies of 2000 particles at various Péclet numbers. The treatment of the solid wall particles is non-trivial. If these particles are not given an equilibrated temperature then there is a temperature jump, which results either in mass migration towards the wall or to the centre. Such a migration is of course unphysical. The problem is resolved by giving the wall particles a fluctuating velocity (but no physical movement) of such intensity that a simulation with no overall flow leads to a more or less constant solidosity profile. This is done by giving these particles the appropriate fluctuating force, while endowing the wall particles with a suitably chosen mass. In the simulations reported here the ratio of the wall particle mass to the bulk particle mass is 34. Thus, wall particles are in fluctuating motion, but not as intensely as the particles in the bulk of the fluid.

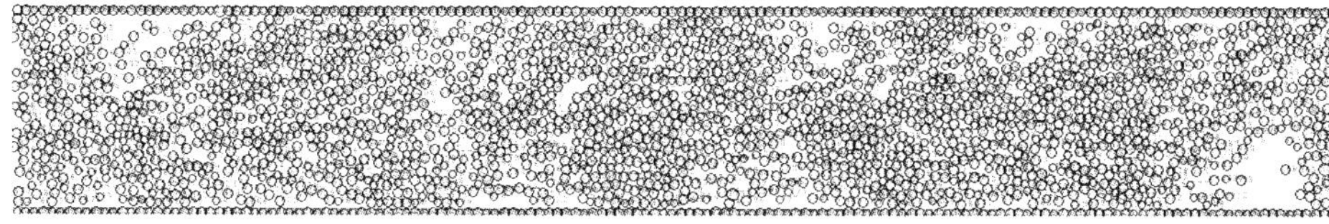

Figure 2 *Simulation after 60,000 time steps at* $Pe \to \infty$

The parameters chosen for the illustrations below are as follows. The flow Reynolds number is 0.78; this is the controlling flow parameter – all other variables are non-dimensionalised. The coefficient of restitution is 0.2; the particle diameter, the fluid viscosity and the particle mass are set to unity; the force on each particle is 0.2. The time step equals 0.001. Data are gathered in strips of some three particles wide along the stream-wise direction and averaged.

At infinite Péclet number (that is, negligible thermal agitation) the migration phenomenon is very clearly visible as illustrated in Figure 2. For this case it takes about 10,000 time steps for the whole assembly to move through the conduit. It is observed that the migration pattern becomes manifest by the creation of mesoscopic regions of low solidosity near the walls.

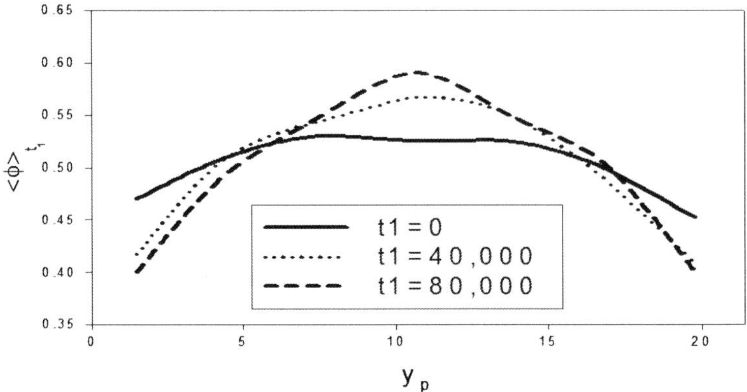

Figure 3 *Development of the migration pattern at $Pe \to \infty$; The time average is taken over 20,000 steps from t_1.*

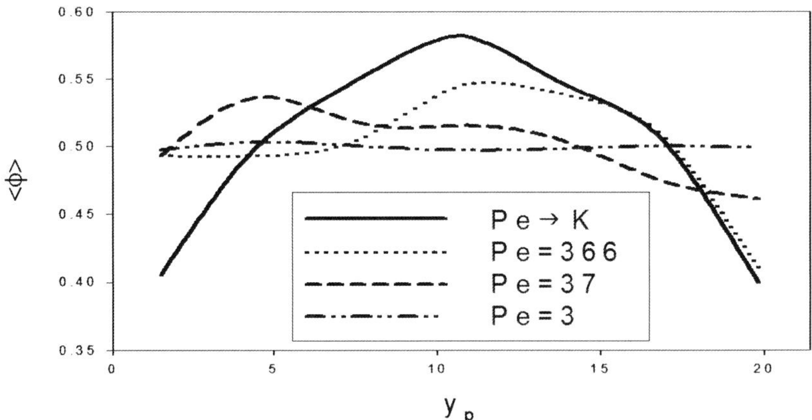

Figure 4 *Average solidosity pattern at various Péclet numbers. The typical dimension of these regions is some five particle diameters.*

The time development of the solidosity pattern is illustrated in Figure 3, where sequences of time averages are shown. The average over 20,000 time step is shown, starting from t_1. Figure 3 is made by using the seven points at which the averages are calculated across the channel and fitting a spline through the points. Each point represents a time average over some 300 particles. Clearly, for the migration pattern to develop fully it needs four or five subsequent passages through the conduit.

Turning now to the effect of Brownian motion in the simulation, the destruction of the migration pattern at higher temperatures (smaller particles, lower shear rates) is illustrated in Figure 4 (the slight asymmetry in the patterns is explained by a remnant discontinuity in a temperature jump at the walls, which leads to a small instability effect). It is observed that the migration effect disappears at smaller Péclet numbers. This is, of course, similar to the effect observed in the experimental literature. These simulations are done in two dimensions (which makes for easy visualisation) and this explains that only a qualitative comparison with the three-dimensional case can be carried out.

4   CONCLUSIONS

It has been shown that the migration phenomena in conduit flow in dense suspensions, as reported in the literature, can be reproduced using a DEM simulation. To achieve this, a fluctuating force that represents the thermal agitation is introduced. This force is calibrated against the properties of dilute suspensions, which is a linear problem. The thermal force is introduced as a pulse at each time step in the DEM simulation.

Acknowledgement

Financial support from the Leverhulme Trust is gratefully acknowledged.

References

1 C.J. Koh and P. Hookham, *J. Fluid Mech.* 1994, **266**, 1.
2 A. Averbakh, A. Shauly, A. Nir and R. Semiat, *Intl J. Multiphase Flow,* 1997, **23**, 409.
3 R.E. Hampton and A.A. Mammoli, *J. Rheol.* 1997, 1997, **41**, 621.
4 M. K. Lyon and L. G. Leal, *J. Fluid Mech.* 1998, **363**, 25.
5 D. Leighton and A. Acrivos, *J. Fluid Mech.* 1987, **181**, 415.
6 R.J. Phillips, R.C. Armstrong, R.A. Brown, A.L. Graham and J.R. Abbott, *Phys. Fluids A*, 1992, **4**, 30.
7 D.F. McTigue and J.T. Jenkins, *Advances in micromechanics of granular materials,* Elsevier, New York, 1992, 381.
8 P.R. Nott and J.F. Brady, *J. Fluid Mech.* 1994, **275**, 157.
9 R.M. Miller and J.F. Morris, *J. Non-Newtonian Fluid Mech.* 2006, **135**, 149.
10 J.J. Stickel and R.L. Powell, *Annual Rev. Fluid Mech.* 2005, **37**, 129.
11 S. Vahid. *Migration and structure formation in sheared slurries*, PhD thesis, Kingston University, UK, 2008.
12 G.K. Batchelor. *An introduction to fluid dynamics*, CUP, Cambridge, 1967.
13 J.T. Jenkins and M.A. Koenders, *Granular Matter*, 2005, **7**, 13.
14 M. Frank, D. Anderson, E.R. Weeks and J.F. Morris, *J. Fluid Mech.* 2003, **493**, 363.
15 A.J.C. Ladd, *J. Fluid Mech.* 1994, **271**, 285.
16 A. Einstein, *Investigations on the Theory of Brownian Movement*, Dover, New York, 1956.
17 C.V. Heer, *Statistical Mechanics, Kinetic Theory and Stochastic Process*, Academic Press, New York, 1972.

FORCE EVALUATION FOR BINGHAM FLUIDS USING MULTIPLE-RELAXATION-TIME LATTICE BOLTZMANN MODEL

S. Chen, Q. Sun and F. Jin

State Key Laboratory of Hydroscience and Engineering, Tsinghua University, China

1 INTRODUCTION

The mechanical behaviours of Bingham fluids are important in both the engineering and physics communities. Force evaluation is a critical variable in studies of fluid-structure interactions. For simple flow fields, analytical methods were provided for the Bingham plastic model. But in some complex cases, especially ones with complicated geometries and boundary conditions, traditional methods, such as finite volume methods[1] and finite element methods,[2] are often of limited use. Therefore, developing an efficient numerical method for Bingham fluids is highly desirable.

The lattice Boltzmann (LB) method has attracted more attention as an alternative approach to traditional methods in recent years. In order to overcome the disadvantage of the single-relaxation-time (BGK) model, in which the numerical calculation was often instable for low viscosity, we employed the multiple-relaxation-time (MRT) LB model for studying dynamic drag force on immersed particles in Bingham fluids. The force calculation methods are briefly introduced firstly, and the validation tests and force on a round particle are presented. The relationship between the drag coefficient and Reynolds number and Bingham number is given finally.

2 FORCE EVALUATION METHOD

Fluid force evaluation on solid particles is crucial to study the behaviours of particle/fluid systems. In LB models, there are two main schemes to calculate the dynamic fluid force: momentum-exchange method and surface stress integration method. In this work, we use the former one, which was firstly proposed by Ladd in the work of evaluating the force on a sphere in suspension flows.[3] As schematically illustrated in Figure 1, solid spherical particles are effectively modelled with a series of stairs (*i.e.* nodes shown in the solid triangles), while the nodes that model the fluid are shown as circles. As the lattice tends smaller, such representation will become more precise. Some of the links between lattice nodes are cut by the boundary surface. Because the halfway bounce-back boundary condition is used, the actual boundary surface in the numerical simulation is located at halfway along the links (as shown in squares). The force on each direction can be obtained according to the momentum exchange between two opposing directions of the neighbouring lattices, as expressed in the below:

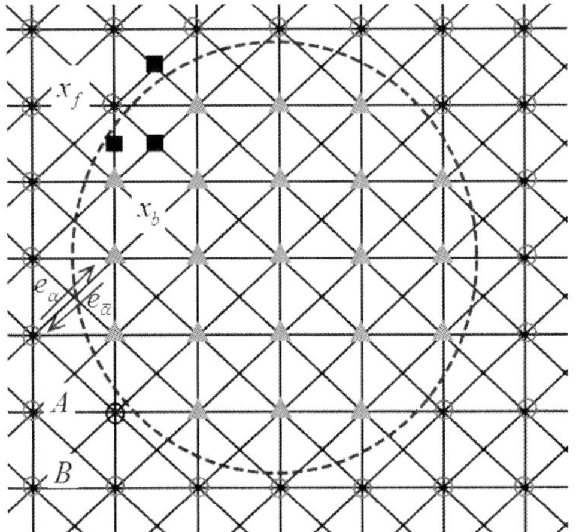

Figure 1 *Layout of the regularly spaced lattices and a spherical particle boundary*

$$F_\alpha = \frac{1}{\delta_t}\left[e_\alpha f_\alpha(x_f) - e_{\bar{\alpha}} f_{\bar{\alpha}}(x_f + e_\alpha \delta_t) \right]$$ (1)

where $e_{\bar{\alpha}}$ is defined as $-e_\alpha$.

In our simulations, two scalar arrays, $flag(i,j)$ and $wall(i,j)$ are introduced in order to implement the momentum exchange method efficiently. A value of 0 is assigned to $flag(i,j)$ for the lattice site (i, j) that are occupied by fluid; a value of 1 is assigned to $flag(i,j)$ for those lattice nodes inside the solid body. The array $wall(i,j)$ is set to be zero, except that the value boundary nodes, whose neighbour nodes in any direction are resided inside the solid body, is equal to 1. For example, in Figure 1 $flag(A) = 0$, $wall(A) = 1$ and when $flag(B) = 0$, $wall(B) = 0$. Over a time step $\delta_t = 1$ the momentum exchange with all possible neighbour solid nodes is given by:

$$F_w = \sum_{\alpha=1}^{8} e_\alpha [f_\alpha^+(x_f) + f_{\bar{\alpha}}(x_f)] flag(x_f + e_\alpha \delta_t)$$ (2)

The total force on the solid particle can be obtained by summing the contribution over all boundary nodes where $wall(i, j)=1$:

$$F = \sum_{wall(i,j)=1} F_w = \sum_{wall(i,j)=1} \sum_{\alpha=1}^{8} e_\alpha [f_\alpha^+(x_f) + f_{\bar{\alpha}}(x_f)] flag(x_f + e_\alpha \delta_t)$$ (3)

where $f_\alpha^+(x_f)$ denotes the density distribution function after collision. The momentum exchange occurs during the subsequent streaming stage when $f_\alpha^+(x_f)$ moves to node x_s in some direction and then bounce back at the halfway of the link. Clearly, the drag force is proportional to the number of the boundary nodes. A dimensionless parameter of drag coefficient C_D is defined to measure the fluid dynamic drag, $C_D = |F_x|/\rho U^2 r$, where U is the average velocity in the channel, ρ is the fluid density and r is the radius of the spherical solid particle.

3 RESULTS AND DISCUSSION

In order to examine the present model, a pressure-driven 2D steady Poiseuille flow is investigated. Based on our knowledge, the domain consisting of 64×128 uniform lattices is fine enough for Bingham fluid flowing in a 2D channel. The pressure boundary conditions are used at inlet and outlet with a constant pressure gradient of $-\nabla P = 5.2 \times 10^{-6}$. The no-slip boundary conditions are applied on both top and bottom boundaries.

The force on the top (or bottom) wall at a given location x is calculated. F_x and F_y are the total shear and normal stresses when δ_x is set to be unit. In a fully developed 2D channel, the analytical expressions are $F_x = (\mathrm{d}p/\mathrm{d}x)H/2$, $F_y = -p$, as listed in Table 1. It can be seen that the error is very small. Therefore, the momentum exchange method is suitable for force evaluation in MRT-LBM model for Bingham fluids proposed in this work.

Table 1 *Comparison of fluid stresses at top and bottom wall in a 2D channel*

y	x/L	F_x_LBM	F_x_analytical	F_y_LBM	F_y_analytical
	0.02	0.0001632	0.000164	-0.33366	-0.33333
	0.17	0.0001632	0.000164	-0.33355	-0.33333
	0.33	0.0001632	0.000164	-0.33345	-0.33333
bottom wall	0.50	0.0001632	0.000164	-0.33334	-0.33333
	0.64	0.0001632	0.000164	-0.33324	-0.33333
	0.80	0.0001631	0.000164	-0.33314	-0.33333
	0.95	0.0001631	0.000164	-0.33303	-0.33333
	0.02	0.0001632	0.000164	0.333656	0.33333
	0.17	0.0001632	0.000164	0.333552	0.33333
	0.33	0.0001632	0.000164	0.333448	0.33333
top wall	0.50	0.0001632	0.000164	0.333344	0.33333
	0.64	0.0001632	0.000164	0.33324	0.33333
	0.80	0.0001631	0.000164	0.333135	0.33333
	0.95	0.0001631	0.000164	0.333031	0.33333

A 2D channel with a spherical solid particle is used to study the force evaluation for the Bingham fluid. The domain consists of 84×440 lattices, and the spherical particle with a radius of 10 is fixed in the centre of the channel as shown in the inset of Figure 2. Higher *Bn* means wider flat plateau, i.e. the thickness of un-yielded zone Y would be larger (see Figure 2), that is to say, the fluid in the core would flow forwards as a rigid body without relative deformation.

In Figure 3, Reynolds number is defined as $Re = \rho U L/\mu_p$, and Bingham number is defined as $Bn = (\tau_0 H)/(\mu_p \bar{U})$, where U is defined as the average velocity of a corresponding Newtonian fluid with viscosity μ_p at the same pressure gradient, and $\bar{U}$ is the average velocity of the Bingham fluid at inlet. An analytical solution of the Bingham fluid is imposed at inlet, while a fully developed velocity boundary condition is used at outlet. No-slip boundary conditions are applied at both top and bottom walls and the solid obstacles. Halfway bounce-back scheme is used for walls. The momentum exchange method presented in §2 is applied. The drag coefficient on the stationary disk was investigated at various Re and Bn values. When $Bn=0$, the fluid is reduced to the Newtonian fluid. The calculated result is shown as solid circles in Figure 3. It can be seen that the correlation between the numerical and experimental results for inertial flows is excellent.[4] However, when $Re<1$ the calculated drag coefficients diverge due to the influence of bounding layer. The same problem has been found in Ref. [5]. For the Bingham fluids ($Bn>0$), the drag coefficient C_D is affected not only by Re, but also by Bn, as shown in Figure 3. The Bn varies from 2.1 to 41.0 and Re from 0.2 to 90.0. The apparent viscosity of the Bingham fluid increases as Bn increases. For the same Re, the drag force is larger at a higher Bn. The wider the rigid body is, the larger drag force will be induced. As far as we know, there are not similar results reported in previous publications.

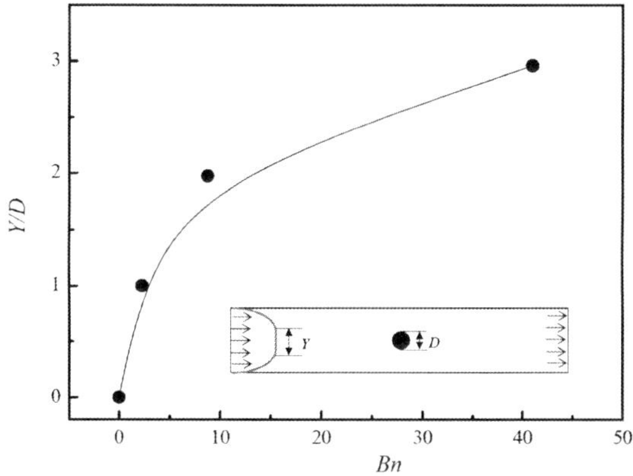

Figure 2 *The ratio of the un-yielded zone thickness Y to particle diameter D versus Bn*

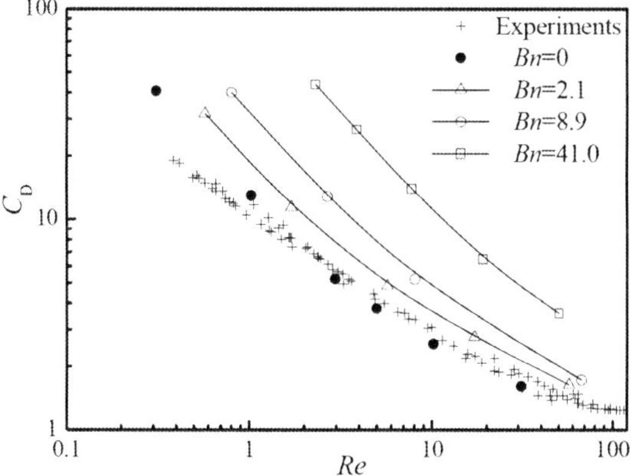

Figure 3 *Drag coefficient C_D for various Re and Bn*

4 OUTLOOK

Our aim is to simulate fresh mortar flows through rock-filled structures using the MTR-LB model, which is of principal significance in hydraulic engineering. The drag coefficient on a stationary cylinder is investigated firstly for a Bingham fluid. It is noted that unlike the Newtonian fluids, the drag coefficient is not only the function of Re, but also Bn, i.e. the fluid force on the cylindrical obstacle grows remarkably as Bn increases. Further work will focus on the fluid-structure interaction for Bingham fluids. The MRT-LB method for three dimensions is being developed.

References

1   P. Neofytou, *Adv. Eng. Software.* 2005, **36**, 664.
2   B. C. Bell and K. S. Surana, *Int. J. Numer. Meth. Engng.* 1994, **37**, 3545.
3   A. J. C. Ladd, *J. Fluid Mech.* 1994, **211**, 285.
4   D. J. Tritton, *J. Fluid Mech.* 1959, **6**, 547.
5   D. R. J. Owen, C. R. Leonardi and Y. T. Feng, *Int. J. Numer. Meth. Engng.* 2011, **87**, 66.

THE EFFECT OF INITIAL BED HEIGHT ON THE BEHAVIOUR OF A SOIL BED DUE TO PIPE LEAKAGE USING THE COUPLED DEM-LBM TECHNIQUE

X. Cui, J. Li, A.H.C. Chan and D. Chapman

School of Civil Engineering, University of Birmingham, Edgbaston, Birmingham, B15 2TT

1 INTRODUCTION

Leakage from underground pipes is a common problem with buried services. Soil particles surrounding the leak could be washed away by the leaking fluid, generating a subsurface cavity. This cavity may develop gradually and suddenly at the end, thus exposing the buried infrastructure to the danger of collapse, and leading to surface subsidence. As it is important to maintain the integrity of the underground environment and to address the associated safety issues, there is a need to understand how soils respond to local leakage.

Compared with field studies and laboratory experiments, numerical simulations provide a more flexible and efficient way in 'visualising' the behaviour of the system in response to a leaking fluid without any field or sample disturbance. As a real challenge is raised attributed to the complicated interactions between the soil and the leaking fluid in the vicinity of the leak, the capability of tracing the fluid-particle interactions at a particle level is desired around the leak. A numerical technique coupling the Discrete Element Method (DEM) and the Lattice Boltzmann Method (LBM)[1,2] is adopted in this study.

Previous study on the pipe leakage model indicates that the coupled DEM-LBM technique is capable of simulating the desired problem in respect of its underlying mechanism and general bed behaviour. This paper presents a parametric study on this problem, by applying the coupled DEM-LBM technique, to explore the effect of initial bed height, i.e. the effective overburden, on the soil response to an underlying leaking pipe.

Therefore, in this paper, a brief introduction to the coupled DEM-LBM technique is firstly given in Section 2. In Section 3, the two-dimensional numerical model used in this study is illustrated, followed by the demonstration and discussion of the simulation results. Finally, conclusions are drawn in Section 4.

2 METHODOLOGY

In the coupled DEM-LBM technique, the soft-sphere DEM approach[3] is used to solve for the particle-particle interactions, while the LBM is employed to model fluid flows at the pore scale. Besides, the Immersed Moving Boundary (IMB) scheme[4] is applied to the LBM framework in order to provide the interface treatment for the fluid-particle interactions. Hence, introductions are given sequentially, in this section, to the DEM approach, the LBM implementation, and the IMB scheme incorporated into the LBM framework.

2.1 Soft-sphere DEM

The coupled DEM-LBM technique is regarded as a promising tool in simulating fluid-particle systems with a large number of moving particles.[5] The DEM analyses the granular soil on a particle level. In DEM, granular material is modelled as an assembly of separate particles. When the soft-sphere approach is used, a slight overlap is allowed between two particles in contact, and the interactions are viewed as a dynamic process in which contact forces accumulate or dissipate over time. Contact forces can be subsequently obtained through the deformation history at the contact. The motion of a single particle is governed by Newton's second law in the form of the following dynamic equations,

$$m\frac{d^2\vec{x}}{dt^2} = \vec{F}_c + \vec{F}_b + \vec{F}_h \tag{1}$$

$$I\frac{d\vec{\omega}}{dt} = \vec{T}_c + \vec{T}_h \tag{2}$$

where $\vec{F}_c$ denotes the total contact force, calculated by summing up the contact forces acting on one particle. In this study, the contact force calculations implemented in the Birmingham DEM code originally developed by Thornton[6,7,8] were directly incorporated into the code developed for this study. $\vec{T}_c$ indicates the total torque generated by the contact force. $\vec{F}_b$ represents the body force, namely, the submerged gravity in this work. $\vec{F}_h$ and $\vec{T}_h$ refer to the force and torque generated by the flowing fluid, respectively. Their values are obtained through the fluid-particle two-way coupling (see §2.3).

With the accelerations computed from Equations (1) and (2) for each particle at each time step, the particle velocities are integrated using a central difference scheme, and the location of each particle is updated for the calculations of next cycle.

2.2 LBM with Large Eddy Simulation (LES)

The LBM has been used as an effective alternative to conventional macroscopic methods for fluid flow simulations.[9] It is a time-stepping procedure, based on microscopic kinetic models. In the classic LBM, the fluid domain is divided into a rectangular in 2D or cubic lattice in 3D with uniform spacing. Fluid is viewed as packets of micro-particles residing on the lattice nodes. During each time step, particles are allowed either to remain in the same node, or to travel to their adjacent nodes with corresponding discrete velocities $\vec{e}_i(i=0,...,n)$. In order to simulate fluid flow at high Reynolds numbers, the Large Eddy Simulation (LES) technique has been implemented with LBM.[10] The governing LBM equation in a turbulence filtered form is given as follows,

$$\tilde{f}_i(\vec{x}+\vec{e}_i\Delta t_{LBM},t+\Delta t_{LBM}) = \tilde{f}_i(\vec{x},t) - \frac{1}{\tau_{total}}[\tilde{f}_i(\vec{x},t)-\tilde{f}_i^{eq}(\vec{x},t)] \quad (i=0,...,n) \tag{3}$$

where $\tilde{f}_i$ are filtered density distribution functions, and $\tilde{f}_i^{eq}$ are a set of filtered equilibrium distribution functions related to the prescribed discrete velocities and the macroscopic variables, including fluid density and velocity. The effect of the unresolved scale, which is principally responsible for energy dissipation through viscous forces, is considered by a relaxation time τ_{total}, taking into account the effect of turbulence viscosity. Full detail of the formulation can be found in Ref. [10].

2.3 Coupling of Solid and Fluid Phases

In order to model the interactions between the fluid and particle phases, the Immersed Moving Boundary (IMB) scheme is adopted in the LBM framework. This scheme treats the particles as moving solid boundaries in the LBM. At each lattice node, the LBM equation is modified using a weighting function depending on the fractional area of a nodal cell covered by a moving solid (see Figure 1). In such a way, the fluid flow influenced by the presence of the moving solid particles can be taken into account. On the other hand, the smooth and accurate hydrodynamic force and torque acting on a moving particle can also be obtained. The detailed implementation of this coupling scheme was illustrated in Ref. [4] and is also given in Ref. [11].

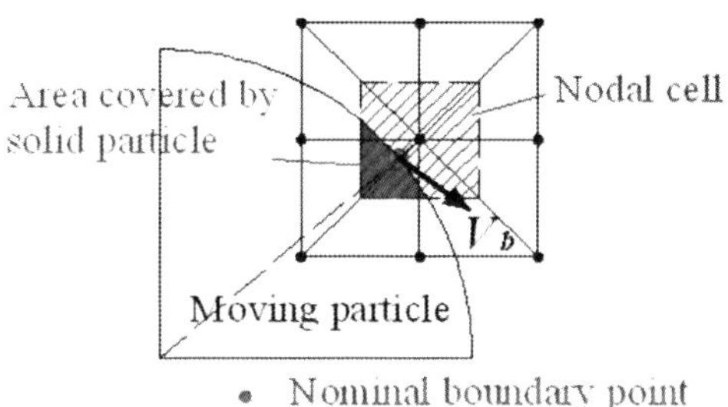

Figure 1 *The nodal cell and the fractional area covered by a moving solid particle*

3 NUMERICAL SIMULATION AND RESULTS

A two-dimensional numerical simulation is carried out using the coupled DEM-LBM technique. A densely-packed bed consisting of different number of circular particles is generated and allowed to settle in a rectangular container (see Figure 2). A particle size distribution ranging from 3.0 mm to 6.0 mm using cumulative beta distribution is employed. In all tests, the soil bed has a fixed width of 600 mm. There is a horizontal fluid pipe underlying the soil bed, and a small orifice with a fixed width of 4.0 mm is opened at the top suface of the pipe simulating the leak. In this study, a same pressure boundary condition is applied both to the pipe inlet (on the left) and outlet (on the right), and fluid flows are driven by the pressure difference between the pipe and the top surface of the soil bed. Initially at the start of the test, the DEM calculation was switched off in order to achieve a well-developed fluid flow in the pipe. After 40,000 time steps, the DEM calculation was then switched on, and the soil bed started to response to the leaking fluid.

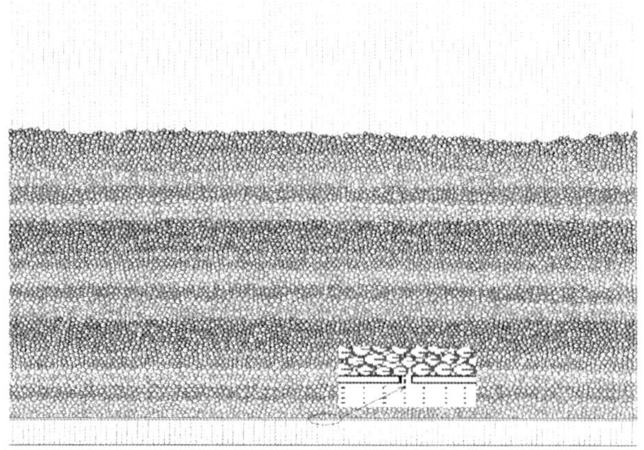

Figure 2 *Numerical setup*

The basic parameters used in the simulations are listed in Table 1.

Table 1 *Basic parameters used in the simulation*

Parameters	Values
Particle density (kg/m^3)	2700
Friction coefficient in the DEM calculation	0.3
Young's modulus (MPa)	69
Poisson's ratio	0.3
DEM time step (s)	2.0×10^{-5}
Fluid density (kg/m^3)	1000
Kinematic viscosity of the fluid (m^2/s)	1.0×10^{-6}
Lattice spacing (m)	1.0×10^{-3}
LBM time step (s)	1.0×10^{-4}
Dimensionless relaxation time	0.5003
Domain size (m × m)	0.6×0.6
Constant relative pressure at the inlet (kPa)	13.33
Constant relative pressure at the outlet (kPa)	13.33

Numerical tests were conducted on a series of initial bed height, and the test arrangement is listed in Table 2.

Table 2 *Test arrangement for various initial bed height*

Test No.	Particle No.	Initial Bed Height (mm)
H1	7238	250
H2	7780	270
H3	8792	300
H4	9348	320
H5	10047	340

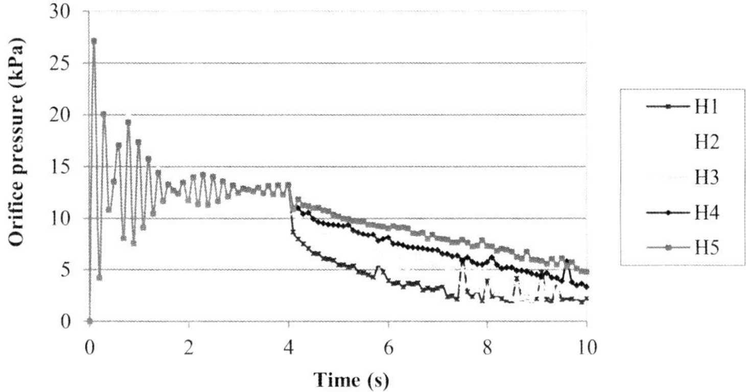

Figure 3 *Time evolution of orifice pressure*

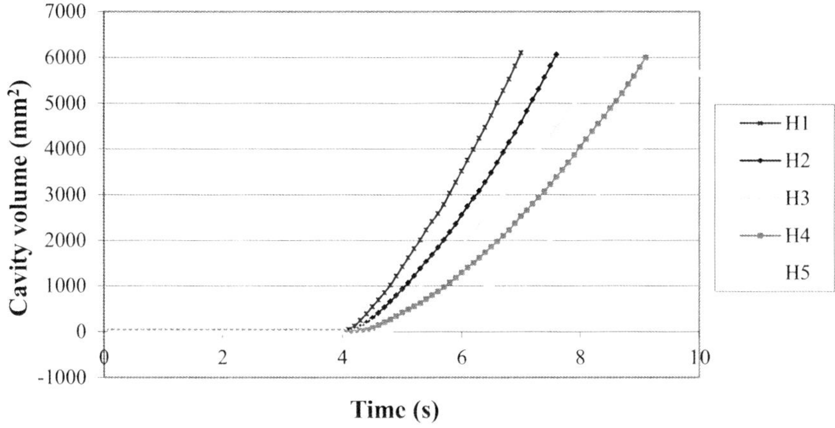

Figure 4 *Time evolution of cavity volume*

Simulation results show that, in all cases, the soil bed undergoes fluidisation due to the leaking fluid. The fluidisation phenomenon is associated with a cavity formed and developing outwards from the leaking area. Time evolution of the orifice pressure and the cavity size are also recorded in Figures 3-5. When fluidisation occurs, a wedge-shaped uplift zone is observed from the results (see Figure 6).

From Figure 3, it can be seen that at the first 4 s when the DEM calculation was switched off, the orifice pressure was oscillating around 13.3 kPa regardless of different initial bed height. The oscillating phenomenon is attributed to the pressure wave movement in the LBM model used in this study. It can be argued that the time-averaged orifice pressure during the first 4 s is related only to the constant pressure values applied to the pipe inlet and outlet. Immediately after the DEM calculation was switched on, the orifice pressure started dropping with time. This suggests that the cavity forms at the orifice, and the 'unblocking' of the surrounding particles subsequently leads to the reduction in pressure. It also suggests that the initial orifice pressure of 13.3 kPa is sufficiently large to initiate the fluidisation by pushing out the wedge-shaped uplift zone. The cavity keeps growing in size and more particles are 'unblocking' to dissipate the injecting energy. Therefore, the orifice pressure continues to decrease. For a deeper bed, the orifice pressure decreases slower, and correspondingly, the cavity develops slower as well (see Figure 4). It is easily understood that for a deeper bed, fluidisation is more difficult due to larger overburden weight. Hence, less energy is available to further 'unblock' the surrounding particles.

It is worth noting that, from Figure 5, the cavity volume has a linear relationship with the initial bed height during the test. And it can be argued that this finding is regarded valuable in predicting the cavity size for a given initial bed height at a particular moment.

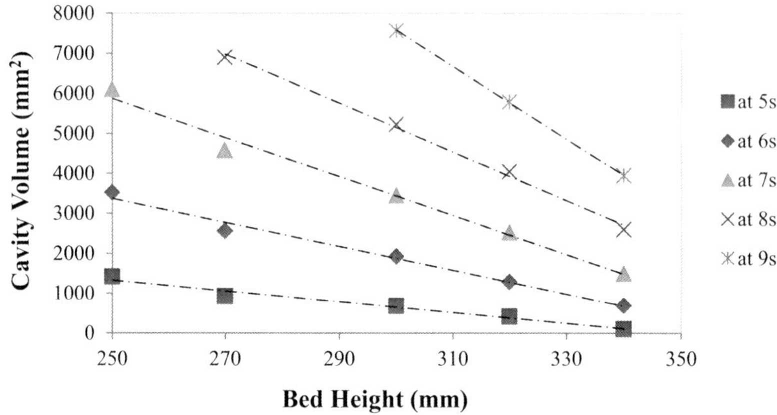

Figure 5 *Time evolution of cavity size with different initial bed height*

In addition, as mentioned above, a wedge-shaped uplift zone is observed when fluidisation occurs. By capturing the snapshots of particle displacement at the onset of fluidisation, the angle of the wedge can be measured. Figure 6 suggests that although different initial bed height is used, the angle of the wedge is kept at a constant value of 62°. This value is comparable with the experimental finding obtained by Alsaydalani,[12] and this can be interpreted as a macroscopic friction angle ϕ' of 34° if the wedge angle is taken as $45^{\circ}+\phi'/2$, which is a reasonable value for a cohesionless material.

4 CONCLUSIONS

In the two-dimensional study presented in this paper, the effect of initial bed height on the soil behaviour in response to a leaking pipe is investigated using the coupled DEM-LBM technique. Simulation results indicate that, with sufficiently large orifice pressure, the soil bed undergoes fluidisation which is characterised by a wedge-shaped uplift zone and a cavity formed at the leaking area. The initial bed height influences neither the initial orifice pressure, nor the angle of the wedge at the onset of fluidisation which should be a property of the soil. However, it can be identified that a soil bed with a deeper initial height leads to a slower decrease in the orifice pressure and a slower developing rate of the cavity. It is also worth noting, from the numerical simulations, that the cavity size at any particular moment relates linearly to the initial bed height. This may be of special value in predicting cavity size at a particular moment with the initial bed height specified. Further validations on this can be carried out against experimental results in the future.

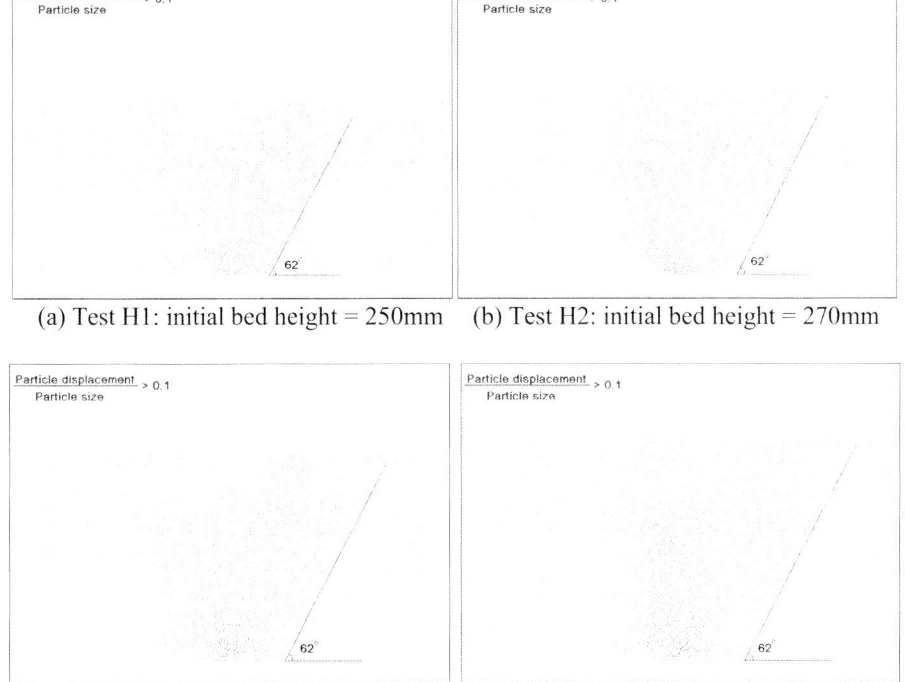

(a) Test H1: initial bed height = 250mm   (b) Test H2: initial bed height = 270mm

(c) Test H3: initial bed height = 300mm   (d) Test H4: initial bed height = 320mm

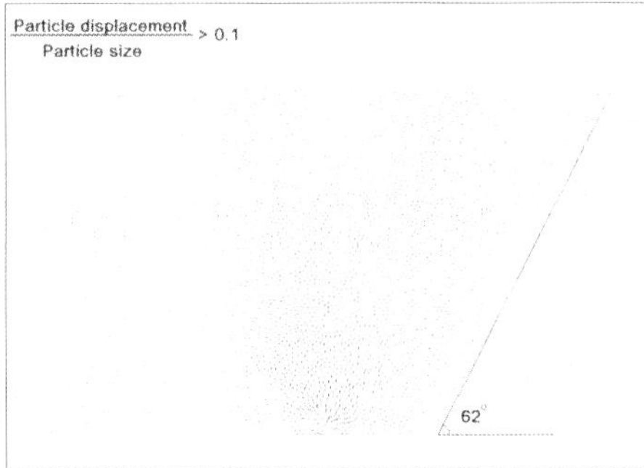

(e) Test H5: initial bed height = 340mm

Figure 6 *Particle displacement and wedge angle plots at the onset of fluidisation*

Acknowledgements

The first author is very grateful for financial support by the Overseas Research Students Awards and Postgraduate Teaching Assistant Scholarship from the University of Birmingham. The computations described in this paper were performed using the

University of Birmingham's BlueBEAR HPC service, which was purchased through HEFCE SRIF-3 funds. See http://www.bear.bham.ac.uk for more details.

References

1 B.K. Cook, D.R. Noble and J.R. Williams, *Engineering Computations*, 2004, **21**, 151.
2 Y.T. Feng, K. Han and D.R.J. Owen, *International Journal for Numerical Methods in Engineering*, 2007, **72**, 1111.
3 N.G. Deen, M. Van Sint Annaland, M.A. Van der Hoef and J.A.M. Kuipers, *Chemical Engineering Science*, 2007, **62**, 28.
4 D. Noble and J. Torczynski, *Int. J. Mod. Phys. C*, 1998, **9**, 1189.
5 Y.T. Feng, K. Han and D.R.J. Owen, *International Journal for Numerical Methods in Engineering*, 2010, **81**, 229.
6 C. Thornton, *Journal of Physics D (Applied Physics)*, 1991, **24**, 1942.
7 C. Thornton and K.K. Yin, *Powder Technology*, 1991, **65**, 153.
8 C. Thornton and Z. Ning, *Powder Technology*, 1998, **99**, 154.
9 S. Chen, D. Martínez and R. Mei, *Phys. Fluids*, 1996, **8**, 2527.
10 K. Han, Y.T. Feng and D.R.J. Owen, *Computers and Structures*, 2007, **85**, 1080.
11 X. Cui, J. Li, A. Chan and D. Chapman, *Particuology*, 2012, **10**, 242.
12 M.O.A. Alsaydalani, *Internal Fluidisation of Granular Material*, unpublished doctoral dissertation, School of Civil Engineering and Environment, University of Southampton, 2010.

GRANULAR FLOWS IN FLUID

K. Kumar[1], K. Soga[1] and J.-Y. Delenne[2]

[1]Department of Engineering, University of Cambridge, Cambridge, CB2 1PZ, UK
[2]LMGC UMR 5508, University of Montpellier 2, Pl E. Bataillon, Montpellier, 34095
Cedex 5, France

1   INTRODUCTION

Avalanches, landslides, and debris flows are geophysical hazards, which involve rapid
mass movement of granular solids, water, and air. Globally, landslides cause billions of
pounds in damage, and thousands of deaths and injuries each year. Hence, it is important to
understand the triggering mechanism and the evolution of flow. The momentum transfer
between the discrete and continuous phases significantly affects the dynamics of the flow
as a whole.[1] Although certain macroscopic models are able to capture simple mechanical
behaviours,[2] the complex physical mechanisms occurring at the grain scale, such as
hydrodynamic instabilities, formation of clusters, collapse, and transport,[1] have largely
been ignored. In particular, when the solid phase reaches a high volume fraction, the strong
heterogeneity arising from the contact forces between the grains, and the hydrodynamic
forces, are difficult to integrate into the homogenization process involving global
averages.[1] In order to describe the mechanism of immersed granular flows, it is important
to consider both the dynamics of the solid phase and the role of the ambient fluid.[3] The
dynamics of the solid phase alone are insufficient to describe the mechanism of granular
flow in a fluid; it is important to consider the effect of hydrodynamic forces that reduce the
weight of the solids inducing a transition from dense-compacted to dense-suspended flows,
and the drag interactions which counteract the movement of the solids.[4] Transient regimes
characterized by change in solid fraction, dilation at the onset of flow and development of
excess pore pressure, result in altering the balance between the stress carried by the fluid
and that carried by the grains, thereby changing the overall behaviour of the flow.[3] In the
present study, 2D Lattice-Boltzmann and Discrete Element Method is adopted to capture
the fluid-soil interactions in underwater avalanches.

2   LBM-DEM Formulation

The Lattice Boltzmann equation Method (LBM) is an alternative approach to the classical
Navier-Stokes solvers for fluid flow and works on an equidistant grid of cells, called lattice
cells, which interact only with their direct neighbors.[5] The fluid domain is divided into a
rectangular grid or lattice, with the same spacing 'h' in both the x- and the y-directions, as
shown in Figure 1. The present study focuses on 2-D problems; hence, the D2Q9

momentum discretization is adopted, where the fluid particles at each node are allowed to move to their eight intermediate neighbors with eight different velocities $\mathbf{e}_i$ ($i = 1, \ldots, 8$).[6]

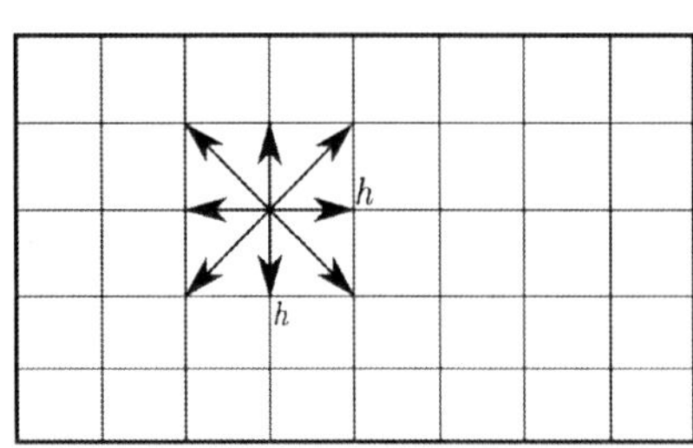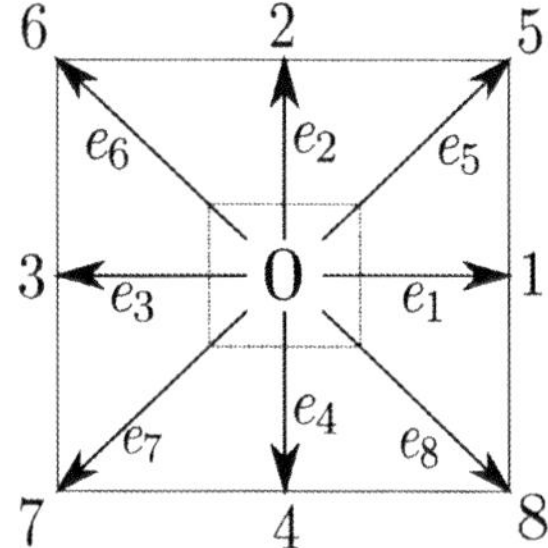

Figure 1 *The Lattice Boltzmann discretization and D2Q9 scheme: (a) a standard LB lattice; (b) D2Q9 model*

The nine discrete velocity vectors are defined as:

$$\begin{cases} e_o = (0,0) \\ e_1 = C(1,0); \ e_2 = C(0,1); \ e_3 = C(-1,0); \ e_4 = C(0,-1); \\ e_5 = C(1,1); \ e_6 = C(-1,1); \ e_7 = C(-1,-1); \ e_8 = C(1,-1); \end{cases} \tag{1}$$

in which C is the lattice speed ($=h/\Delta t$). Where, Δt is the discrete time step. The primary variables in the Lattice Boltzmann formulation are called the fluid density distribution functions, f_i, each relating the portable amount of fluid particles moving with the velocity e_i along the ith direction at each node. The macroscopic variables can be obtained from the particle distribution functions, according to:

$$\rho = \sum_{i=0}^{\beta-1} f_i \quad \text{(macroscopic fluid density) and } \mathbf{u} = \frac{1}{\rho}\sum_{i=0}^{\beta-1} f_i e_i \quad \text{(macroscopic velocity)} \tag{2}$$

where, $i \in [0, \beta-1]$ is an index spanning the discretized momentum space. There are nine fluid density distribution functions, f_i ($i = 0, \ldots, 8$), associated with each node in the D2Q9 model. This simple equation allows us to pass from the discrete microscopic velocities that comprise the LBM back to a continuum of macroscopic velocities representing the fluid's motion. The BGK (Bhatnagar-Gross-Krook) approximation is used to describe the streaming and the collision of the particles. Streaming and collision (i.e., relaxation to local equilibrium) is defined as:

$$f_i(\mathbf{x}+\mathbf{e}_i\Delta t, t+\Delta t) = f_i(\mathbf{x},t) - \frac{1}{\tau}[f_i(\mathbf{x},t) - f_i^{eq}(\mathbf{x},t)] \quad (i = 0,\ldots,8) \tag{3}$$

where, $f_i(\mathbf{x}+\mathbf{e}_i\Delta t, t+\Delta t) = f_i(\mathbf{x},t)$ describes the streaming part and the collision part, which brings the system to local equilibrium is described by $\frac{1}{\tau}[f_i(\mathbf{x},t) - f_i^{eq}(\mathbf{x},t)]$. τ is a non-dimensional relaxation time parameter, which is related to the fluid viscosity.

2.1 LBM-DEM Coupling

Lattice Boltzmann approach can accommodate large grain sizes and the interaction between the fluid and the moving grains can be modelled through relatively simple fluid – grain interface treatments. Further, employing the Discrete Element Method (DEM) to account for the grain – grain interaction naturally leads to a combined LB – DEM procedure.[7] The Eulerian nature of the LBM formulation, together with the common explicit time step scheme of both LBM and DEM makes this coupling strategy an efficient numerical procedure for the simulation of grain – fluid systems. Such a coupled methodology is used in simulating grain – fluid systems dominated by grain – fluid and grain – grain interactions. To capture the actual physical behavior of the fluid – grain system, it is essential to model the boundary condition between the fluid and the grain as a non-slip boundary condition, i.e. the fluid near the grain should have similar velocity as the grain boundary. The solid grains inside the fluid are represented by lattice nodes. The discrete nature of lattice, results in a stepwise representation of the surfaces, which are circular, hence sufficiently small lattice spacing is adopted.

2.2 Permeability

In DEM, the grain – grain interaction is described based on the contact interactions. In a 3D granular assembly, the pore spaces between grains are interconnected, whereas in 2-D assembly, the grains are in contact with each other that result in a non-interconnected pore-fluid space. This results in a no flow condition in a 2-D case. In order to overcome this difficulty, a reduction in radius is assumed only during LBM computations (fluid and fluid – solid interaction). The reduction in radius allows interconnected pore space through which the surrounding fluid can flow (reduced R=0.7r to 0.95r, 'r' is grain radius). The reduction in radius is assumed only during LBM computations, hence this technique has no effect on the grain – grain interactions computed using DEM. Different permeability can be obtained, for any given initial packing, by varying the reduction in the radius for the grains, without changing the actual granular packing. Cumulative β method is adopted to generate a randomly packed granular assembly with a polydispersity of 1.8 (see Figure 2).[8] The size of the grains varies between 1.25mm to 2.2mm.

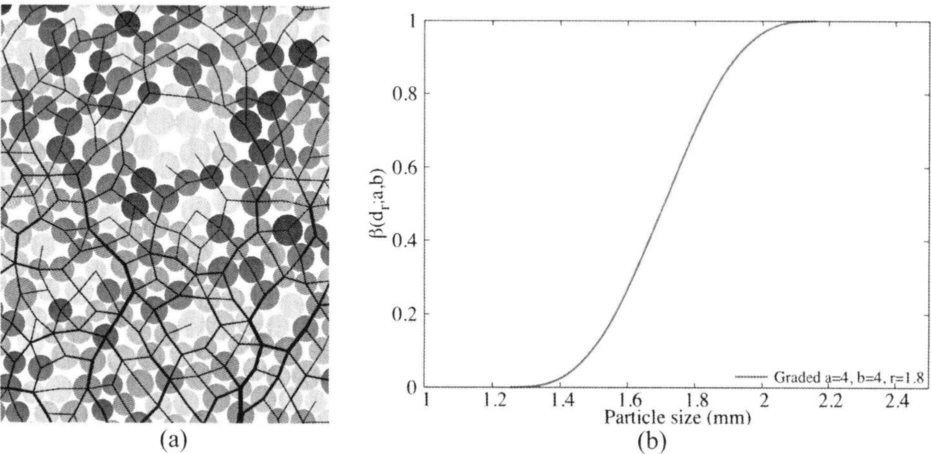

Figure 2 *(a) Randomly packed sample generated by cumulative β method; (b) PSD Curve*

A square sample of 50 mm x 50 mm is used to determine the transverse permeability. Dirichlet boundary condition,[9] i.e. pressure/density constrain is applied at the left and the right boundaries. A small increment (10^{-4} ρ_0) in density is applied at the left hand side boundary and a constant density is maintained at the right hand side boundary. This results in a gradient of pressure causing the fluid in the domain to flow. The mean velocity of flow (v) is determined and the permeability of the sample (k) is computed as:

$$k = v \cdot \mu \cdot \frac{\Delta x}{\Delta P} \tag{4}$$

where μ is the dynamic viscosity of the fluid (Pa.s), Δx is the thickness of the bed of porous medium (m), and ΔP is the applied pressure difference (Pa). In the present study, the radius is varied from 0.7 to 0.95 to obtain a wide range of permeability for the sample. Increase in the size of the grain from 0.7 to 0.95 reduces the porosity from 0.60 to 0.27. The permeability computed from LB – DEM method is verified by comparing it with an analytical solution. One of the widely used analytical solutions for permeability is the Carman – Kozeny equation (i.e. the CK Model), which is based on Poiseuille flow through pipe and is mainly used for 3D, homogenous, isotropic, granular porous media at moderate porosities. In the present study, a modified Carman – Kozeny equation that takes into account the microstructure of the fibers and is valid in a wide range of porosities is adopted.[10] The normalized permeability is defined as:

$$\frac{k}{d^2} = \frac{\varepsilon^3}{\psi_{CK}(1-\varepsilon)^2} \tag{5}$$

In the CK model, the hydraulic diameter D_h, is expressed as a function of measurable quantities porosity and specific surface area.

$$D_h = \frac{4\varepsilon V}{S_v} = \frac{\varepsilon d}{(1-\varepsilon)}; \text{with a}_v = \frac{\text{particle surface}}{\text{particle volume}} = \frac{S_v}{(1-\varepsilon)V} = \frac{4}{d} \tag{6}$$

With the total wetted surface, S_v, and the specific surface area, a_v. The above value of a_v is for circles (cylinders) – for spheres $a_v = 6/d$. ψ_{CK} is the empirically measured CK factor, which represents both the shape factor and the deviation of flow direction from that in a duct. It is approximated for randomly packed beds of spherical grains. The variation in the flow rate for different reduction in radius is presented in Figure 3a. The normalized permeability for different porosity obtained by varying the radius from 0.7 to 0.95 is presented in Figure 3b.

It can be observed from the figure that the permeability decreases drastically as the radius is varied from 0.7r to 0.95r. The granular assembly is almost impermeable for a radius of 0.95r. The normalized permeability is found to match the qualitative trend of the Carman-Kozeny equations. The LB – DEM permeability curve lies between the permeability curves for spherical and cylindrical grain arrangements implying a better approximation of permeability in 2D granular assembly by reducing the radius during LBM computations.

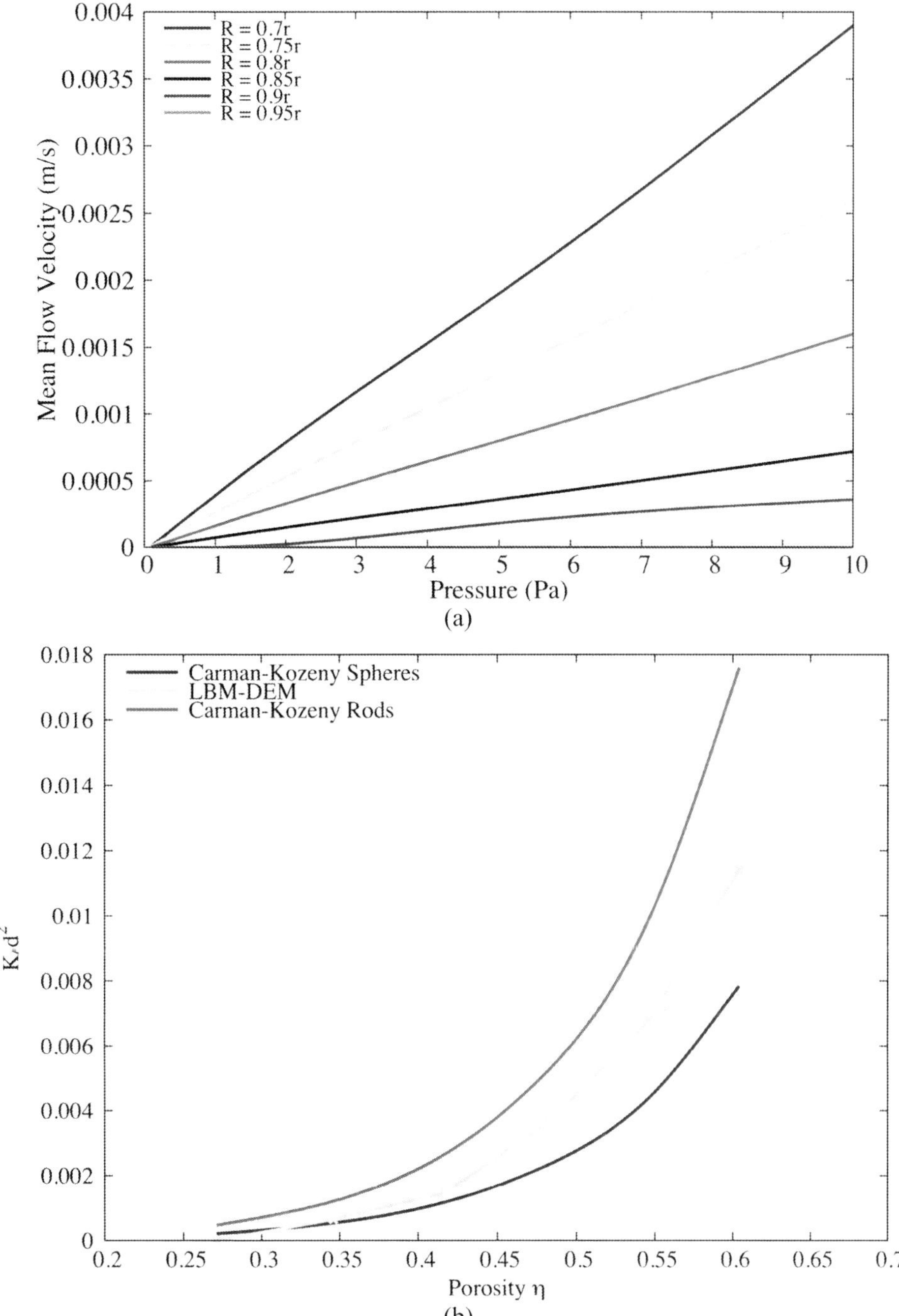

Figure 3 *(a) Variation of flow rate for different reduction in radius, (b) Permeability vs. porosity*

3   GRANULAR AVALANCHES IN FLUID

In this section, we study the behavior of immersed granular avalanches for different permeability. We consider 2D poly-disperse system (d_{max}/d_{min}=1.8) of circular discs in fluid. The simulations were carried out with 1000 grains of density 2650 kg/m^3 and a contact friction angle of 26°. The collapse of the column was simulated inside a fluid with a density of 1000 kg/m^3 and a kinematic viscosity of 1×10^{-6} m^2/s. The choice of a 2D geometry has the advantage of cheaper computational effort than a 3D case, making it feasible to simulate very large systems with an important number of nodes for a reasonable computing time. A granular column of aspect ratio 'a' of 0.8 was considered. Radius of the grains was varied from 0.7r to 0.95r during LBM computations. Dry and buoyant analyses were done to compare the effect of hydrodynamic forces on the run-out distance.

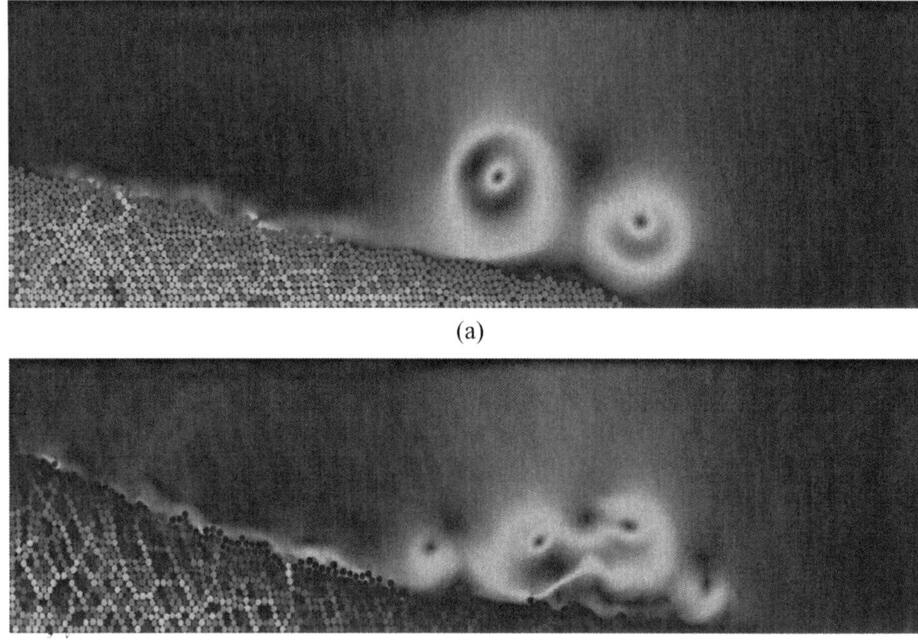

(a)

(b)

Figure 4 *Final run-out profile for reduction in radius of (a) 0.7r and (b) 0.95r*

The final run-out profile for a reduced grain size of 0.7r and 0.95r are presented in Figure 4. The less permeable column collapses further, while the highly permeable column entraps more water and has smaller run-out distance. The rheology of the flow tends to change, with change in permeability. Figure 5a presents the evolution of normalized run-out distance for different permeability conditions. Normalized time is defined as $\sqrt{H/g}$,[11] where, H is the height of the granular column and g is acceleration due to gravity. It can be observed from the figure that the dry and buoyant columns run farther and evolve quicker than the immersed case. The run-out is found to increase with decrease in permeability. The time required for the flow to initialize increases with decrease in permeability, i.e. a highly permeable column collapse quicker in comparison to a less permeable granular column. Three distinct regimes can be observed in the immersed avalanche, similar to the observations in dry granular column collapse:[11] (1) a vertical-fall regime, where the grains located at the top undergo vertical falls and the grains at the bottom are ejected horizontally by the fluid; (2) a heap regime, where the grains at the top move along an inclined stationary deposit; (3) a horizontal regime, where the grains move essentially horizontally. In this last regime, fluid re-circulations were observed leading to the formation of vortex flows, which moves towards the surface (see Figure 4). Figure 5b depicts the evolution of kinetic energy for low (R=0.95r and η=0.25) and high (R=0.7r and η=0.6) permeable granular columns. The collapse of granular pile in fluid takes longer to initialize in comparison with dry and buoyant collapse. Similar to the dry granular column collapse, the run-out distance for collapse in fluid is proportional to $\sqrt{KE_{max}}$. Where, KE_{max} is the maximum kinetic energy observed during the flow.

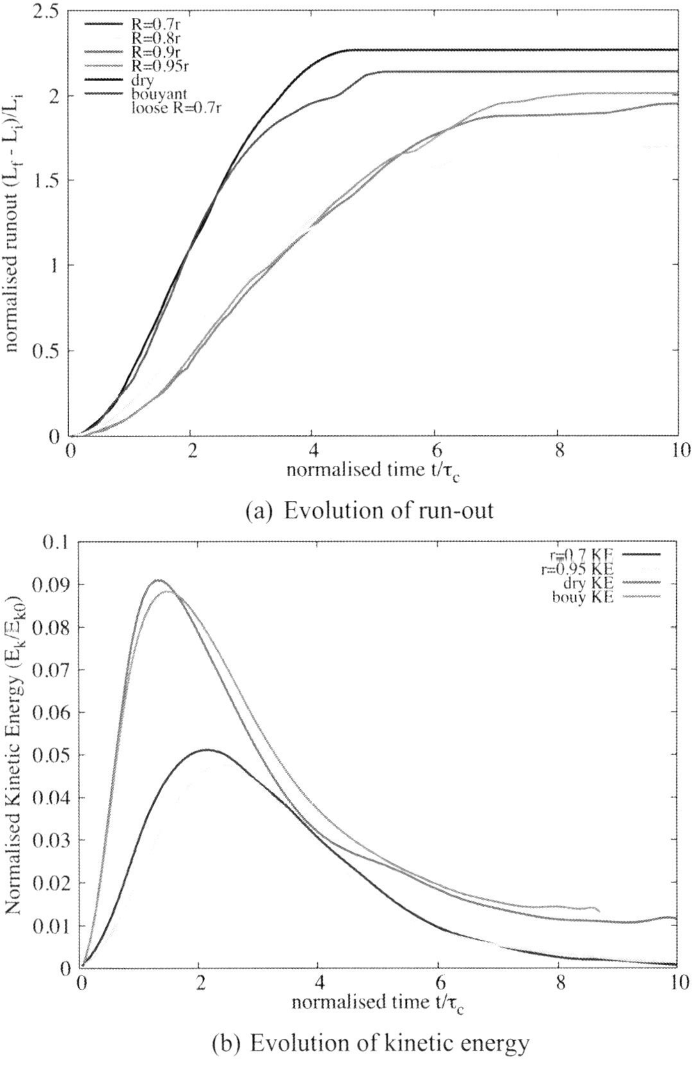

Figure 5 *Effect of permeability on the flow dynamics*

4   SUMMARY

In this work, we have studied the behavior of immersed granular avalanches using LB – DEM approach. In 2D immersed column collapse simulation, the radius of the grains is reduced during LBM computations to allow fluid to flow through the pore-space, which is otherwise not interconnected. By reducing the radius of grains during LBM computations, different permeability can be achieved for the same initial packing. Immersed granular avalanches are significantly affected by hydrodynamic forces. Development of a negative pore pressure postpones the onset of avalanche. The run-out distance is found to be significantly affected by the permeability, the run-out distance is found to increase with decrease in permeability. Further research will be carried out to study the effect of relative density on the evolution of immersed granular avalanches.

Acknowledgement

The authors would like to thank F. Radjai, LMGC, Montpellier, France for stimulating discussions regarding this work.

References

1. V. Topin, F. Dubois, Y. Monerie, F. Perales and Wachs, *J. Non-Newtonian Fluid Mechanics,* 2011, **166**, 63.
2. S. M. Peker and S. S. Helvacı, in *Solid-liquid two phase flow*, Elsevier, 2007.
3. R. P. Denlinger and R. M. Iverson, *J. Geophys. Res*, 2011, **106**, 553.
4. C. Meruane, A. Tamburrino and O. Roche, *J. Fluid Mechanics*, 2010, **648**, 381.
5. S. Chen and G. D. Doolen, *Annual review of fluid mechanics*, 1998, **30**, 329.
6. K. Han, Y. Feng and D. Owen, *Computers & structures*, 2007, **85**, 1080.
7. M. Mansouri, J.Y. Delenne, M. S. El Youssoufi and A. Seridi, *Comptes Rendus Mécanique*, 2009, **337**, 675.
8. C. Voivret, F. Radjai, J. Y. Delenne and M. S. El Youssoufi, *Physical Review E*, 2007, **76**, 021301.
9. X. He, Q. Zou, L. S. Luo and M. Dembo, *J. Statistical Physics*, 1997, **87**, 115.
10. K. Yazdchi, S. Srivastava and S. Luding, *Intl. J. Multiphase Flow*, 2011, **37**, 956.
11. L. Staron, and E. J. Hinch, *J. Fluid Mechanics*, 2005, **545**, 1.

Cohesive Systems

A STUDY OF THE INFLUENCE OF SURFACE ENERGY ON THE MECHANICAL PROPERTIES OF LUNAR SOIL USING DEM

C. Modenese[1], S. Utili[2] and G.T. Houlsby[1]

[1]Department of Engineering Science, University of Oxford, Parks Road, OX1 3PJ, UK
[2]School of Engineering, University of Warwick, Coventry, CV4 7AL, UK

1   INTRODUCTION

A deep understanding of the geotechnical behaviour of Lunar Soil is still lacking in the geotechnical literature. However, the knowledge of this is crucial to guarantee the success of any future operation on the Moon, be it manned or robotic. Although decades have passed from the last human mission, the Moon still represents a very attractive object, as confirmed by the current Chinese Lunar Exploration Program (CLEP). Its attractiveness is built not only on the potential to provide us with resources otherwise rare on Earth, but also because a permanent outpost on the Moon would offer the opportunity to explore the Solar system further towards more terrestrial planets, like Mars.

Whether the range of problems that lunar scientists will soon be facing will include the design of a foundation, the stability of the slope or the interaction between a rigid wheel and the soil, the knowledge of Lunar Soil shear strength is required. Although some estimates of Lunar Soil strength properties have been provided, thanks to direct observations and in-situ experiments,[1] little is known about the unusual Lunar Soil cohesion and its origin. Lunar Soil cohesion plays a significant role on the behaviour of the soil on the Moon, as Figure 1 illustrates. The origin of Lunar Soil cohesion was attributed in the past to the presence of high surface energy forces due to the characteristic features of the Lunar environment.[2-4] However, no conclusion has yet been reached on what causes Lunar Soil cohesion, calling for further research.

The work presented in this paper aims to shed light on this unresolved issue by exploring the relationship between surface energy and the macroscopic cohesion of the soil. The problem is tackled by the Discrete Element Method (DEM), which is considered a powerful tool in the modelling of granular materials. To this end, the JKR contact model[5] is implemented into the open source DEM code Yade[6] to model the effect of surface energy at the inter-particle level. Realistic properties of Lunar Soil materials are assigned to the grains at the microscopic level. Macroscopic observations will be compared with and without surface energy.

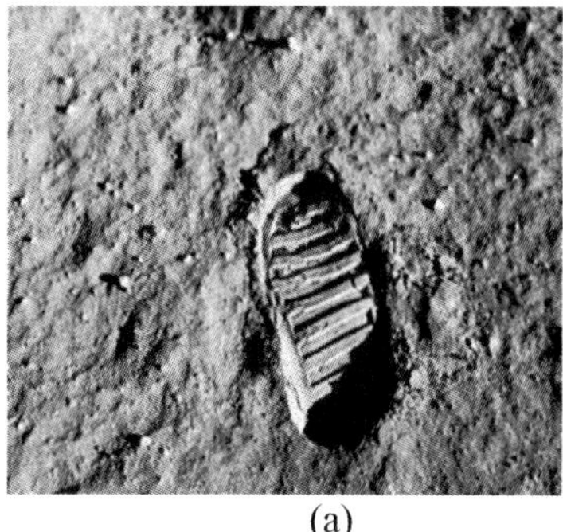

(a)   (b)

Figure 1   *Evidence of the cohesiveness of Lunar Soil. (a) The footprint of an astronaut (Photo ID: AS11-40-5878, NASA/courtesy of nasaimages.org.) and (b) the landing gear and scoop of the Surveyor 3 spacecraft on the lunar surface (Photo ID: AS12-48-7110, NASA/courtesy of nasaimages.org.)*

2 SURFACE ENERGY OF LUNAR SOIL

Several factors lead to think that surface energy forces might be not negligible on the Moon. First of all, Lunar Soil is a very fine material, with an average particle size around 70 μm. Second, gravitational forces are rather low in the lunar environment (g = 1.63 m/s^2) if compared to terrestrial conditions. Finally, atmospheric pressure is almost absent on the Moon. The combination of all these factors results in attractive inter-particle van der Waals forces being dominant. The van der Waals forces considered in this paper act at a very short range and are intrinsically related to the surface energy of the material. The latter is defined as half of the energy necessary to separate two surfaces from contact to infinity for two media of the same kind.[7] When two surfaces come into contact the value of surface energy can be expressed as:

$$\gamma = \frac{A}{24\pi D^2} \tag{1}$$

where A is the Hamaker constant and D the distance between the two surfaces. Unfortunately, no measurements of surface energy for real Lunar Soil have been made so far. However, such knowledge is crucial to any quantitative study of the effect of surface energy on the bulk behaviour. Nevertheless, important qualitative results can still be provided numerically. In this study a value of surface energy equal to 0.07 J/m^2 is taken as representative for Lunar Soil. This value was already employed by other authors studying the rheology of Lunar material.[4].

3   THE NUMERICAL MODEL

3.1 The Particle Size Distribution

Despite the variability of the lunar environment, the particle size distribution (PSD) of Lunar Soil samples collected at different sites on the Moon showed to be very similar.[8] A representative lunar size distribution was identified by Carrier.[8] Therefore, in our DEM analyses, the PSD for all the samples was generated by taking a number of discrete points from the representative curve obtained by Carrier[8] for Lunar Soil as shown in Figure 2. Due to computational reasons, only a constrained discrete PSD was considered as indicated in Figure 2. In particular, the grading curve was restricted to a ratio of maximum to minimum particle size equal to $R_D = D_{100}/D_0 = 8$. The largest particles were not included in the PSD since their contribution to the strength of the material is negligible.[9] On the other hand, it is important to capture the effect of grading since this may influence the critical state of the soil, as numerically shown by Wood and Maeda.[10] However, the smallest particles were not considered in the simulations to keep the computational times affordable.

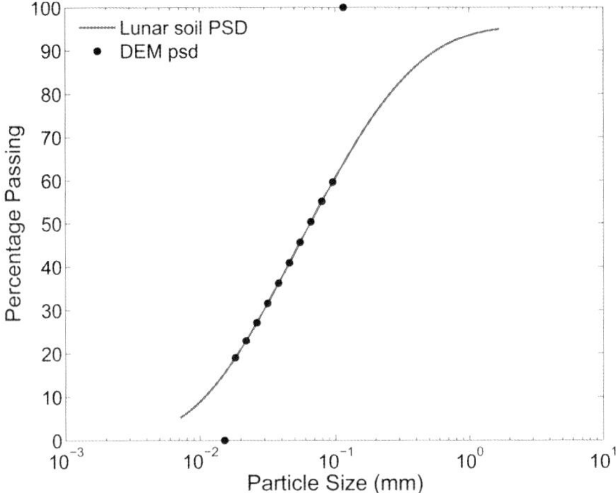

Figure 2 *Particle size distribution of Lunar Soil and of the DEM model*

3.2 The Contact Model

The aim of this research is to examine the origin of Lunar Soil cohesion by investigating the effect of surface energy on the soil shear strength. Surface energy was introduced in the numerical model at the microscopic scale, following the so-called JKR analytical solution.[5] The JKR solution represents an approximation of the contact behaviour of two elastic bodies in the presence of adhesion under quasi-static loading. The model accounts for the effect of adhesion inside the contact region. The maximum adhesion force predicted by JKR is equal to $P_{ad} = -3\pi R^* \gamma$ and is independent of the applied pressure.

A modified version of the JKR model was proposed by Thornton and Ning,[11] which also includes the effect of plasticity. On the other hand, plastic deformations are expected to be negligible at very small pressures. Therefore, only the elastic-adhesive JKR solution was employed in this study. In the absence of adhesion, the contact behaviour follows the classical Hertzian solution. A comparison of the contact behaviour between JKR and Hertz models is presented in Figure 3.

In the shear direction, the behaviour is approximated by the Mindlin and Deresiewicz[12] partial-slip solution both with and without adhesion. In addiction, the Mohr-Coulomb (MC) force limits the contact shear force. However, a modified MC sliding criterion is employed for the adhesive case.[13] Further details of the implementation can be found in Modenese *et al.*[14]

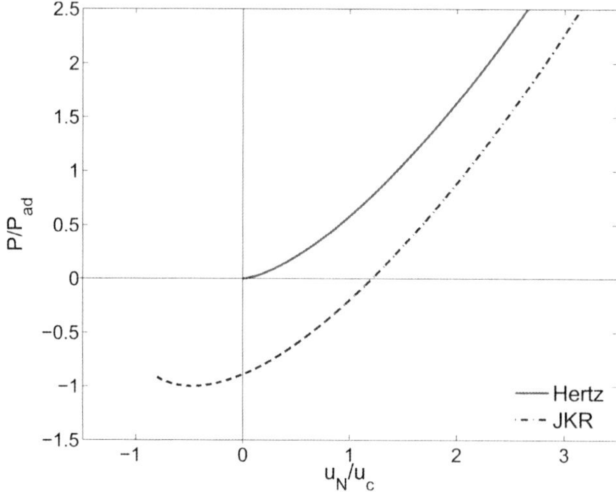

Figure 3   *Comparison between JKR and Hertz contact models. Relationships between normal force versus normal overlapping ($u_c = 3/2[(\pi^2\gamma^2 R^*)/(2E^*)]^{1/3}$)*

3.3 Sample Preparation

Standard triaxial numerical simulations were run in a 3D periodic cell under extremely small confining pressures, ranging from 1 kPa to 10 kPa. Due to the low lunar gravity, 1.63 m/s^2, these stress levels correspond to a depth on the Moon ranging from 0.40 to 4 m. To perform experimental tests at such small stress levels is still a challenge to be overcome. The advantage of numerical experiments lies in the fact that any boundary condition can be easily imposed. However, an attentive calibration of the numerical parameters, ruling the stress-strain servo-control algorithm, was required in order to impose triaxial stress paths under such small external loads.

Table 1   *Input DEM parameters of the tests*

Number of particles	N	10,000	
Young's modulus	E	70	GPa
Poisson's ratio	v	0.25	
Average particle size	R	72	μm
Particle density	ρ	2,650	kg/m^3
Strain rate	$\dot{\varepsilon}$	0.1	1/s
Friction coefficient	μ	0.6	
Surface energy	γ	0.07	J/m^2

Each sample was made of 10,000 spherical particles. The size of the sample was then selected so as to generate a representative elementary volume (REV) and at the same time to limit the fluctuations of the mechanical response. The generation procedure of very dense samples ($D_R = 100\%$) can be summarized as follows:
- Isotropic compression under zero friction angle until the pressure of 1kPa was reached and quasi-static conditions were attained;

- Gradual increase of the inter-particle friction angle until a realistic value was achieved under quasi-static conditions;
- Isotropic compression until the desired cell pressure was attained.

A summary of the parameters employed in the DEM model is given in Table 1. All the parameters are realistic except for the strain rate that was scaled up of a few orders of magnitude in comparison with the value adopted in laboratory experiments in order to reduce the computation time. However, care was taken throughout the simulations to maintain samples under quasi-static equilibrium.

4   RESULTS

Triaxial compression tests at constant mean pressures were run for very small confining stresses, in the range between 1 kPa and 1 MPa. Stresses were measured in the usual manner, i.e. by considering the contribution of all the contact forces within the periodic cell and taking the average of those over the total cell volume. The stress tensor, under quasi-static conditions, reads as[15]

$$\sigma_{ij} = \frac{1}{V} \sum_p \sum_c l_i^c f_j^c \tag{2}$$

where l_i is the branch vector at the contact, f_j is the total contact force and V is the volume of the periodic cell. The two stress invariants, output from the analysis, are the deviatoric stress, computed as $q = \sigma_1 - 0.5(\sigma_2 + \sigma_3)$, i.e. as the difference between the axial principal stress and the mean lateral pressure, and the mean cell pressure equal to $p = (\sigma_1 + \sigma_2 + \sigma_3)/3$. Finally, strains were computed directly from the deformations of the periodic cell.

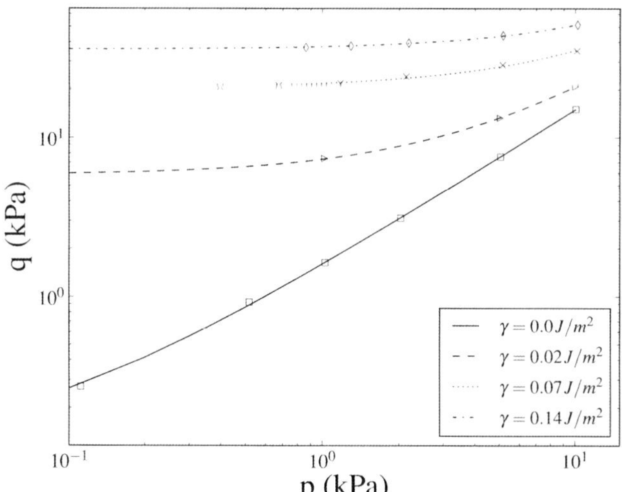

Figure 4 *Peak strength envelopes for different values of surface energy*

The triaxial tests were carried out until very large strains, i.e. up to 0.30 of deviatoric strain, in order to achieve the condition of critical state. Tests were run for the Hertz-Mindlin and JKR-Mindlin contact models. Three values of surface energy were used, such as 0.02, 0.07 and 0.14 J/m^2. The results on the q, p plane at both the peak and critical states are shown in Figures 4 and 5. In order to make it easier to see the amount of cohesion in the sample, a log-log scale was preferred.

Some remarkable observations can be made as follows:
- The macroscopic friction angle is almost unaffected by the introduction of surface energy;
- The presence of surface energy increases the macroscopic cohesion of the soil in a non linear manner;
- The effect of inter-particle adhesion is more significant at the peak rather than the critical state;
- The effect of surface energy decreases with increasing pressure, which is expected since the external loading becomes more significant compared to inter-particle adhesive forces.

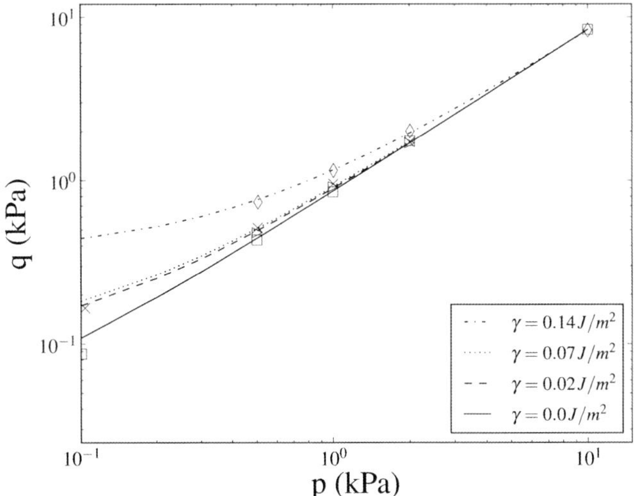

Figure 5 *Critical strength envelopes for different values of surface energy*

Table 2 *Microscopic cohesion and friction angles at the peak and critical state for different levels of surface energy*

γ (J/m^2)	Φ_{ps} ($°$)	Φ_{cs} ($°$)	c_{ps} (kPa)	c_{cs} (kPa)
0.00	37	22	0.00	0.00
0.02	37	21	2.89	0.04
0.07	37	21	10.18	0.05
0.14	37	21	17.85	0.14

Table 2 summarizes the results in terms of macroscopic cohesion and friction at the peak and critical states. The results were obtained without any tuning of the input parameters, which were selected to match realistic value for Lunar Soil. However, the limitation of the model to spherical particles lead to friction angles which are much lower than the expected values for real Lunar Soil and Lunar Soil simulants, in the range between 40 and 60 degrees, increasing with increasing density. Nevertheless, the focus of this study was to investigate the effect of van der Waals forces on the macroscopic properties of the soil, although different quantitative results may be predicted when particle shape effects are taken into account.

5   CONCLUSIONS

This paper investigated the influence of surface energy on the macroscopic mechanical properties of an assembly of spheres mimicking Lunar regolith. To this end, the JKR model was employed at the particle contacts. It emerged that the presence of inter-particle adhesion produced some macroscopic cohesion. This result shows that a physical link between microscopic adhesion and true cohesion at the macroscopic level exists. Moreover, the influence wasì more significant at the peak than at the critical state. On the other hand, no influence on the friction angle was observed, either at the peak or at the critical state. Finally, the influence of surface energy becomes negligible with increasing pressure, due to the fact that inter-particle van der Waals attractive forces become negligible in comparison with the applied pressure. This is consistent with the fact that the maximum adhesion force is independent of the applied confining pressure since plastic deformations of the grains are excluded from the adopted contact model.

Further work is currently underway on the mutual influence of particle shape effects and inter-particle adhesion.

References

1   W.D. Carrier, Research note, *Lunar Geotech. Institute*, 2005.
2   H.A. Perko, J.D. Nelson and W.Z. Sadeh, *J. Geotech. Geoenviron.*, 2001, **127**, 371.
3   C.S. Chang and P.Y. Hicher, *J. Aerospace Eng.*, 2009, **22**, 43.
4   O. Baran, A. De Gennaro, and R.A. Wilkinson, *Powders and Grains*, 2009.
5   K.L. Jonhson, K. Kendall and A.D. Roberts, *Proc. R. Soc. A*, 1971, **324**, 301.
6   V. Šmilauer, E. Catalano, B. Chareyre, S. Dorofeenko, J. Duriez, A. Gladky, J. Kozicki, C. Modenese, L. Scholtès, L. Sibille, J. Stránský and K. Thoeni, *Yade Reference Documentation*, ed. V. Šmilauer, 2010.
7   J. Israelachvili, *Intermolecular and Surface Forces*, ed. Academic Press, 1992.
8   W.D. Carrier, *J. Geotech. Geoenviron.*, 2003, **129**, 956.
9   T. Ueda, T. Matsushima and Y. Yamada, *Granul. Matter*, 2011, **13**, 731.
10 D.M. Wood and K. Maeda, *Acta Geotech.*, 2008, **3**, 3.
11 C. Thornton and Z. Ning, *Powder Technol.*, 1998, **99**, 154.
12 R.D. Mindlin and H. Deresiewicz, *J. Appl. Mech.*, 1953, **16**, 259.
13 K. Kendall, *Nature*, 1986, **319**, 203.
14 C. Modenese, S. Utili, and G.T. Houlsby, *Proc. Earth and Space ASCE Conf.*, 2012.
15 C. Thornton, *Géotechnique*, 2000, **50**, 43.

MODELLING OF THE CONTACT BEHAVIOUR BETWEEN FINE ADHESIVE PARTICLES WITH VISCOUS DAMPING

K. Mader and J. Tomas

Mechanical Process Engineering, Department of Process Engineering, Otto-von- Guericke-University Magdeburg, P.O. Box 4120, D-39106 Magdeburg, Germany

1 INTRODUCTION

Cohesive and compressible powders consist of fine (d < 100 µm), ultrafine (d < 10 µm) or nanosized particles (d < 1 µm). These powders show a list of flow problems in processing apparatuses or machines, in product handling equipment or in storage and transportation containers.[1,2] Most of the problems are caused by undesired adhesion or sticking. Otherwise desired adhesion effects are very useful to agglomerate, formulate or coat these particles to improve the product quality. The micromechanics of particle adhesion and macroscopic cohesive powder flow behaviour are very essential to assess the product quality and to improve the process performance in particle technology.

This means that the interparticular adhesive forces, especially the van der Waals attraction forces, exceed the gravitational forces of the particles by several orders of magnitude (see Table 1). Therefore, the undisturbed handling of these products represents a scientific challenge and is particularly important for many industries.

Table 1 *Evaluation of the adhesiveness of fine, dry particles*[3,4,5]

	Particle size d (µm)	Ratio F_{H0}/F_G	Evaluation
F_{H0}	10 - 100	$1 - 10^2$	Slightly adhesive
	1 - 10	$10^2 - 10^4$	Adhesive
F_G	0.01 - 1	$10^4 - 10^8$	Very adhesive

The key issue is the understanding of the approach, contact and detachment of fine, dry and adhesive particles.[3,4,6,7] This requires the derivation and modelling of the force-displacement and torsional moment-angle functions of fine adhesive particles for the six mechanical degrees of freedom of translation and rotation including the van der Waals forces (Figure 1). Furthermore, the inelastic contact deformation needs to be combined with the intensification of the van der Waals force within the flattened contact zone of fine particles.

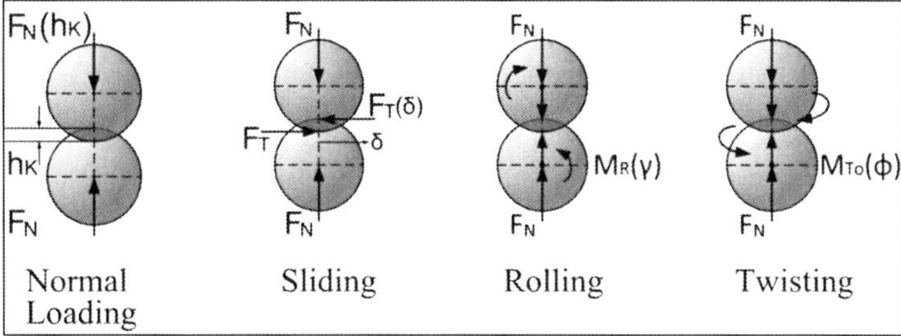

Figure 1 *Overview of the six degrees of freedom of the mechanical stress of a particle contact through translation and rotation, including the desired cause-response functions: normal force-displacement function, tangential force-displacement function, rolling moment-rolling angle function and torsional moment-angle function.*

2 MODELS FOR 'STIFF PARTICLES WITH SOFT CONTACTS'

In the model, the external forces and short-range adhesion forces immediately generate a localized contact deformation as a response. The generated contact surfaces are small in comparison to the particle size. For the model, it is not necessary to consider the nano-scale deformations outside of the contact zone. Furthermore, the model is characteristic for the particle contact with variable adhesion. This includes the adhesive, elastic-plastic and viscous, elastic-plastic contact deformation with the unloading and reloading hysteresis as well as the energy absorption. With this load dependent adhesion force, we obtained the micromechanical models for elastic and frictional sliding, rolling and twisting.

To evaluate the model, we selected reference particle systems. Because of their different material properties we used stiff glass and soft titania (Figure 2). The corresponding values were taken from the literature and were back-calculated using shear tests.

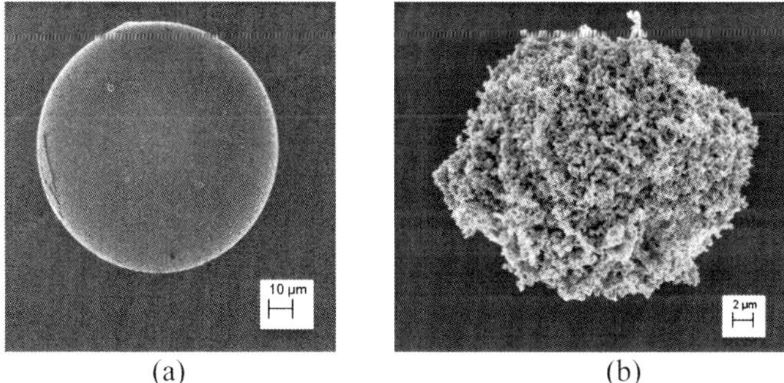

(a)   (b)

Figure 2 *SEM images of (a) A Spheriglass particle (F_{H0}/F_G: $1.3\cdot10^3$; contact consolidation coefficient κ: 0.1); (b) A Titania particle (F_{H0}/F_G: $2.5\cdot10^5$; κ: 1.1)*

We can see that the micro glass particles are adhesive because the ratio between the attraction force and the gravitational force (F_{H0}/F_G) is high. This ratio for the micro glass particles is lower than that of titania. When we compare the contact consolidation coefficients κ we can see that the glass is stiffer compared to the extremely soft titania.

2.1 Normal Force-Displacement Function with Variable Adhesion

The first fundamental of the elastic contact compression under normal loading was developed by Hertz.[8] With the approach of the elliptical stress distribution within the contact surface, Hertz derived the nonlinear elastic normal force-displacement law for isotropic smooth particles. The principal stresses within and outside the contact area was introduced by Huber.[9] Examples for analytical and numerical models of the spatial stress distribution within a sphere were given by Lurje[10] and Chen *et al.*[11] An additional contribution to the normal force by a constant adhesion of the elastic contact was observed by Sperling,[12] Derjaguin,[13, 14] Dahneke,[15, 16] Johnson,[17, 18] Greenwood [19] and Peukert.[20, 21, 22] This is only a short excerpt from the numerous existing contact models. However, the variable adhesion by the production and handling of fine particles is not adequately considered in these models. In addition, there are significant deficits in the realistic modelling of the mechanical contact behaviour of fine particles at the nano-scale. The interactions between the plastic and viscoplastic contact deformation and the increasing adhesion are underestimated and undervalued in the literature. Therefore, an analytical function of the normal force-displacement behaviour of the contact compression of continuous transitions from the elastic, elastic-plastic and fully plastic contact behaviour with a constant flow pressure (yield limit) and a variable adhesion force, which depends on the contact deformation (flattening), was derived. The contact of two isotropic and smooth spheres, as a typical part of a packed bed or a particle packing, was considered under an applied load F_N. Figure 3a shows the normal force-displacement behaviour of a particle-particle contact, according to the new contact model.[15]

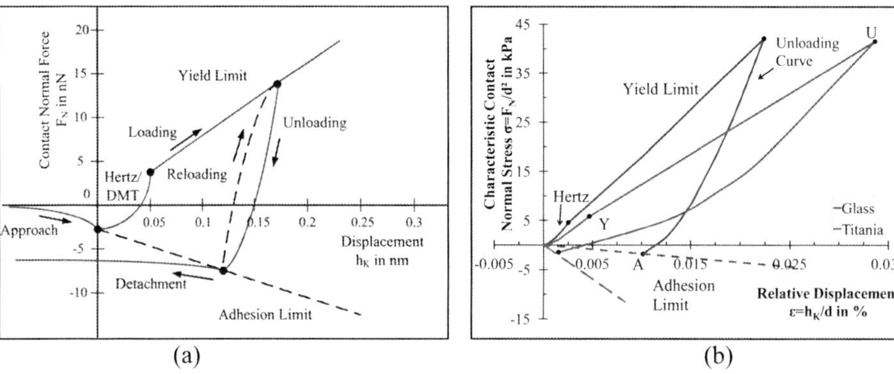

(a)   (b)

Figure 3 *(a) Characteristic normal force-displacement curve of smooth isotropic particles, (b) Contact normal stress-strain-relation for glass and titania*

After the approach of the two particles from an infinite distance to a minimum separation, a smooth sphere-sphere contact without any contact deformation by the short range attractive adhesion force is formed. This is the so called jump-in. The zero point of the y-axis, where the displacement is equal to zero, is equivalent to the characteristic adhesion separation of a direct contact. If the contact may be loaded, from the point of the adhesion force to the yield point, it is elastically deformed with an approximated circular contact area according to Hertz[8] and Derjaguin[13, 14] (with adhesion). When the Hertz curve intersects the abscissa, the total force equilibrium within the self equilibrating contact is obtained. By increasing the external normal load this soft contact transits at the pressure of the micro-yield strength to a plastic yielding at this point. There is the transition from elastic to elastic-plastic deformation.[23, 24, 25] The elastic domain is located between the elastic plastic yield limit and the adhesion limit. Any normal loading yields an increasing displacement. But if one would be unloaded, beginning at an arbitrary point, the elastically deformed, annular contact zone would be recovered along a parabolic curve. The reloading curve would run along an equivalent curve from the adhesion limit to the elastic plastic yield limit. If the adhesion limit is reached, the contact planes detach with an increasing distance. This particle separation can be considered for the calculation by means of a long range hyperbolic adhesion force curve.

The comparison of the materials is shown in Figure 3b. For a better illustration, we set the calculated values into a relation, $\sigma = F_N / d^2 = f(\varepsilon = h_K / d^2)$, and recorded the characteristic contact normal stress versus the relative displacement. To compare both materials with their different physical properties (Table 2), it is important to consider the behaviour of the elastic-plastic yield limit. One evaluation criterion is the stiffness κ of the contact and the other one is the plastic repulsion coefficient κ_p. The stiffness of glass is 2200 N/m and the repulsion coefficient is 0.07. These values show the stiff contact behaviour of glass. The behaviour of titania is opposite because of the stiffness k=165 N/m and the repulsion coefficient κ_p=0.44.

Table 2 *Material properties of glass and titania*

Properties	Glass	Titania
Particle size (µm)	5.8	0.6
Elasticity modulus (kN/mm²)	100	50
Charact. adhesion force (nN)	1.74	0.54
Micro-Yield strength (MPa)	300	400
Contact friction coefficient	0.8	0.64

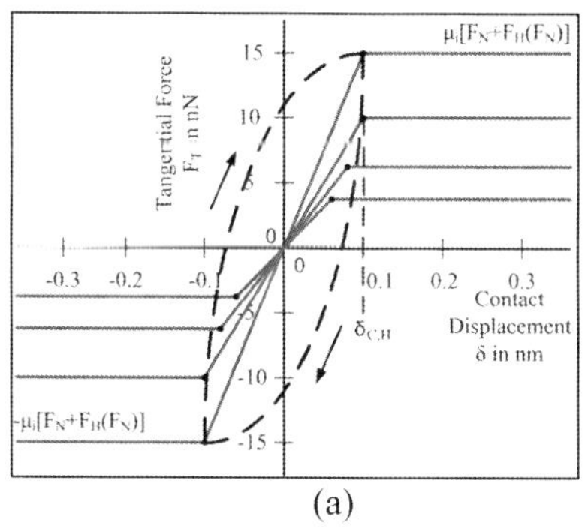

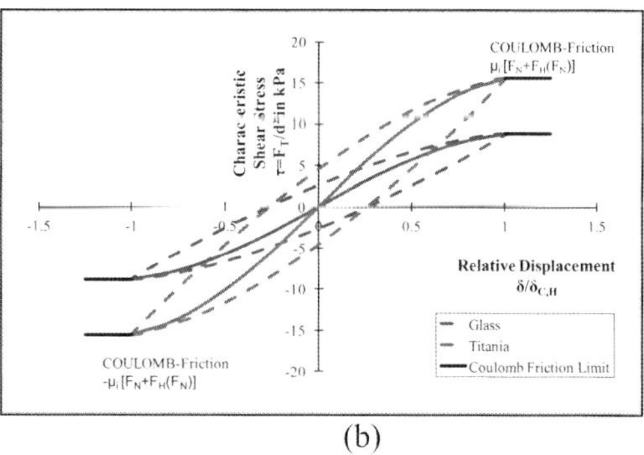

(a)   (b)

Figure 4 *(a) Characteristic tangential force-displacement curve of smooth isotropic particles, (b) Contact shear stress-strain-relation for glass and titania*

2.2 Tangential Force-Displacement Function with Variable Adhesion

In addition to the classical, linear elastic tangential force-displacement law of Hook,[26] Fromm,[26] Cattaneo,[27] Föppl,[28] Mindlin,[29] Sonntag[30] and Mindlin/Deresiewicz[31] modelled the nonlinear frictional contact behaviour. In all publications the adhesion of monodisperse, fine particles is not sufficiently considered. Therefore, the load-dependent adhesion force and its influence is presented and discussed in Figure 4a.

We modified Mindlins[29] non-linear model for the first tangential loading of a pre-consolidated or flattened elastic-plastic contact. The coulomb friction limit of the tangential force is described by the coefficient of internal friction μ_i. It depends on the elastic-plastic contact consolidation. If the contact loses its elastic tangential stiffness and completely passes over into contact sliding, it obtains the friction limit of displacement under load-dependent adhesion force. Both friction limits are geometrically similar and increase with the increasing load normal force. In addition to the loading curves, the cyclic parabolic unloading and reloading curves are indicated. In Figure 4b we show the soft contact behaviour of titania, because the shear stress is much larger than a stiff glass-glass contact. In all cases the consolidation stress was 15 kPa.

2.3 Rolling Resistance Force-Rolling Angle Function with Variable Adhesion

Micro-roughness, which occurs because of the contact deformation and partial adhesion under normal load, is the reason for the rolling resistance of spheres. This was described by Fromm,[26] Sonntag,[30] Johnson[17] and Iwashita.[32] For this reason, the new rolling resistance force-rolling angle function includes a nonlinear elastic component for partial load dependent adhesion within a rolling contact surface, see Figure 5a.

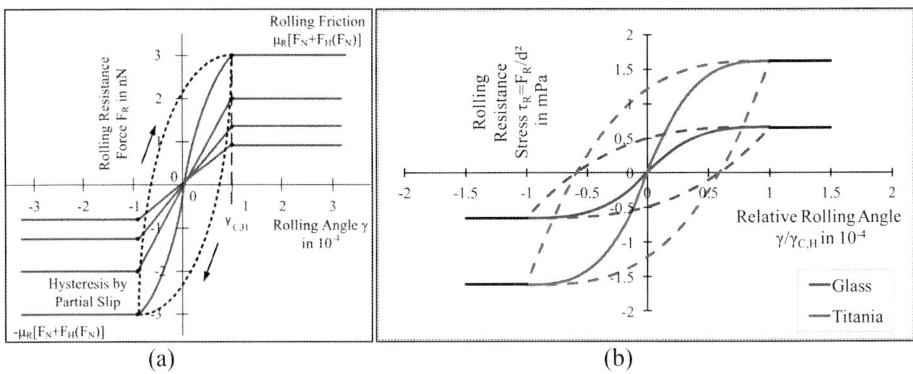

(a)	(b)

Figure 5 *(a) Characteristic rolling resistance force-rolling angle curve of smooth isotropic particles, (b) Contact rolling resistance stress-strain-relation for glass and titania*

Johnsons creep model[17] for elastic frictional behaviour without adhesion is the basis for the extension to the elastic-plastic contact flattening. The critical rolling resistance results from the elastic-plastic contact displacement in normal direction and the load-dependent adhesion force. When the contact loses its elastic rolling stiffness the rolling friction limit with respect to the critical rolling angle is obtained. The friction limit is geometrically similar to the tangent limit calculated by ratio of the force limit and the initial stiffness. By integrating the equation of rolling resistance stiffness the force rolling angle function for the first loading of the flattened elastic-plastic contact is obtained. Similarly to the constitutive tangential force-displacement model, the functions for unloading and reloading are obtained. In Figure 5b, we observe the same behaviour as in sliding. The titania has a much larger rolling resistance stress because of its soft contact behaviour.

2.4 Torsional Moment-Rotation Angle Function with Variable Adhesion

When a particle contact rotates around its own major axis, it is the loading of the torsion. Mindlin,[29] Cattaneo,[27] Deresiewicz[33] and Johnson[17] derived the radial torsional distribution of the elastically deformed contact circle as a function of the rotation angle. Despite the coupling of the four types of stresses[34] the elastic-plastic contact deformation and the load-dependent adhesion were neglected. The rotation includes a non-linear part of the partial load-dependent adhesion in the normal direction in the elastic-plastic deformed contact area (Figure 6a).

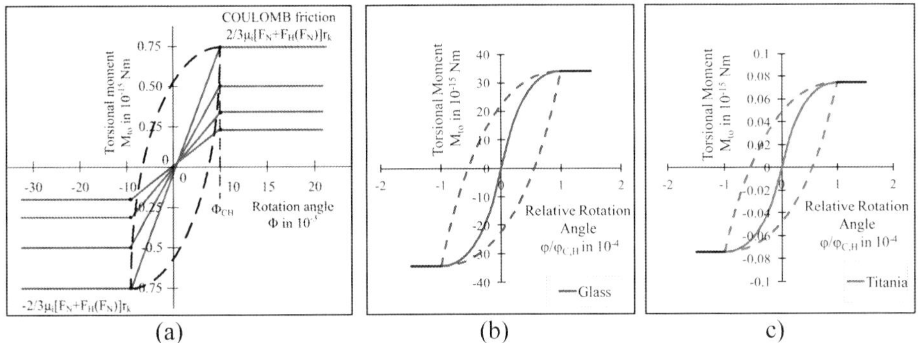

Figure 6 *(a) Characteristic torsional moment-rotation angle curve of smooth isotropic particles, (b) Contact torsional moment-relative rotation angle-relation for glass, (c) Contact torsional moment-relative rotation angle-relation for titania*

The torsional stiffness for the elastic-plastic contact without any adhesion force can be derived according to Deresiewicz[33] as a function of the torsional moment. When the contact loses its elastic torsional stiffness and completely passes over into frictional contact rotation, it obtains the friction limit of moment under load-dependent adhesion force. The critical rotation angle results from the completely mobilized frictional contact rotation of the moment. This friction limit is geometrically similar to the tangent limit calculated by the ratio of moment limit and initial stiffness. The value does not depend on the load normal force but only on the particle properties. The torsional moment results in a function of rotation angle for the first torsional loading of the flattened elastic plastic contact.

In Figure 6b and 6c, the abscissa is set into a relation, (φ / φ_{CH}). We can see that the torsion of the glass is 500 times larger than the torsion of titania. The reasons are in the different material properties. For example, when we consider the modulus of elasticity of glass (100 kN/mm²), we can see that the shear modulus is the twice of titania.

3 COMBINATION OF THE VELOCITY-DEPENDENT VISCOUS DAMPING AND THE NORMAL LOADING

If the balance of forces is established the sum of the inertia force, the elastic and elastic-plastic contact force and the viscous damping force has to be equal to zero (Eq. 1).

$$F_{Inertia} = F_{elastic} + F_{elastic-plastic} + F_{damping} \tag{1}$$

The approach of the contact model was the impact of two spherical particles. The equations were established for each of elastic, elastic-plastic and viscous contact. Hereby the contact is characterized by a spring, dashpot and plasticity. The modelling was carried out using Matlab with the parameters of the reference particle systems. For our model the damping approach of Tsuji *et al.*[35] was chosen (see Equation 2).

$$F_{damping} = \eta \cdot \dot{h}_K = \alpha_{Viscous} \cdot \sqrt{m_{1,2} \cdot k_{normalelastic}} \cdot h_K^{\frac{1}{4}} \cdot \dot{h}_K \qquad (2)$$

The varied parameters were the damping coefficient and initial velocity. In Figure 7 the numerical results are shown.

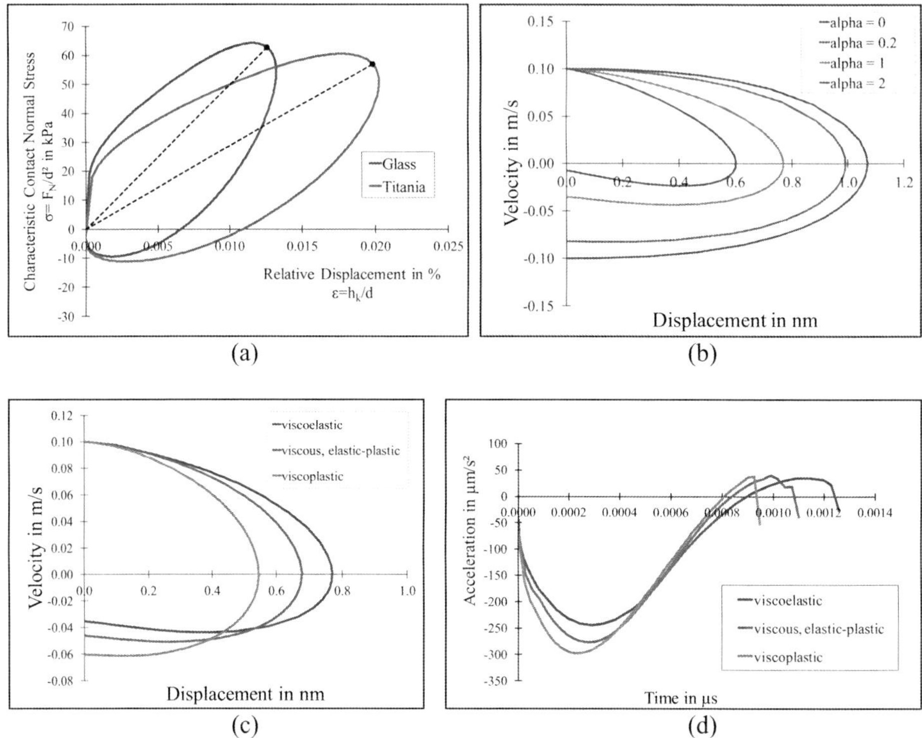

Figure 7 *(a) Contact normal stress-strain relation for glass and titania, (b) Viscoelastic contact of glass; comparison of different damping coefficients; $v_0=0.1$ m/s, (c) Impact of two glass particles; comparison of viscous types; $\alpha_{damping}=1$, (d) Acceleration versus material properties at impact; $\alpha_{damping}=1$; $v_0=0.1$ m/s*

Figure 7a shows the contact normal stress-strain relation for glass and titania. Therefore the viscoelastic contact with an impact velocity of 0.1 m/s and a damping coefficient of 1 were chosen. Because of its stiff material properties, the displacement of the glass particles is smaller than the soft titania particle-particle contact. Figure 7b shows the results for velocity versus displacement. The modelling was performed with the parameters of glass particles at an initial velocity of 0.1 m/s and a damping coefficient from 0 to 2. We consider a single particle contact, which starts when both particles contact each other and finishes if both particles detach from each other. If the damping coefficient is zero, the damping term is eliminated. When no damping is present, the particles contact each other with the same velocity as they detach. This velocity corresponds to the given velocity of 0.1 m/s. When the damping coefficient is larger the influence of the damping term increases too. This means that an increasing damping coefficient decreases the velocity. Figure 7c also shows the velocity versus displacement. In this figure the different kinds of viscosity were distinguished. The damping coefficient was 1 and the initial velocity was 0.1 m/s. The largest displacement was observed by the viscoelastic impact and the smallest one by the viscoplastic impact. The reason for this behaviour is the relationship between surface flattening, which can be considered as a circle, and the necessary force, which had to be applied for the different deformation types. The viscous, elastic-plastic deformation is located in the middle, because it is viscous, elastically and plastically deformed. Figure 7d shows the particle acceleration versus the material properties at impact. When we consider the strength of the dragging of the particles, that means the negative acceleration, it follows theoretically that the plastic deformation must be the largest one. This fact is confirmed in the figure. The point of maximum contact flattening when the acceleration has its minimum is also shown for the viscoplastic contact. The viscoelastic deformation on the other hand has the smallest acceleration.

4 CONCLUSIONS

In this work the model for 'stiff particles with soft contact' was developed. The derivation and calculation for the four types of loading: normal loading, sliding, rolling and twisting with a constant micro-yield strength p_f were shown. These were evaluated and compared using micro glass spheres and titania as reference particle systems. It is observed that the contact of glass-glass is stiffer compared to that of soft titania. For the normal loading, the velocity-dependent viscous damping was modelled using Matlab. The damping approach proposed by Tsuji *et al.*[35] was chosen. For the evaluation, parameters such as damping coefficient, type of contact (viscoelastic, viscous elastic-plastic and viscoplastic) and the velocities at the detachment were examined.

List of Symbols

$\alpha_{viscous}$	[-]	damping coefficient
d	[μm]	particle diameter
E	[kN/mm²]	modulus of elasticity
ε	[%]	strain
$F_{damping}$	[nN]	damping force
$F_{elastic}$	[nN]	elastic contact force
$F_{elastic-plastic}$	[nN]	elastic-plastic contact force
F_G	[nN]	gravitational force
F_{H0}	[nN]	adhesion force
$F_{inertia}$	[nN]	inertia force
h_K	[nm]	displacement
k	[N/m]	stiffness
κ	[-]	contact consolidation coefficient
κ_p	[-]	plastic repulsion coefficient
$m_{1,2}$	[kg]	mean particle mass
μ_i	[-]	internal friction

η	[kg/s]	damping term
φ	[°]	rotation angle
φ_{CH}	[°]	critical rotation angle
σ	[kPa]	stress
v_0	[m/s]	impact velocity

References

1   K. Borho, R. Polke, K. Wintermantel, H. Schubert, K. Sommer, *Chemie Ingenieur Technik*, 1991, **63**, 792.
2   H. Rumpf, *Chemie Ingenieur Technik*, 1958, **30**, 144.
3   J. Tomas, S. Kleinschmidt, *Chem. Eng. Technol.*, 2009, **32**.
4   J. Tomas, S. Kleinschmidt, *Chemie Ingenieur Technik*, 2009, **81**, 717.
5   J. Tomas, *Abhandlungen der Sächsischen Akademie der Wissenschaften zu Leipzig, Technikwissenschaftliche Klasse*, 2009, **1**, 1.
6   C. Thornton, *J. Phys. D.: Appl. Phys.*, 1991, **24**, 1942.
7   C. Thornton, Z. Ning, *Powder Technology*, 1998, **99**, 154.
8   H. Hertz, *Journal reine und angewandte Mathematik*, 1882, **92**, 156.
9   M.T. Huber, *Annalen der Physik*, 1904, **14**, 153.
10 A.I. Lurje, in *Räumliche Probleme der Elastizitätstheorie*, Berlin: Akademieverlag, 1963.
11 Y. Chen, A. Best, T. Haschke, W. Wiechert, H.-J. Butt, *Journal of Applied Physics*, 2007, **101**.
12 G. Sperling, in *Eine Theorie der Haftung von Feststoffteilchen an Festkörpern*, TH Karlsruhe: Dissertation, 1964.
13 B.V. Derjaguin, *Kolloid Zeitschrift*, 1934, **69**, 155.
14 B.V. Derjaguin, V.M. Muller, U.P. Toporov, *Journal of Colloid and Interface Science*, 1975, **53**, 314.
15 B. Dahneke, *Journal of Colloid and Interface Science*, 1975, **51**, 58.
16 B. Dahneke, *Journal of Colloid Interface Science*, 1972, **40**, 1.
17 K.L. Johnson, in *Contact Mechanics*, Cambridge University Press, 1985.
18 K.L. Johnson, K. Kendall, A.D. Roberts, *Proceedings of Royal Society*, 1971, **A324**, 301.
19 J.A. Greenwood, *Proceedings of Royal Society*, 1997, **A453**, 1277.
20 M. Götzinger, W. Peukert, *Langmuir*, 2004, **20**, 5298.
21 Q. Li, V. Rudolph, W. Peukert, *Powder Technology*, 2006, 248.
22 H. Zhou, W. Peukert, *Langmuir*, 2008, **24**, 1459.
23 E.M. Lifshitz, *Soviet Physics (JETP)*, 1956, **2**, 73.
24 H.C. Hamaker, *Physica*, 1937, **4**, 1058.
25 J. Tomas, *Chemical Engineering Science*, 2007, **62**, 1997.
26 G. Fromm, *Zeitschrift für Angewandte Mathematik und Mechanik*, 1927, **7**, 27.
27 C. Cattaneo, *Academia Nationale dei Lincei, Rendiconti*, 1938, **27**, 342.
28 L. Föppl, in *Die strenge Lösung für die rollende Reibung*, München: Leibnitz Verlag, 1947.
29 R.D. Mindlin, *Transactions of American Society Mechanical Engineers, Journal of Applied Mechanics*, 1949, **16**, 259.
30 G. Sonntag, *Zeitschrift für Angewandte Mathematik und Mechanik*, 1950, **30**, 73.
31 R.D. Mindlin, H. Deresiewicz, *Transactions of American Society Mechanical Engineers, Journal of Applied Mechanics*, 1953, **20**, 327.
32 K. Iwashita, M. Oda, *Powder Technology*, 2000, **109**, 192.

33 H. Deresiewicz, *Transactions of American Society Mechanical Engineers, Journal of Applied Mechanics*, 1954, **21**, 52.
34 H. Göldner, in *Lehrbuch höhere Festigkeitslehre*, vol. 1 & 2, Leipzig: Fachbuchverlag, 1992.
35 Y. Tsuji, T. Tanaka, T. Ishida, *Powder Technology*, 1992, **71**, 239.

REBOUND OF A PARTICLE FROM A SOLID SURFACE WITH A VISCOUS OR NONLINEAR VISCOELASTIC LIQUID FILM IN THE CONTACT ZONE

J. Bowen[1], D. Cheneler[2], J.W. Andrews[1], C-Y. Wu[1], M.C.L. Ward[2] and M.J. Adams[1]

[1] School of Chemical Engineering, The University of Birmingham, Edgbaston, Birmingham, B15 2TT, UK
[2] School of Mechanical Engineering, The University of Birmingham, Edgbaston, Birmingham, B15 2TT, UK

1 INTRODUCTION

The importance of the physical properties of liquid binders in wet agglomeration processes is well established and has been an area of significant research interest for many years. Under static conditions, the interparticle binding mechanism arises from capillary attraction while viscous, and possibly viscoelastic forces for some fluids, will be developed when there is relative motion between the particles. Closed-form approximations have been derived to compute the capillary and viscous forces for interparticle pendular liquid bridges.[1] They have been incorporated in discrete element modelling (DEM) simulations to examine the effect of such bridges on the behaviour of wet agglomerates during collisions.[2] The cohesive effect of liquid bridges has subsequently been included in simulations of fluidized beds,[3] vibrated granular beds,[4,5] soils[6] and high shear mixers.[7] There has also been considerable work examining other factors that influence the liquid bridge forces including, the separation-distance dependence of the force between two rigid spherical bodies,[8] elastohydrodynamic collisions of spherical particles,[9] the influence of liquid bridge volume[10] and particle wetting hysteresis.[11]

Organic polymers, either in solution or as a melt, are commonly used for liquid binders.[12-16] Such polymers are generally composed of uncrosslinked flexible macromolecules of high molecular weight. Under quiescent conditions, the polymer chains will exist as random coils but during flow they will elongate and align in the direction of the flow field. Consequently, polymeric binders can exhibit non-Newtonian rheological properties. For relatively small molecular weights it is possible to approximate their flow characteristics by a power law fluid. In this case, a closed-form approximation has been developed to determine the viscous forces.[17] However, if the molecular weight and concentration exceeds a critical value, such binders will exhibit viscoelastic behaviour. This arises because the polymer chains become entangled during the transition from a random coil to stretched state. The entanglements behave like temporary cross-links that result in elastic forces being generated.[18]

Recently Bowen et al.[19] reported the use of colloid probe atomic force microscope (AFM) for measuring the peak cohesive forces at small separation distances for thin films of a range of Newtonian and viscoelastic poly(dimethylsiloxane) (PDMS) liquids deposited on Si wafers. The data were interpreted using existing models of the capillary and viscous forces and, for molecular weights greater than the critical value, a closed-form approximation was derived to describe the viscoelastic force. The mean fluid strains are relatively large when a particle separates from a surface so that it was necessary to develop a nonlinear model. Viscoelastic stress overshoot was also considered but was not a contributing factor for the separation velocities available with an AFM. This work could form the basis of computing the forces between particles in DEM simulations involving polymeric liquid binders

A brief overview of the AFM work is presented here, with the equations of motion for models that include capillary forces and (i) viscous forces, and (ii) nonlinear viscoelastic forces. These equations are then used to calculate whether a particle will rebound or stick when it has undergone a collision in the presence of (i) a nonlinear viscoelastic liquid binder, and (ii) a viscous binder of having a similar zero-shear rate viscosity.

2 THEORY

As described in the previous section, theoretical models for calculating the cohesive force as a function of separation distance have been derived previously,[19] based on the system geometry shown in Figure 1, which is a schematic diagram of a spherical particle in contact with a liquid film supported by a planar half-space. The liquid contact radius, a, is approximated by ignoring the meniscus. The axial and radial coordinates, (z, r), are also shown with an origin at the planar surface. The gap between the two solid bodies, $s(r)$, has a minimum value $s(0) = S$. It is assumed that the cohesive force is the sum of the capillary attraction and that associated with shear flow in the lubrication limit.

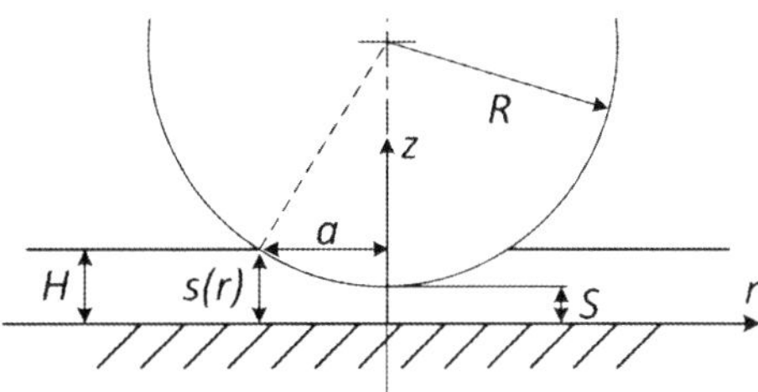

Figure 1 *Sphere-on-flat geometry for a film thickness H*

2.1 Capillary-viscous model

The total force, F, acting on a sphere of radius R, when partially immersed in a perfectly wetting thin film of a Newtonian liquid and with a separation velocity, $\dot{S}$, is the sum of the capillary, F_C, and viscous forces, F_V, thus:[20]

$$F = F_C + F_V \tag{1}$$

where

$$F_C = 4\pi\,\sigma\,R\left(1 - \frac{S}{H}\right) \tag{2}$$

$$F_V = 6\pi\eta R^2 \left(1 - \frac{S}{H}\right)^2 \frac{\dot{S}}{S} \tag{3}$$

where σ is the liquid surface tension, H is the film thickness, and η is the liquid viscosity. Equation (3) assumes a no slip boundary condition and that the lubrication approximation is valid.

A force balance leads to the following relationship between the capillary and viscous forces and the inertia of an AFM beam assuming that the effect of fluid inertia is negligibly small:

$$\left(1 - \frac{S}{H}\right) + \frac{3\,\eta R}{2\,\sigma}\left(1 - \frac{S}{H}\right)^2 \frac{\dot{S}}{S} = \frac{m_b\,\ddot{S}}{4\pi\sigma R} \tag{4}$$

where m_b is the total effective mass of the cantilever and $\ddot{S}$ is the acceleration of the probe.

2.2 Capillary-nonlinear viscoelastic model

The non-dimensional equation of motion incorporating the nonlinear viscoelastic force, which is analogous to (4), neglecting stress overshoot,[21] is given by:

$$\left(1 - \frac{S}{H}\right) + \frac{3}{2\sigma}R\left(1 - \frac{S}{H}\right)^2 \frac{\dot{S}}{S}\left(\frac{1}{1 + \alpha\,\gamma(t)^\beta}\right)\sum_{i=1}^{N}\eta_i\left(1 - e^{-t/\lambda_i}\right) = \frac{m_b\,\ddot{S}}{4\pi\sigma\,R} \tag{5}$$

where the strain, γ, and frequency, ω, dependent viscoelastic modulus of the liquid, $G(\omega,\gamma)$, is described by a damping function, $h(\gamma)$, of the following form:[22]

$$h(\gamma) = \frac{1}{1 + \alpha\,\gamma^\beta} \tag{6}$$

where α and β are parameters fitted to the measured strain-dependent behaviour of the liquid, and N is the number of Maxwellian spring-dashpot elements in parallel; the modulus of each element, G_i, is given by:

$$G_i = \frac{\eta_i}{\lambda_i} \tag{7}$$

where η_i and λ_i are the viscosity and relaxation time of each element.[23]

The appropriate measure of strain in (6) is the mean value, γ_{av}, at the surface of the particle as discussed in Ref. [19]. The corresponding mean strain rate, $\dot{\gamma}_{av}$, is given by:

$$\dot{\gamma}_{av} = \frac{1}{\pi a^2}\int_0^a\int_0^{2\pi}\dot{\gamma}[s(r)]\,r\,\mathrm{d}\theta\,\mathrm{d}r = \frac{3\dot{S}}{(H-S)}\left[\frac{\sqrt{2R}}{\sqrt{S}}\tan^{-1}\left(\frac{a}{\sqrt{2SR}}\right) - \frac{a}{H}\right] \tag{8}$$

where θ is the azimuthal angle and the liquid contact radius, a, is given by:

$$a = \left[2R(H-S)\right]^{0.5} \tag{9}$$

The mean strain, γ_{av}, is obtained by integrating Equation (8) at every instant over all previous times given that $\dot{S}$, S and a are functions of time calculated at each prior time step. This result can then be substituted into the equation of motion, Equation (5).

3 COMPARISON OF THEORY WITH AFM MEASUREMENTS

3.1 Summary of AFM results

PDMS films with thicknesses in the range 0.2-2 µm were deposited onto highly polished Si wafers using a spin coater. The film thicknesses were measured using an ellipsometer. Linear PDMS liquids with zero shear rate viscosities in the range 0.1-1,124 Pa.s were used; shear rate dependent viscosities, frequency dependent properties, and strain dependent behaviour were measured using cone-and-plate rheometry. AFM force measurements were performed using rectangular Si cantilevers with attached spherical SiO_2 colloid probes of radii 2.5-6 µm. Cantilever spring constants were in the range 10-30 N/m and were calibrated according to the method described by Bowen *et al.*[24] Measurements were performed for a drive velocity range of 0.1-50 µm/s. Calculations were performed using Matlab, solving Equations (4) and (5) iteratively for S as a function of time using a numerical ordinary differential equation solver, ODE15, to integrate the equation of motion.

3.2 Peak cohesive forces

Figure 2 shows the measured and calculated values of the normalised peak cohesive force, F^{*}_{peak}, as a function of the Capillary number, Ca. These parameters are defined as follows:

$$F^{*}_{peak} = \frac{F_{peak}}{4\pi\sigma R} \tag{10}$$

$$Ca = \frac{\eta_0 \dot{S}_{peak}}{\sigma} = \frac{\eta_0 V}{\sigma} \tag{11}$$

where $\dot{S}_{peak}$ is the velocity corresponding to the maximum cohesive force when it is equal to that of the drive velocity, V, during retraction of the AFM cantilever.

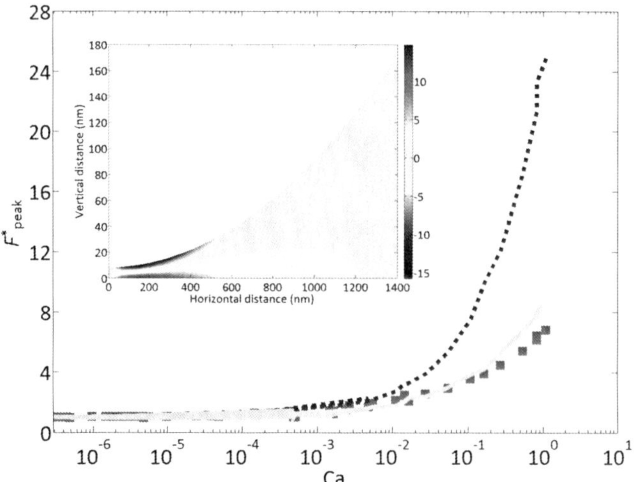

Figure 2 *The measured values of the non-dimensional cohesive force (■), F^*_{peak}, as a function of the capillary number, Ca, together with the values calculated using the capillary-viscous model (continuous, Equation (4)), and capillary-nonlinear viscoelastic model (dotted, Equation (5)). For Ca < 10^{-4} the experimental and calculated data are approximately equal. Inset shows the strain field for Ca = 1 where η_0 = 1,124 Pa.s, H = 186 nm, S = 10 nm, S_0 = 5.6 nm, R = 6 μm, V = 10 μm/s.*

Figure 2 shows that the measured value of F^*_{peak} is ~ 1 for Ca < 10^{-3} and begins to increase gradually for 10^{-3} < Ca < 10^{-2}, up to F^*_{peak} ~ 9 for Ca ~ 1. This represents a transition from the capillary plateau region where F^*_{peak} is independent of velocity to a region where the viscous or viscoelastic interactions are important. At values of Ca > 10^{-2}, the capillary-viscous model (Equation (4)) leads to an increasing over-estimation of the measured forces, *e.g.* for Ca ~ 1, the calculated value is approximately a factor of four greater than that measured. In comparison, the capillary-nonlinear viscoelastic model (Equation (5)) displays excellent agreement with the measured values for Ca < 0.1 but with some divergence for Ca > 0.1. For example, at Ca ~ 1, the measured value of F^*_{peak} is ~ 6.8, while the calculated value using the nonlinear viscoelastic model is ~ 8.3. The inset in Figure (2) is the calculated strain field in the fluid as the particle is separated from the surface, corresponding to Ca = 1. It can be seen that the maximum strain occurs at approximately 200 nm from the axis of symmetry, reaching a value of 14 at the wall.

4 PARTICLE COLLISION SIMULATIONS

4.1 Collisions and elastic rebound in the presence of a thin liquid film

Here the models developed in §2.1 and §2.2 will be applied to the collisions of spherical elastic particles. It is assumed that the compliance of the particles is sufficiently small that the deformation of the solid bodies in the contact region does not significantly affect the flow field of the fluid i.e. elastohydrodynamic coupling is ignored. The collision between two spherical elastic spheres in the absence of a fluid is described by the Hertz equations.[25] The effective Young's modulus, E^*, for two particles is given by:

$$\frac{1}{E^*} = \frac{1-v_1^2}{E_1} + \frac{1-v_2^2}{E_2} \tag{12}$$

where v_i is the Poisson's ratio and E_i is the Young's modulus for particle i. The effective radius, R^* is defined by:

$$\frac{1}{R^*} = \frac{1}{R_1} + \frac{1}{R_2} \tag{13}$$

where R_i is the radius of particle i. The effective mass, m^*, is defined as:

$$\frac{1}{m^*} = \frac{1}{m_1} + \frac{1}{m_2} \tag{14}$$

where m_i is the mass of particle i. Here the collision between a particle ($i = 1$) and a half-space ($i = 2$) is considered for which $1/R_2 = 1/m_2 = 0$ and with $E_1 = E_2$.

It has been shown that the elastohydrodynamic collisions of spherical elastic particles involving a Newtonian fluid depend on two dimensionless parameters.[26] The first is the elasticity parameter, ε, which determines the tendency of the particles to deform.

$$\varepsilon = \frac{4\eta \, \dot{S} R^{*3/2}}{E^* S_0^{5/2}} \tag{15}$$

where η is the fluid viscosity, and $\dot{S}$ is the particle velocity at an initial separation distance S_0. The second is the Stokes number:

$$St = \frac{m^* \dot{S}}{6\pi\eta \, R^{*2}} \tag{16}$$

For $\varepsilon \ll 1$ and $St < 5$, particle deformations are negligible and the rigid sphere approximation considered here is valid.

A critical Stokes number exists, St_c, above which the particles will rebound rather than stick.[9, 27]

$$St_c = b + c \ln\left(\frac{1}{\varepsilon}\right) \tag{17}$$

where b and c are system-specific parameters that depend on the energy dissipated within the impacting bodies, or through deviations due to vortex shedding for high Reynolds number systems.[28] Here the Reynolds number is defined as:

$$Re = \frac{\rho_L R^* \dot{S}}{\eta} \tag{18}$$

where ρ_L is the density of the liquid.

Typical particle radii range from 1–100 µm for pharmaceutical and soil mechanics applications,[29] whereas spherical particles of 12.7 mm radius have recently been employed in the construction of a Stokes cradle,[27] which was used to study wetted particle collisions in a pendulum arrangement similar to that of a Newton's cradle.

4.2 Simulations of particle rebound in the presence of a thin liquid film

For Newtonian liquids, the value of η used in Equations (14), (15) and (18) can be obtained from rheological measurements. For nonlinear viscoelastic liquids, more complex non-dimensional groups would need to be specified due to the strain and strain rate dependence although the zero shear rate viscosity can be used as an upper bound for η.

Here we present example simulations of the normal collisions of spherical SiO_2 particles against a planar half-space, using (i) the capillary-viscous model (Equation (4)) and (ii) the capillary-nonlinear viscoelastic model (Equation (5)). The constitutive parameters for the liquids are given in the figure legend of Figure 3; both liquids have similar zero shear rate viscosities. Particles with $R = 1$ mm, a liquid film of thickness $H = 200$ nm, an initial separation distance $S_0 = 10$ nm, at given initial rebound velocities $\dot{S}_0 = 1$ m/s and 5 m/s are considered. The influence of gravity was ignored in the calculations. In the case of the Newtonian fluid, for example, when $\dot{S}_0 = 5$ m/s and $t = 0$, $\varepsilon = 4.3 \times 10^{-15}$ and $St = 2.3 \times 10^{-2}$. The upper bounds for the viscoelastic cases also satisfied the rigid particle approximation of $\varepsilon \ll 1$ and $St < 5$.

Assuming a coefficient of restitution of unity, in the absence of a liquid film, the maximum compression corresponding to highest velocity, δ_z, would be 7.2 µm, using the following expression:[25]

$$\delta_z = \left(\frac{15 m^* \dot{S}_0^2}{16 R^{0.5} E^*} \right)^{0.4} \tag{19}$$

Thus the deviation from a spherical geometry under the conditions considered here is negligibly small and moreover the actual compression would be less due to the presence of the liquid film.

The results are shown in Figure 3 and for both initial velocities considered, stick occurred for the viscous liquid and rebound occurred for the viscoelastic liquid. For the viscous fluid, the maximum separation distance was greater for the higher velocity as would be expected but the values are small compared to the film thickness. At longer times, the capillary forces cause the particles to return to the initial separation thickness. In the case of the viscoelastic liquid, the rebound velocity corresponds to a constant velocity equal to $\dot{S}_0$. In reality a particle would accelerate from the minimum separation due to the elastic recovery but this has been ignored in the current work.

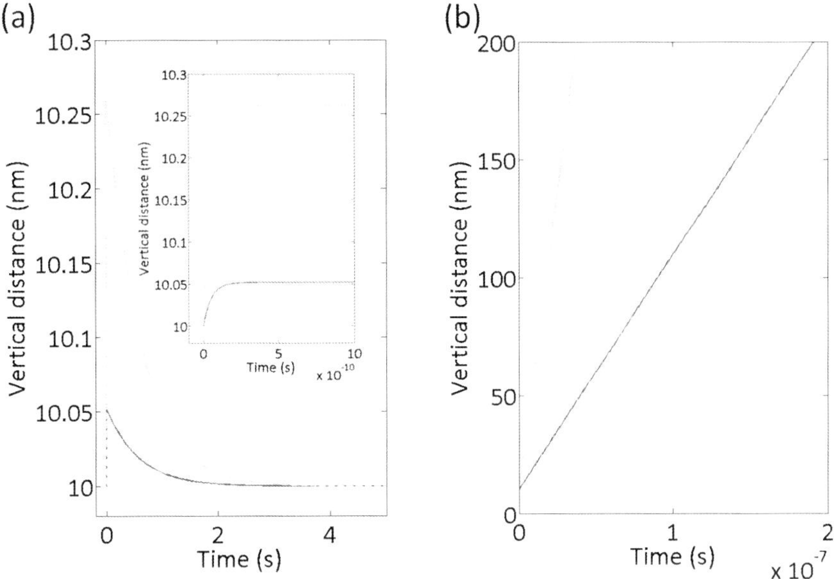

Figure 3 *Calculated trajectories of particles with R = 1 mm in a liquid film of H = 200 nm, at an initial separation distance S_0 = 10 nm, at an initial velocity $\dot{S}_0$ = 1 m/s (solid line) and $\dot{S}_0$ = 5 m/s (dashed line). The liquid film properties are (a) η_0 = 105 Pa.s, and (b) η_1 = 90.77 Pa.s, η_2 = 49.46 Pa.s, η_3 = 63.3 Pa.s, η_4 = 94.99 Pa.s, λ_1 = 56 ms, λ_2 = 84.2 ms, λ_3 = 1 ms, λ_4 = 21.4 ms, α = 0.02, β = 1.9.*

5 CONCLUSIONS

A first order approximation of the impact of particles demonstrates that while a viscous liquid film in the contact region may result in stick, it is probable that a viscoelastic fluid with a similar zero shear rate viscosity will lead to rebound. That is, the current data show that the use of a zero-shear rate viscosity to model polymeric liquid binders is not appropriate since it is likely to overestimate the cohesive strength that they impart to particles. This arises because the maximum strains, corresponding to the peak cohesive force, are sufficiently large that the elastic stresses have been significantly dissipated. At the molecular length scale this may be regarded as a flow-induced untangling of the polymer chains. Nevertheless, the strains are still finite and not sufficient to be treated as a steady state flow condition. An alternative approach is to assign the reduction in the cohesive force to a decrease in the steady flow viscosity with increasing strain rate. Such shear thinning is due to the increasing alignment of the polymer chains with increasing strain rate, which can be described by the Cross model.[30] However, when such a model was investigated, the results were not consistent with the experimental values measured using an AFM because of the unrealistically large strains.[31]

Acknowledgements

The atomic force microscope used in this research was obtained, through Birmingham Science City: Innovative Uses for Advanced Materials in the Modern World (West Midlands Centre for Advanced Materials Project 2), with support from Advantage West Midlands (AWM) and part funded by the European Regional Development Fund (ERDF).

References

1. C. Thornton, G. Lian and M.J. Adams, in *Proc. 2nd Int. Conf. on Discrete Element Methods*, J.R. Williams and G.W. Mustoe (Eds), *IESL Publications*, **1993**, 177.
2. G. Lian, C. Thornton and M.J. Adams, in *Powders and Grains 93*, C. Thornton (Ed), *Balkema*, **1993**, 59.
3. T. Mikami, H. Kamiya and M. Horio, *Chem. Eng. Sci.*, 53, **1998**, 1927.
4. S.C. Yang and S.S. Hsiau, *Chem. Eng. Sci.*, 56, **2001**, 6837.
5. S.S. Hsiau and S.C. Yang, *Chem. Eng. Sci.*, 58, **2003**, 339.
6. R. Zhang and J. Li, *J. Terramech.*, 43, **2006**, 303.
7. H. Zhai, S. Li, D.S. Jones, G.M. Walker and G.P. Andrews, *Chem. Eng. J.*, 164, **2010**, 275.
8. G. Lian, C. Thornton and M.J. Adams, *J. Colloid. Interface Sci.*, 161, **1993**, 138.
9. G. Lian, M.J. Adams and C. Thornton, *J. Fluid Mech.*, 311, **1996**, 141.
10. C.D. Willett , M.J. Adams, S.A. Johnson and J.P.K. Seville, *Langmuir*, 16, **2000**, 9396.
11. C.D. Willett, M.J. Adams, S.A. Johnson and J.P.K. Seville, *Powd. Tech.*, 130, **2003**, 63.
12. P.J.T. Mills, J.P.K. Seville, P.C. Knight and M.J. Adams, *Powd. Tech.*, 113, **2000**, 140.
13. S.J.R. Simons and R.J. Fairbrother, *Powd. Tech.*, 110, **2000**, 44.
14. M.D. Parker, P. York and R.C. Rowe, *Int. J. Pharmaceutics*, 64, **1990**, 207.
15. M.D. Parker, P. York and R.C. Rowe, *Int. J. Pharmaceutics*, 72, **1991**, 243.
16. W. Pietsch, Agglomeration in Industry Volume 2, *Wiley-VCH*, **2005**.
17. G. Lian, Y. Xu, W. Huang and M.J. Adams, *J. Non-Newt. Fluid Mech.*, 100, **2001**, 151.
18. J.D. Ferry, Viscoelastic Properties of Polymers 3rd Ed., *John Wiley & Sons*, **1980**.
19. J. Bowen, D. Cheneler, J.W. Andrews, A.R. Avery, Z. Zhang, M.C.L. Ward and M.J. Adams, *Langmuir*, 27, **2011**, 11489.
20. M.J. Matthewson, *Phil. Mag. A*, 57, **1988**, 207.
21. P.J. Leider and R.B. Bird, *Ind. Eng. Chem. Fundam.*, 13, **1974**, 336.
22. C.W. Macosko, Rheology, *Wiley-VCH*, **1994**.
23. R.B. Bird, R.C. Armstrong and O. Hassager, *Dynamics of Polymeric Liquids Volume 1: Fluid Mechanics*, 2nd Edition, John Wiley & Sons, **1987**.
24. J. Bowen, D. Cheneler, D. Walliman, S.G. Arkless, Z. Zhang, M.C.L. Ward and M.J. Adams, *Meas. Sci. Technol.*, 21, **2010**, 115106.
25. K.L. Johnson, *Contact Mechanics*, Cambridge University Press (Cambridge), **1985**.
26. R.H. Davis, J.M. Serayssol and E.J. Hinch, *J. Fluid Mech.*, 163, **1986**, 479.
27. C.M. Donahue, C.M. Hrenya and R.H. Davis, *Phys. Rev. Lett.*, 105, **2010**, 034501.
28. E. Achenbach, *J. Fluid Mech.*, 62, **1974**, 209.
29. D.W. Taylor, *Fundamentals of Soil Mechanics*, Wiley, **1948**.
30. M.M. Cross, *J. Colloid Sci.*, 20, **1965**, 417.
31. J. Bowen, D. Cheneler, J.W. Andrews, A.R. Avery, Z. Zhang, M.C.L. Ward and M.J. Adams, Unpublished results.

EFFECT OF THE PENDULAR STATE ON THE COLLAPSE OF GRANULAR COLUMNS

R. Artoni[1], F. Gabrieli[2], A. Santomaso[1] and S. Cola[2]

[1] Dept. of Industrial Engineering, University of Padova, via Marzolo 9, 35131 Padova, Italy; e-mail: riccardo.artoni@unipd.it
[2] Dept. of Civil, Architecture and Environmental Engineering (ICEA), University of Padova, via Ognissanti 39, 35129 Padova, Italy.

1 INTRODUCTION

Due to the complex behaviour of granular materials, the study of the transition between static and dynamic conditions becomes an important challenge for the scientific investigation. Several experimental and numerical analyses of both quasi-static and fully developed flows at steady state conditions were carried out in the past. Many of these studies examined flows on chutes,[1] in ring-shear cells,[2] or the deposition on a slope with the natural repose angle.[3]

On the other hand, three most famous tests incorporating transition phases (flowing and stopping/jamming) are the rotating drum test, the silo discharge and the column collapse. The latter was explored in many conditions: 2D or 3D, with cylindrical or rectangular initial geometry, with different aspect ratios or with various granular materials, etc.[4, 5]

It is well known that the presence of water strongly affects the behaviour of a granular material especially in the transition between static and dynamic conditions in which the attractive forces between particles become lower than the inertial ones and vice versa. The water effect is important especially in the pendular state (for saturation degree lower than about 5%) in many engineering applications: in geotechnics (e.g.. in slope stability problems), in many chemical processes (e.g. wet agglomeration) or granular transport and handling (discharge from silos, clogging, etc.).

In comparison with other experiments the collapse test permits to better focus on phenomena involved in transition phases without introducing the effects of other variables, such as angular velocity and axial dispersion of particles occurring in rotating drums, or orifice geometry in case of discharge from silos.

The Discrete Element Method (DEM) is particularly suitable to simulate these kinds of tests providing that a capillary law is introduced in the model and a redistribution function of the liquid volumes is implemented, as it will be demonstrated in this work. The numerical analyses performed with this improved tool enhance the comprehension of the mechanical behaviour of granular materials with different grain sizes, wetting concentrations, liquid surface tensions and contact angles.

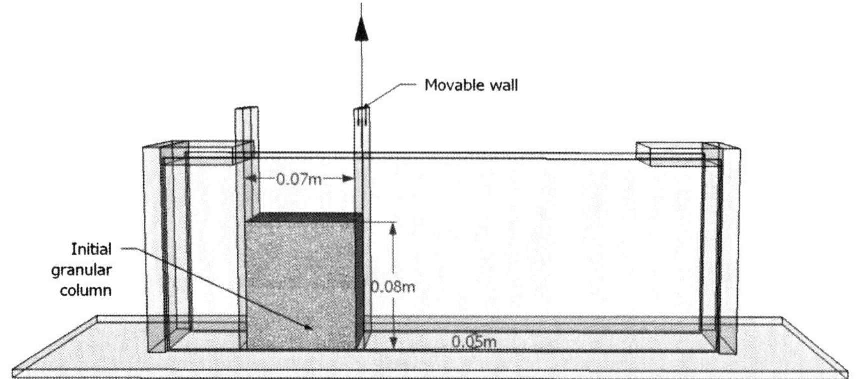

Figure 1 *Layout of the collapse test device.*

2 COLUMN COLLAPSE EXPERIMENTS

2.1 Materials and methods

The experimental program consists of some collapse tests with an initial cuboid geometry (size 7 x 8 x 5 cm) inside the glass box as sketched in Figure 1. Glass ballotini with different grain sizes (2, 3 and 5 mm in diameter) were used in various wetting conditions (liquid contents equal to 0, 0.5, 1, 2 and 4% of the weight of the dry materials). Two liquids were employed for wetting: distilled water (surface tension γ = 72.75 mN/m; density ρ = 1 g/cm^3; viscosity 1 mPa s) and water added with a fluorinated surfactant (γ = 17.2 mN/m; ρ = 1 g/ml; viscosity 1 mPa s).

The initial granular column was formed by pouring the wetted materials with a funnel, sustaining the column with a movable glass lateral wall. To start the test the wall was lifted by a weight/pulley system while its movement was constrained in the vertical direction.

Each test was captured using a semi-professional high-speed camera (Casio EX-F1). The analysis of the image sequence enabled us to obtain the time evolution of the granular mass and to measure the run-out length, the height of the pile and the angles of the slope at each time step. These experimental data were compared with DEM simulations to evaluate the accuracy of the numerical analysis.

2.2 Experimental Results

Figure 2 showed the normalized run-out length obtained in all the experimental tests. It is evident that the collapse behaviour depends on both water content and grain size especially for finer graded materials. The dry materials stopped approximately with the same slope profiles and run-out lengths irrespective of the grain size. The same results are obtained for the coarsest material with all the moisture contents (from 0 to 4%) and surface tensions considered, indicating that for coarse materials the effect of liquid is negligible.

On the other hand, small particles (D = 2 mm) show a dependence on the liquid content with lower values of run-out length for higher liquid content, and a weak dependence on the surface tension of the wetting phase. The intermediate size particles (D = 3 mm) are sensitive to both liquid content and surface tension in a weak manner.

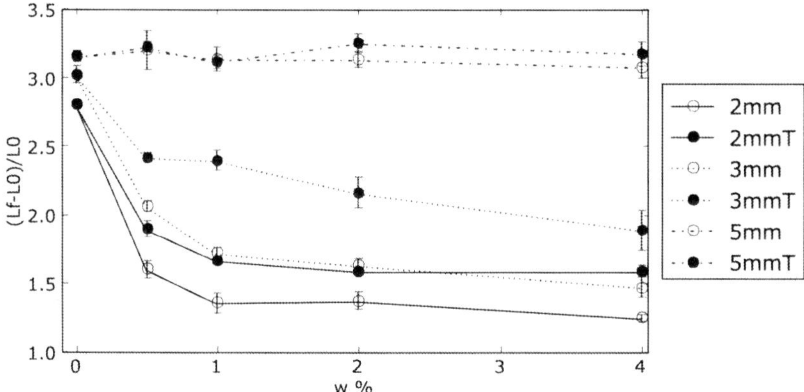

Figure 2 *Normalized run-out length with different water contents, grain sizes and wetting fluids. The suffix "T" in the legend refers to tests with water-surfactant solution. Lf is the final run-out value and L0 the initial width of the column (=7 cm).*

3   THE CAPILLARY FORCE ALGORITHM

When a liquid is present bewteen two particles an attraction force exists. Experimental analyses to quantify the intensity of this force were extensively carried out in the past, but its measure is difficult when the capillary volume or the particle distance is very small.

The capillary force could be taken into account in numerical algorithms using two different theoretical approaches: the approach based on the classical Young-Laplace equation[6,7] or the minimum energy approach.[8,9] Some authors demonstrated that the two methods are equivalent in some simple cases,[9] the minimum energy approach, for which Lambert et al.[9] have recently developed an approximate theoretical solution, is hence adapted in this study.

In the model developed by Lambert et al.[9], the capillary force variation is related in a simple way to particle distance and capillary volume, which can also be used to analyse the effect of water evaporation.[10] The capillary force between two spherical particles is given as:[9]

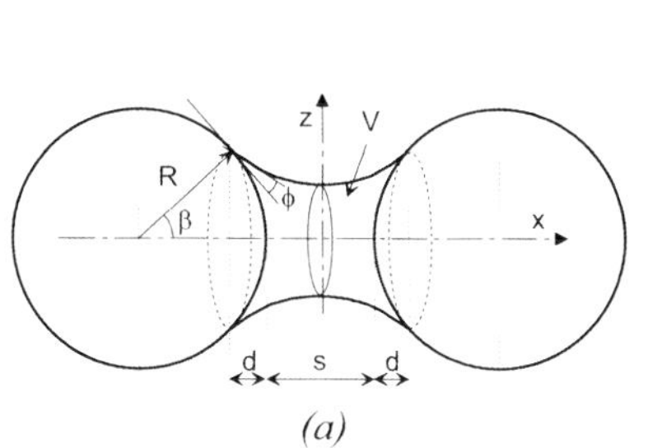

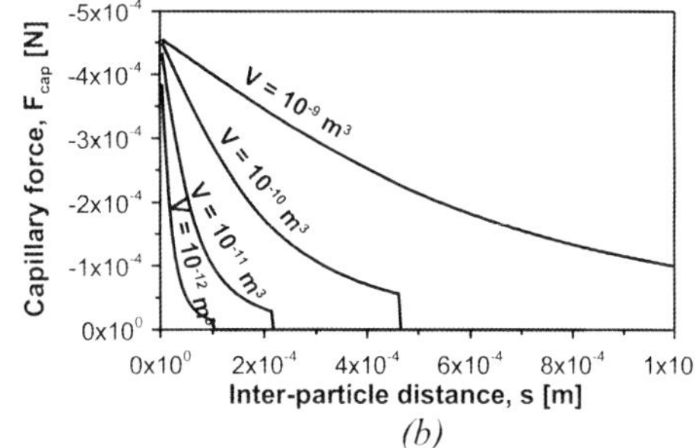

(a)   (b)

Figure 3 *(a) Sketch of the capillary bridge geometry; (b) capillary force functions for different capillary volumes (R = 1mm; γ = 72.75 mN/m).*

$$F_{cap}(s,V) = -\frac{2\pi R\gamma\cos\phi}{1+[s/2d(s,V)]} \tag{1}$$

where R is the particle radius, s the gap, V the capillary volume (Figure 3a). The parameter d is the wet spherical segment on grain surface that depends on V and s:

$$d(s,V) = \frac{s}{2}\left[-1+\sqrt{1+\frac{2V}{\pi Rs^2}}\right] \tag{2}$$

During the collapse of the wet granular column, the net of capillary bridges deforms and changes continuously, breaking at some points as well as reforming when the water volumes of two nearby particles come in contact.

The distance at which the capillary bridge fails is called critical distance and it can be computed using the relationship proposed by Lian et al.[7]

$$s_c = \left(1+\frac{\phi}{2}\right)V^{1/3} \tag{3}$$

where ϕ is the contact angle (Figure 3a).

It was assumed that when a capillary bridge breaks the volume splits into two equal drops on the two particles. The drops are fixed to the particles surface and their position is updated at each time step to follow the particle rotation in space. As two drops approach, they can form a new capillary bridge if the following two conditions are simultaneously met:

i) the distance between the two particles is smaller than:[11]

$$s_{form} = \frac{12V}{\pi\left(1+\dfrac{1}{1-\cos\phi}\right)} \tag{4}$$

with $s_{form} < s_c$.

ii) the angle between the directions of the two drops is lower than a certain threshold θ. In the following study we adopted $\theta = 45°$, which is a reasonable value considering that the drops have a finite volume and the glass surface is not completely hydrophilic.

If a new capillary bridge is formed, it assumes that its volume is the sum of the volumes of the linked drops.

4 SIMULATIONS AND RESULTS

The experimental tests were reproduced with 1:1 scale simulations: the discrete model used 2253 particles when 5 mm diameter glass ballotini is used and 37540 particles for 2 mm diameter spheres.

The initial distribution of liquid volume in the column mass was assumed homogeneous. The total water volume was subdivided among all the potential contacts (i.e. contacts with distance satisfying Equation (3)) starting from a trial value of capillary bridge volume and iteratively verifying the total amount of water. In this way, it was possible to find a stable particle configuration and the assembly could relax without artificial localized tension.

In dry condition, the contact between two particles was simulated using a linear spring for the normal contact and a linear spring plus a slider in tangential direction.[12] A viscous dashpot was applied at the normal and tangential contact. The elastic normal and tangential stiffness (k_n and k_t) of the springs and the contact friction coefficient ($\tan\phi$) of slider were calibrated comparing the results of a triaxial test performed on glass ballotini at a confinement stress σ_3 equal to 100 kPa with a cubic true triaxial test simulation.[10] The same parameters (stiffness and friction) were applied to the ball-wall contact. Table 1 lists the values of all the contact parameters chosen in the DEM analyses.

In the analyses the wall was lifted up imposing an acceleration equal to the gravitational acceleration, which was measured in the laboratory tests.

Table 1 *Contact parameters used in DEM simulations.*

Normal stiffness k_n [kN/m]	Tangential stiffness k_s [kN/m]	Contact friction coeff. $\tan\phi$	Normal and tangential viscous damping coeff. n
400	100	0.62	0.1

4.1 Effect of water contents

The variation of the water content in a granular material in the pendular regime leads to a variation of the volume of capillary bridges and then to a change in the capillary forces. In Figure 3*b* capillary forces for different capillary volumes are presented. It can be noted that for $s = 0$ (at the contact) the capillary force is independent of the liquid volume. This is consistent with the fact that for two particles in contact the capillary force is at its maximum, equal to $2\pi R\gamma\cos\phi$.

In the collapse column tests different water contents produce different final slope profiles. However, the differences observed are not significant as it might be expected (Figure 4). Remarkable differences were observed when comparing dry (Figure 4*b*) with wet conditions (Figures 4*c*, 4*d*, 4*e* and 4*f*), but the concentration of water in the granular mass slightly affects the run-out length and slope angle. Similar results was also obtained in other studies considering different experiments: unconfined axial compression[13] or shear tests.[14] The presence of water seems to act like a switch (on or off) rather than a gentle regulator of the dependence on the water concentration.

This is partially consistent with the previous information: for gaps $s = 0$, no variation of the capillary force is possible for any variations of the capillary volume. If the material is dense (spherical particles are prone to dense state than others) in the granular mass there are many capillary volumes with zero gap.

4.2 Effect of liquid surface tension

In collapse tests the effect of a lower surface tension reflects on a slight sticky behaviour of the material, a larger run-out and a final profile with lower angles as it can be seen from the Figure 5. This behaviour was only observed using small particles (2 and 3 mm) while it disappeared for the larger ones.

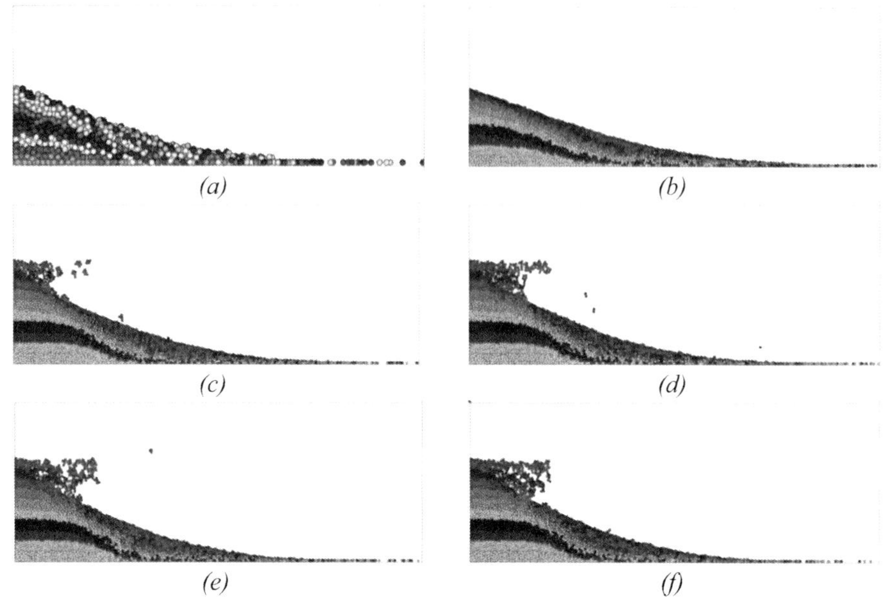

Figure 4 *Final slope profiles with dry particles of (a) D = 5 mm and (b) D= 2 mm and wet particles of D = 2 mm and a water content of (c) 0.5%, (d) 1%, (e) 2% and (f) 4%.*

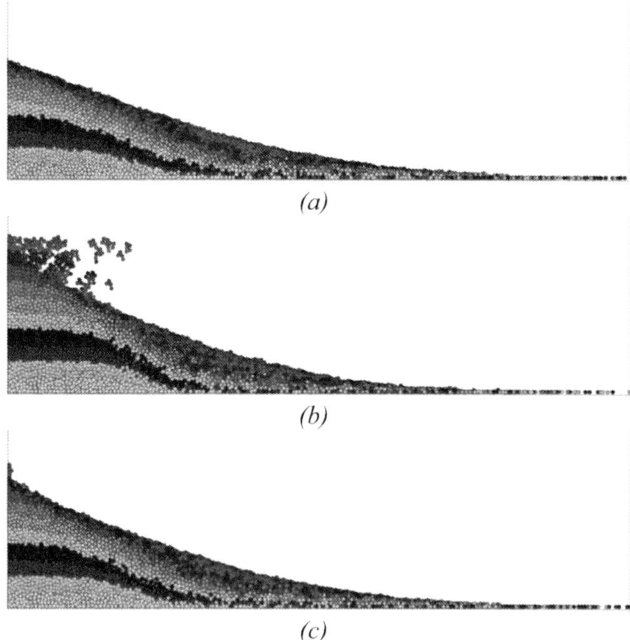

Figure 5 *Final profiles for D = 2mm in (a) dry case, (b) wet case with 1% of water or (c) with 1% of water and surfactant.*

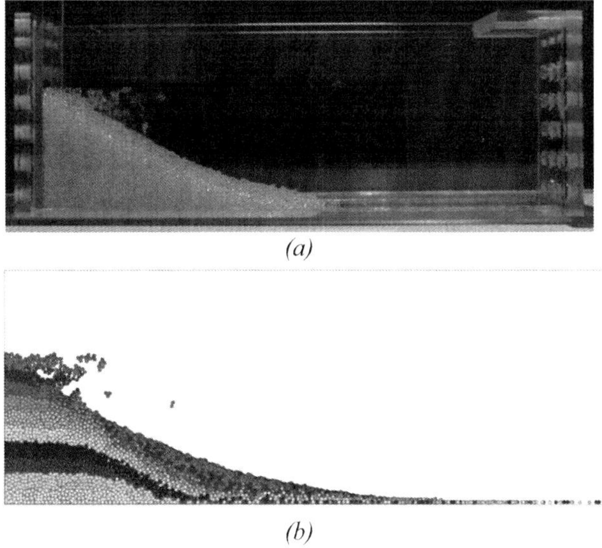

Figure 6 *Final slope profiles for D = 2mm and w = 0.5% obtained from (a) experimental and (b) numerical column collapse tests.*

4.3 Experiments vs. simulations

Analysing the toe of the final slope profiles in numerical simulations it is evident that the run-out length is overestimated resulting in gentler toe angles compared to the experimental results (Figure 6).

A close examination of the mechanical behaviour of particles at the slope toe in the simulations reveals that there are particle rotations despite of the presence of many capillary bridges. This fact is supposed to be not real since the liquid bridge between two moving surface should exert some rotational constraints that are mainly due to the hysteretic behaviour of the contact angle.

5 CONCLUSIONS

Discrete element simulations of column collapse tests were successfully performed to mimic 1:1 scale experiments. The effect of the water was recognized comparing dry with wet conditions while the variation in final slope profile with different liquid concentrations (0.5, 1, 2 and 4%) seemed to be very low both in simulations and in experiments.

Generally, greater differences were experienced in the rear of the slope where static and quasi-static phenomena prevail. The final angle of repose at the toe is not so different for different water contents, which is probably due to large inertial forces that drive the whole phenomena at the toe.

The run-out lengths of tests in wet cases are quite different in comparison with that obtained in simulations. In particular, a local large spread of the toe in the run-out direction was observed in all the wet simulations. This deviation could be probably corrected by adding a rolling resistance to the particles sharing a capillary bridge due to the hysteretic behaviour of the contact angles. This model improvement is currently ongoing.

References

1. L. Silbert, G. Grest, T. Halsey, D. Levine and S. Plimpton, *Phys. Rev. E*, 2001, **64**, 051302, 1.
2. S. Luding, *Particulate Science and Technology*, 2008, **26**, 33.
3. A. Samadani and A. Kudrolli, *Phys. Rev. E*, 2001, **64**, 051301.
4. G. Lube, H. Huppert, R. Sparks and A. Freundt, *Phys. Rev. E*, 2005, **72**, 1.
5. E. Lajeunesse, J. Monnier and G. Homsy, *Phys. Fluids*, 2005, **17**, 103302.
6. R.A. Fisher, *J. Agricult. Sci.*, 1926, **16**, 492.
7. G. Lian, C. Thornton and M. Adams, *J. Colloid Interface Sci.*, 1993, **161**, 138.
8. J.N. Israelachvili, *Intermolecular and Surface Forces*. Academic Press, London, 1992, 224.
9. P. Lambert, A. Chau, A. Delchambre and S. Régnier, *Langmuir*, 2008, **24**, 3157.
10. F. Gabrieli, P. Lambert, S. Cola and F. Calvetti, *Int. J. Num. Anal. Meth. Geomech.*, 2012, (in press), doi: 10.1002/nag.1038.
11. Y.I. Rabinovich, M.S. Esayanur and B.M. Moudgil, *Langmuir*, 2005, **21**, 10992.
12. P.A. Cundall and O.D.L. Strack, *Geotechnique*, 1979, **29**, 47.
13. F. Soulié, F. Cherblanc, M. El Youssoufi and C. Saix, *Int. J. Num. Anal. Meth. Geomech.*, 2006, **30**, 213.
14. V. Richefeu, M.S. El Youssoufi and F. Radjaï, *Phys. Rev. E*, 2006, **73**, 051304.

INVESTIGATION OF DYNAMIC BEHAVIOUR OF A PARTICLE-LOADED SINGLE FIBRE USING DISCRETE ELEMENT METHODS

M. Yang[1], S.Q Li[1], G. Liu[1] and J. S. Marshall[2]

[1] Key Laboratory for Thermal Science and Power Engineering of Ministry of Education, Department of Thermal Engineering, Tsinghua University, Beijing, 100084, China
[2] School of Engineering, The University of Vermont, Burlington, VT 05405, USA

1 INTRODUCTION

The capture of fine particles, as an essential environment issue, has attracted significant attention with the recognition of their serious effects to the environment, climate change, and public health. The fibrous filtration, with the advantages of high removal efficiency of fine particles than other de-dust devices, has been widely used in power plants, mining engineering, cement industries and even life areas. A fibrous filter is usually made up of many slender fibres, arranged more or less normal to the direction of the fluid flow. During the filtration process, particles traveling with the fluid through the regions between fibres are removed by their collisions with the fibre or the particles that already deposited.[1] As the accumulation of the deposited particles, the fibre blocking area dramatically increases, resulting in a notable change of capture efficiency and pressure drop. This kind of enhancement effect of fibre particle-loading on the capture efficiency should be carefully considered.

Several literatures discussed the change in capture efficiency caused by particle loading. Billings[2] observed an increase of the single fibre efficiency as a linear function of loaded particle number. Kasper *et al.*[3] reported a power-law relationship between the efficiency and the loaded mass of particles. However, the microscopic mechanism beyond the experimental observations has not been understood yet. There are several phenomenological models to account for the capture efficiency of a particle-loaded single fibre. However, those models hardly considered the particle distribution on the fibre and the dynamic variation of particle dendrites.

In this paper, based on JKR adhesive contact theory, a novel discrete element method (DEM) approach is used to investigate the mechanism of the efficiency change during the filtration of the particle-loaded single fibre, and to detailedly examine the influence of the particle dendrite structure on the filtration.

2 THE DEM MODEL

The DEM approach tracks the motion of every individual particle by using Newton's second law, both transitionally and rotationally. The governing equations are given as:

$$m\frac{d\mathbf{v}}{dt} = \mathbf{F}_F + \mathbf{F}_A, \quad I\frac{d\mathbf{\Omega}}{dt} = \mathbf{M}_F + \mathbf{M}_A, \tag{1}$$

where m denotes the particle mass, and $I = (2/5)mr_p^2$ is the particle moment of inertia with r_p as the particle radius. The forces acting on the particle are mainly the fluid force $\mathbf{F}_F$, the contact force and the van der Waals (VDW) adhesive force. The contact force and VDW force both acting during the contact of particles are coupled, so in the model they are calculated together and here denoted by $\mathbf{F}_A$. $\mathbf{M}_F$ and $\mathbf{M}_A$ give the fluid torque and the coupled collision and van der Waals adhesion torque respectively, for the equation of particle rotational motion. A brief view of DEM model is given here, and readers are referred to Marshall[4] and Li *et al.*[5] for more details.

2.1 The Fluid Force and Torque

The dominated fluid force on the particles is the viscous drag, considering the particles are heavy compared to the fluid ($\chi \equiv \rho_f/\rho_p < 1$). In the model the force is approximated by a modified form of the Stokes drag law,

$$\mathbf{F}_F = -3\pi\mu d_p\left(\mathbf{v} - \mathbf{u}\right)f, \tag{2}$$

where $\mathbf{v}$ and $\mathbf{u}$ are the particle velocity and the local fluid velocity. The friction factor f is a correction for local particle crowding effect. The form proposed by Di Felice is used in the model with f as a function of particle concentration c and the particle Reynold number $\mathrm{Re}_p = \rho_f|\mathbf{v} - \mathbf{u}|d_p/\mu$ at the local position.[6]

$$f = \left(1-c\right)^{-\beta}, \quad \beta = 3.7 - 0.65\exp\left(-\frac{1}{2}\left[1.5 - \ln\left(\mathrm{Re}_p\right)\right]^2\right), \tag{3}$$

when accounting for the isolated particle, the factor takes on the value $f = 1$, and the equation recovers to the form of classic solution. The dominated fluid torque on particles is caused by the difference in rotation rate of particle and corresponding fluid region. The torque is given as,

$$\mathbf{M}_F = -\pi\mu d_p^3\left(\mathbf{\Omega} - \frac{1}{2}\boldsymbol{\omega}\right), \tag{4}$$

where $\mathbf{\Omega}$ is the particle rotation rate and $\boldsymbol{\omega}$ is the fluid vorticity vector. Other forces including lift force, added mass force, Magus force and reduced gravitational force are also induced by fluid, which are also considered in the model.

2.2 The Collision, Adhesive Force and Torque

Particle collisions are simulated using the soft-sphere model, also accounting for the contact force and a variety of adhesive forces.[5,7] The contact force and torque reflect the elastic and dissipation resistance during the normal-approaching, sliding, rolling and twisting processes of two particles. The presence of particle adhesion mainly causes a change in the normal and sliding resistance, and also introduces a rolling resistance torque. Coupled with the van der Waals adhesion, the collision force and torque on particle i can be written as,

$$\mathbf{F}_A = F_n \mathbf{n} + F_s \mathbf{t}_s, \qquad \mathbf{M}_A = r_i F_s \left(\mathbf{t}_s \times \mathbf{n} \right) + M_r \left(\mathbf{t}_r \times \mathbf{n} \right) + M_t \mathbf{n}, \tag{5}$$

where $\mathbf{n}$, $\mathbf{t}_s$ and $\mathbf{t}_r$ are respectively the normal, sliding, and rolling direction unit. F_n and F_s are the normal and the sliding contact force, while M_r and M_t are the rolling and twisting torques. All these forces and torques consist of an elastic component and a dissipative component.

For spherical particles, the normal contact force acts along the line connecting two particle centres, the direction unit of which is given by $\mathbf{n} = \left(\mathbf{x}_j - \mathbf{x}_i \right) / \left| \mathbf{x}_j - \mathbf{x}_i \right|$. The combined normal elastic and adhesive force is modelled with the John-Kendall-Roberts (JKR) theory. In order to simplify construction of a DEM simulation, Chokshi et al. rearranged the results of JKR theory to express F_{ne} in terms of the contact region radius $a(t)$ and the normal overlap δ_N as,[8]

$$\frac{F_{ne}}{F_C} = 4 \left(\frac{a}{a_0} \right)^3 - 4 \left(\frac{a}{a_0} \right)^{3/2}, \qquad \frac{\delta_{ij}}{\delta_C} = 6^{1/3} \left[2 \left(\frac{a}{a_0} \right)^2 - \frac{4}{3} \left(\frac{a}{a_0} \right)^{1/2} \right]. \tag{6}$$

In these equations, a is the current contact radius, and δ_{ij} is the corresponding overlap of two particles. The critical contact force F_C and overlap δ_C, and the equilibrium contact radius a_0 are determined by,

$$F_C = 3 \pi \gamma R_{eff}, \qquad \delta_C = \frac{a_0^2}{2(6)^{1/3} R}, \qquad a_0 = \left(\frac{9 \pi \gamma R^2}{E} \right)^{1/3}, \tag{7}$$

in which γ is the surface energy, while R_{eff} and E are respectively the equivalent radius and elastic moduli of the two collision particles. The dissipative normal term is counted in with the expression derived by Tsuji *et al.*[9] as,

$$F_{nd} = -\eta_N \mathbf{v}_R \cdot \mathbf{n}, \qquad \mathbf{v}_R = \mathbf{v}_{Ci} - \mathbf{v}_{Cj}, \tag{8}$$

where η_N coefficient of damping, and $\mathbf{v}_R$ is the relative velocity at the contact point.

The sliding resistance can also be decomposed into the elastic and dissipative components. It is given as,

$$F_s = -k_T \left(\int_{t_0}^t \mathbf{v}_s \left(\tau \right) d\tau \right) - \eta_T \mathbf{v}_s \cdot \mathbf{t}_s, \qquad \mathbf{v}_s = \mathbf{v}_R - \left(\mathbf{v}_R \cdot \mathbf{n} \right) \mathbf{n} \tag{9}$$

where k_T and η_T are respectively the tangential dissipation and damping coefficient, and $\mathbf{v}_s$ is the sliding velocity.

The twisting torque can be seen as the iteration of torques introduced by local sliding with the relative velocity $v_R = r\Omega_T$ between the contact surfaces. So it can also be given in the form of equation (9),

$$M_t = -k_Q \left(\int_{t_0}^{t} \Omega_T(\tau)\,d\tau \right) - \eta_Q \Omega_T, \quad \Omega_T = \left(\Omega_i - \Omega_j \right) \cdot \mathbf{n} \tag{10}$$

where k_Q and η_Q are respectively the torsional dissipation and damping coefficient, which could be derived from k_T and η_T.

Bagi and Kuhn[10] defined an objective expression for rolling displacement, with the description of rolling velocity independent of reference frame. Using the definition of rolling velocity, the rolling resistance can be similarly given as sliding and twisting,[11]

$$\mathbf{v}_L = -R_{eff}\left(\Omega_i - \Omega_j \right) \times \mathbf{n} - \frac{1}{2}\frac{r_j - r_i}{r_j + r_i}\mathbf{v}_R, \quad \mathbf{t}_R = \mathbf{v}_L / |\mathbf{v}_L|,$$

$$M_r = -k_R \left(\int_{t_0}^{t} \mathbf{v}_L(\tau)\,d\tau \right) \cdot \mathbf{t}_R - \eta_R \mathbf{v}_L \cdot \mathbf{t}_R, \tag{11}$$

where k_R and η_R are respectively the torsional dissipation and damping coefficient.

The sliding, twisting and rolling resistance will reach critical values as the deformation exceeds a certain limit, after which the resistance keeps constant and the particles start to slide or spin. In presence of van der Waals adhesion, the critical values are given as,

$$F_{r,crit} = \mu_f \left| F_{ne} + 2F_C \right|,$$

$$M_{t,crit} = \frac{2}{3}\mu_f a \left| F_{ne} + 2F_C \right|, \tag{12}$$

$$M_{r,crit} = -4F_C \left(a / a_0 \right)^{3/2} \theta_{crit} R.$$

Here μ_f is the friction coefficient.

2.3 Multiple Time Steps Framework

The ultra-small time scale of particle adhesive contact makes it difficult and challenging to simulate long period and large particle number process. An algorithm reported in Marshall[4] or Li et al.[5] is used to overcome the problem. The algorithm is based on three time steps— a fluid time scale $T_F = O(L/U)$, a particle time $T_P = O(\min(T_{CP}, T_{AP}))$ and a collision time scale $T_C = O\left(d_p \left(\rho_p^2 / E_p^2 U \right)^{1/5} \right)$, where $T_{CP} = d_p / U$ is the particle convection time scale, and $T_{CP} = St T_F$ is the particle aerodynamic time scale. The stokes number is given by $St = \rho_p d_p^2 U / 18\mu L$, where L and U are the characteristic length and velocity of the fluid field. For simulation of fine particles we can generally acquire the relation between the three time scales as $T_F \gg T_P \gg T_C$. Three corresponding time steps ($\Delta t_F = f_1 T_F, \Delta t_P = f_2 T_P, \Delta t_C = f_3 T_C$) are defined on the basis of these time scales, with f_1, f_2, f_3 much less than unit. Different computational tasks are assigned to each of the time steps. On the fluid time step, a particle 'local list' is formed to identify the adjacent particles, and particle forces are fed back for fluid field updating. On the particle time step, forces introduced by fluid are affirmed, and collision of particles is identified. The particles baring adhesive and collision forces are evolved on the collision time step.

3 SIMULATION SETUP

3.1 Simulation Parameters

Focusing on the behaviour of the particle-loaded isolated fibre, 3D DEM Simulations are conducted. We used polystyrene (PSL) spheres as the material in this work, with the density of 1050 kg/m^3. The work of adhesion for PSL particles are assumed to be 20 mJ/m^2 that is similar to the value reported in Ding et al.[12] The size of particles used in this work is 2.6 μm. The air is used as filtration streams (the density and viscosity are respectively 1.205 kg/m^3 and 1.79×10^{-5} $Pa \cdot s$). The filtration gas velocity is 1.2 m/s. The fibre diameter is set at 30 μm.

3.2 Simulation Domain and Fluid Field

The fluid field is acquired by the CFD method numerically with the flow variables normalized by the characteristic length L and velocity scale U_0.

$$u^* = u/U_0, x^* = x/L, t^* = t/(L/U), p^* = p/(\rho_f U_0^2) \tag{13}$$

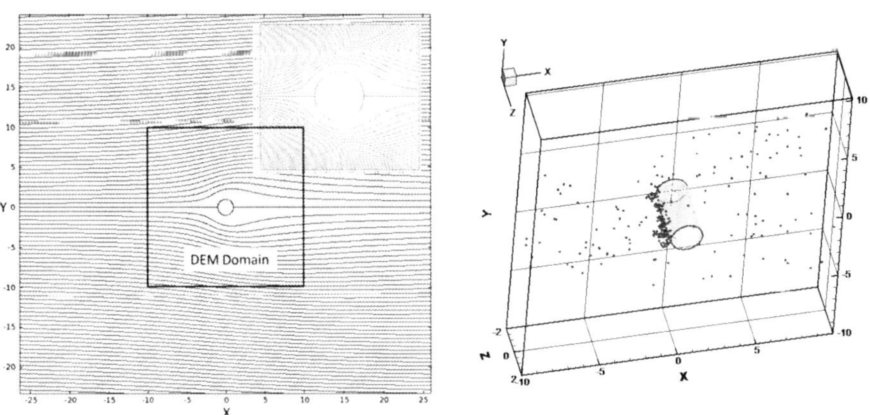

Figure 1 *The fluid field calculation and DEM domain*

The characteristic length L is set as the fibre radius ($L = r_f = 15\ \mu m$), and the capital velocity is 1.2 m/s. In order to acquire the fluid field around an isolated fibre, the domain length applied in fluid field computation is much larger than the fibre radius ($l >> 1$). The particle calculation is conducted in the select region ($-10 \leq X \leq 10$, $-10 \leq Y \leq 10$) as shown in Figure 1. On the third dimension, the simulation domain size is 4 ($-2 \leq Z \leq 2$).

The left boundary ($X = -10$) is the inlet for fluid and particles which are injected in the region $-4 \leq Y \leq 4$, $-2 \leq Z \leq 2$. There is no need to inject particle from the full length of Y, which will be shown later. The right, top and bottom boundaries are outlet conditions for fluid and particles. The front and back boundaries are periodic conditions, which represents the infinite length of the fibre. The boundary of the cylinder, denoting the fibre, is of the wall condition.

3.3 The Statistics Methods

Two methods are used to calculate the capture efficiency of the particle-loaded fibre. As for the first one, the capture efficiency of the N-particles loaded fibre is defined by the fraction of the next n (n=100) captured particles to the total particle injected into the fibre projecting domain. Then, the second method is to carry on another simulation for the N-particles state, where the states of these particles are held, and newly captured particles are removed immediately after their deposition. The efficiency is given by fraction of the particles removed by both the fibre and the dendrites to the total particles that are injected in. The two methods are compared in the following section.

The third statistics method is also applied to investigate the blocking effect of the fibre and the deposited particle dendrites. In this work, the states of selected deposited particles are also held or "frozen" and the newly captured particles are removed. The number of particles passing through the surface at X=0 is counted, and distributed along the Y axis.

4 RESULTS AND DISSCUSSION

4.1 The Capture Efficiency

Figure 2 shows the dynamic variation of single-fibre capture efficiency with the increment of the loaded particle number on the fibre. Two statistical methods used for predicting capture efficiency are almost consistent with each other. The particle capture process of a single isolated fibre can generally be divided into two stage—the bare stage and the particle-loaded stage. The bare stage starts initially, and ends up no longer than 100 particles that have been loaded. It equals about 2.17% of total volume of deposited particles to the fibre volume. In this initial stage of the bare fibre, particles loading on the fibre are relatively few, and are well distributed along the fibre surface. In this stage, it is also found that the short particles' dendrites provide little blocking area in contrast to fibre itself. So in the bare stage, the capture efficiency changes little with the number of loaded particles.

For the second stage of the loaded fibre, the particles are nearly captured by the dendrites of deposited particles. These particle dendrites (or aggregates) provide the major blocking area for coming particles. It is concluded that the changes of capture efficiency along the time behaves a power-law with the number of loaded particles. The left subplot of Figure 2 also shows a comparison between the DEM simulation result and the analytical results predicted by Kasper *et al.*[3] The capture efficiency of a loaded fibre is normalized by the initial efficiency. Except the apparent difference between experiments and DEM simulations at the initial stage, good agreement is obtained when more than 400 particles are loaded.

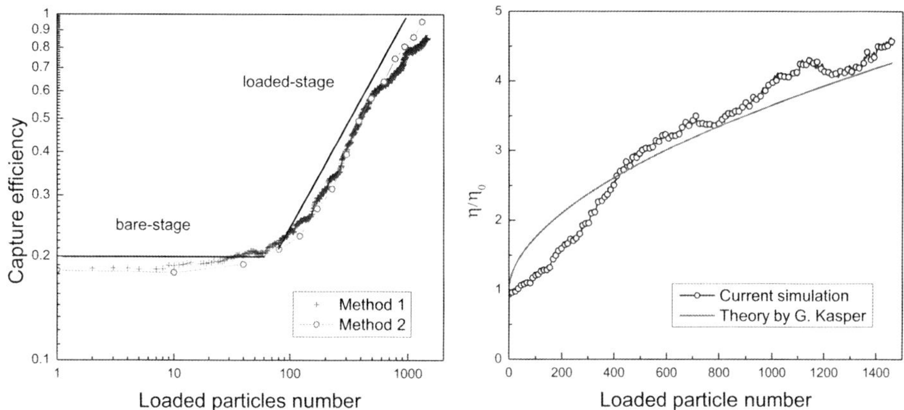

Figure 2 *The capture efficiency v.s. the loaded particle number*

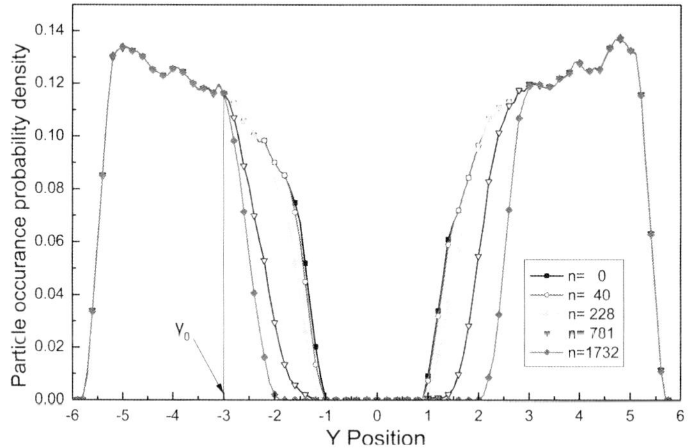

Figure 3 *The blocking effect of the fibre and loaded particles*

4.2 The Blocking Effect

The change of capture efficiency with the deposit accumulation is mainly due to the increment of blocking area of both fibre and loaded particles. The particle aggregates on the fibre are dendrite-like, branching out into the fluid field, providing much more particle blocking area than the bare fibre alone. Figure 3 shows the blocking effect of different loaded particle number. The abscissa is the normalized Y position, while the ordinate is the probability density of particle occurrence. The statistics method is given in Section 3.3.

According to the definition of the axes, the area below the line denotes the penetration ($1-\eta'$, $\eta'=\eta/4$). So the area within two lines gives the efficiency increase due to the particles accumulation. The position where the line meets the line of n=0 reflects the maximum of blocking range.

At the bare-fibre stage, the blocking effect changes little with particles loaded. When the particle number exceeds a certain value (about 100 for this case), the blocking of loaded particles affects. As the particle number increases, the blocking range extends. There seems to be a maximum for the blocking ([-3, 3] this case). It reflects the limit of the particle dendrite length in the Y direction. Longer dendrites will be collapsed or even blown off by strong fluid shear. As the loaded particle number increases, the dendrites crowd and intercept more particles, leading to a lower penetration.

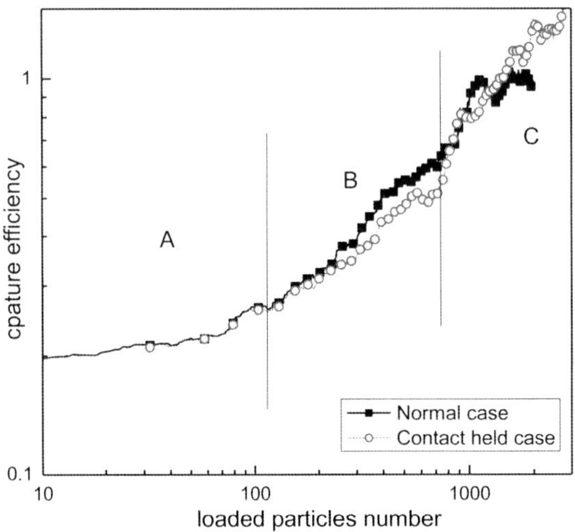

Figure 4 *Comparison between the conventional case (■) with dendrite's dynamic variation and the "frozen" contact case (O)*

4.3 The Behaviour of Particle Dendrite

The loaded particles on the fibre cause the increase in the total blocking area and then the increase in capture efficiency. The change in capture efficiency depends not only on the accumulated number but also on the state in which these particles exist. The formed dendrites of particles undergo dynamic variation during the capture process. Bending, collapsing and blowing-off of the dendrites are observed, which are consistent with the previous reports.[13, 14] In this work, particularly a comparison between the conventional case and the "frozen" contact case is conducted to closely examine the effect of dendrite's dynamic variation. In the initial stage, particles on the fibre are mainly isolated or in small aggregates, which rarely undergo variations. So at the initial stage (stage A in Figure 4), the two cases predict similar efficiencies. In the second stage B of the loaded fibre, the bending and collapsing happen. It causes the dendrites to spread from the front to side of the fibre. The particle dendrite on fibre side provides more area than the front ones under the same particles number. So the conventional case generates a high efficiency than the "frozen" contact case. In the stage C, when particles accumulate to a certain mount, the dynamic variation of the dendrites will cause the dendrite length to a limit and hence the blocking area, while the area continuously increases for the "frozen" contact case. So efficiency for the "frozen" contact case increases more rapidly than the conventional case, and catch up with it after a certain time period.

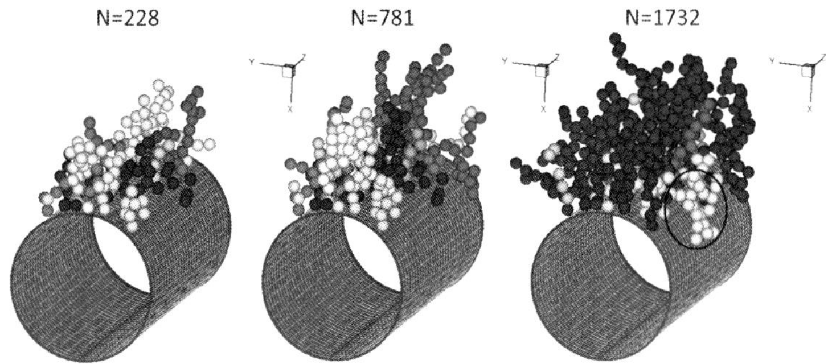

Figure 5 *Evolution of particle dendrites during the filtration*

In the simulation, we can identify the dendrites and inspect the variant of a selected particle dendrite. Figure 5 shows the progressive evolution of all particle dendrites, from which we choose one to examine the dynamic variation, as presented in Figure 6.

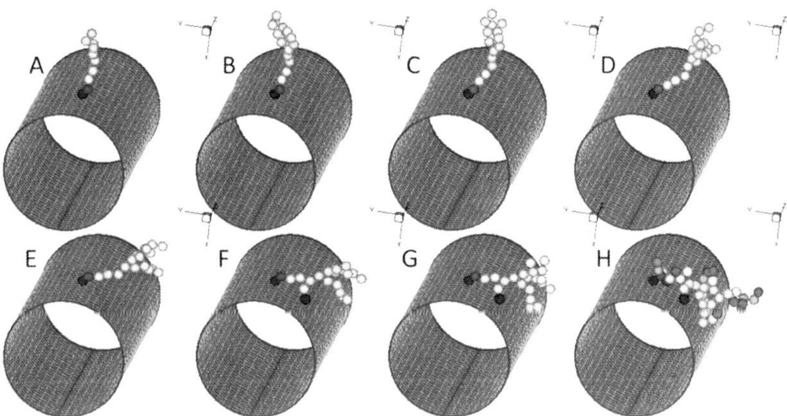

Figure 6 *Evolution of a single particle dendrite*

Figure 6 shows the growth and dynamic variation of a single particle dendrite. During stage A-B, particles are captured by the linear chain, forming a dendrite-like aggregate. As the number of particles on the aggregate top increase, the bottom of the trunk bares a growing torque caused by drag force on the above particles. The aggregate collapses (C-F), when the torque reaches a certain value. The collapse stops when the aggregate touches another one, forming a second support. More particles are captured, and bending happen at the end of the aggregate (F-H). The capture efficiencies of state C and E are acquired with the second statistics method in Section 3.3. The efficiency is defined for the fibre segment of the same Z range with the aggregate. The two efficiencies are 0.3042 and 0.3552 for C and E, respectively, with the relationship C < E.

5 CONCLUSIONS

In this paper, a novel discrete element method, based on JKR theory of adhesive contact, is used to investigate the dynamic behaviour of a particle-loaded fibre. In the initial stage, the capture efficiency is independent of the particle number, while the efficiency changes with loaded particle number in a power-law relationship in the second fibre-loaded stage. The change is due to the increase in the blocking area caused by the formation of particle dendrites. The dynamic variations of dendrites, mainly bending, collapsing and breaking off, affect the efficiency by spreading the particle dendrites to the fibre side and limiting the blocking area in the heavy-loaded stage.

References

1   S.J. Dunnett and C.F. Clement, *Journal of Aerosol Science*, 2006, **37**, 1116.
2   C.E. Billings, *Effects of particle accumulation in aerosol filtration*, Caltech, 1966.
3   G. Kasper, S. Schollmeier, J. Meyer and J. Hoferer, *Journal of Aerosol Science*, 2009, **40**, 993.
4   J.S. Marshall, *Journal of Computational Physics*, 2009, **228**, 1541.
5   S. Q. Li, J. S. Marshall, G. Q. Liu and Q. Yao, *Progress in Energy and Combustion Science*, 2011, **37**, 633.
6   R. Di Felice, *International Journal of Multiphase Flow*, 1994, **20**, 153.
7   C. Thornton and K. K. Yin, *Powder Technology*, 1991, **65**, 153.
8   A. Chokshi, A.G.G.M. Tielens and D. Hollenbach, *Astrophysical Journal,* 1993, **407**, 806.
9   Y. Tsuji, T. Kawaguchi and T. Tanaka, *Powder Technology*, 1993, **77**, 79.
10 K. Bagi and M. R. Kuhn, *ASME Journal of Applied Mechanics*, 2004, **71**, 493.
11 K. Iwashita and M. Oda, *Journal of Engineering Mechanics*, 1998, **124**, 285.
12 W. Ding, H. Zhang and C. Cetinkaya, *Journal of Adhesion*, 2008, **84**, 996
13 Li, S.Q. and J.S. Marshall, *Journal of Aerosol Science*, 2007, **38**, 1031.
14 B. Huang, Q. Yao, S.Q. Li, H.L. Zhao, Q. Song and C.F. You, *Powder Technology*, 2006, **163**, 125.

MODELLING OF THE FILTRATION BEHAVIOUR USING COUPLED DEM AND CFD

S. Stein and J. Tomas

Faculty for Process and Systems Engineering, Institute of Process Engineering, Otto-von-Guericke-University Magdeburg, Germany

1 INTRODUCTION

The filtration with cake formation is an effective method to drain a particle suspension. In order to decrease the moisture content after the filtration process, the cake can be consolidated, which is usually realised in the so called press-filters.[1] The generated cake consists of fine particles which are more or less compressible depending mainly on the particle contact stiffness and packing structure. The liquid flows through a compressed particle packing during the filtration process and therefore, the flow rate difference between solid and liquid has to be taken into account.[2] The dynamics of cake formation is responsible for the liquid drag force between the particles. It also leads to an axial particle pressure p_s transferred in the contacts within the packing. The filter cake resistance increases with increasing filtration time and cake height resulting in a decrease in the filtration rate.

One possibility to improve the filtration process is the formation of flocs, which deposit faster and produce a filter cake with large pores and high porosities. This decreases filter cake resistances and increases permeabilities. The flocculation of the particles can be done by changing the interparticular interaction forces like van der Waals attraction and electrostatic repulsion.

The aim of this work is to predict the filtration behaviour and the effects of flocculation, and to simulate the flow rate during generation and compression of the porous ultrafine particle packing by coupling Discrete Element Methods with Computational Fluid Dynamics in two dimensions within the micron range. A Press-Shear-Cell was used to experimentally validate the simulation.

The fluid is modelled by implementation of a "fixed course-grid fluid scheme" in the DEM-software. The scheme solves the locally averaged, two-phase (fluid and solid) mass and momentum equations for the fluid velocities and pressures,[3] which can be considered as a generalized form of the Navier-Stokes equations for a fluid interacting with a solid phase. The fluid solver uses the SIMPLE scheme[4] for incompressible viscous or inviscid flow on a fixed rectangular geometry aligned with the Cartesian axes. The internal discretization is fixed and regular. The underlying coupling formulation assumes that the particle radius is small compared to the length of a single fluid element. There is no interaction with walls, and no turbulence terms are added at any Reynolds number. The earliest work with this method of particle/fluid interaction was proposed by Tsuji *et al.*[5,6] and was also discussed in Shimizu *et al.*[7]

2 DEM MODELS

The discrete element method was introduced by Cundall[8] for the analysis of rock mechanics problems and then applied to soils by Cundall and Struck in 1979.[9] The model is composed of discrete particles that displace independently of each other, and interact only at contacts or interfaces between the particles. The particles are assumed to be rigid and the behaviour of the contacts is characterized using a soft contact approach, in which a finite normal stiffness is taken to represent the measurable stiffness that exists at a contact. The mechanical behaviour of such a system is described in terms of the movement of each particle and the interparticle forces acting at each contact point. Force and moment balances provide the fundamental relationship between individual particles.[10] In DEM, a particle possesses the translational and rotation motion that can be described by:

$$m_i \frac{dv_i}{dt} = F_i \tag{1}$$

$$I_i \frac{d\omega_i}{dt} = T_i \tag{2}$$

where F_i is the force and T_i is the momentum applied to particle i.

In order to simulate the flocculation of the ultrafine particle suspensions an attractive force according to the DLVO-theory[11,12] was introduced.

To couple the DEM with the fluid dynamics the so called fixed coarse-grid fluid flow scheme is incorporated into the DEM software. As aforementioned the scheme solves the locally averaged, two-phase mass and momentum equations for the fluid velocities and pressures, which can be considered a generalized form of the Navier-Stokes equation for a fluid interacting with a solid phase.

The Navier-Stokes equations for incompressible viscous or laminar flow can be modified to include the effect of a particulate solid phase mixed into the fluid. The average effects over many particles can be characterized in terms of porosity, ε.[3]

$$\rho_f \frac{\partial \varepsilon \vec{u}}{\partial t} + \rho f \vec{v} \cdot \nabla \left(\varepsilon u \right) = -\varepsilon \nabla p + \mu \nabla^2 \left(\varepsilon \vec{u} \right) + \vec{f_b} \tag{3}$$

$$\frac{\partial \varepsilon}{\partial t} + \nabla \cdot \left(\varepsilon \vec{u} \right) = 0 \tag{4}$$

where f_b is the drag force per unit volume. The fluid velocity is denoted as $\vec{u}$.[10] The drag force of fluid flow around a particle is given as:

$$\vec{f_b} = \beta \left(\vec{u} - \vec{v} \right) \tag{5}$$

where $\vec{v}$ is the averaged velocity of all particles in a given fluid element. The coefficient β is calculated based on the porosity ε of the fluid element. For low values of porosity ($\varepsilon<0.8$), the ERGUN relation is used, *i.e.*[13]

$$\beta = \frac{(1-\varepsilon)}{d_{50}^{2}\varepsilon^{2}}\left(150(1-\varepsilon)\mu + 1.75\rho_{f}d_{50}\,|\,\vec{u}-\vec{v}\,|\right) \qquad \varepsilon < 0.8 \qquad (6)$$

For higher values of solid fraction ($\varphi_{s} > 0.8$), β is used from the corrected nonlinear drag force exerted on a spherical particle by a fluid:[14]

$$\beta = \frac{4}{3}C_{d}\frac{|\,\vec{u}-\vec{v}\,|\,\rho_{f}(1-\varepsilon)}{d_{50}\varepsilon^{1.7}} \qquad \varepsilon > 0.8 \qquad (7)$$

where C_{d} is a turbulent drag coefficient defined in terms of the particle Reynolds number:

$$C_{d} = \begin{cases} \dfrac{24\left(1+0.15\,\mathrm{Re}_{p}^{0.687}\right)}{\mathrm{Re}_{p}} & \mathrm{Re}_{p} < 1000 \\[2ex] 0.44 & \mathrm{Re}_{p} > 1000 \end{cases} \qquad (8)$$

where

$$\mathrm{Re}_{p} = \frac{|\,\vec{U}\,|\,\varepsilon\rho_{f}d_{50}}{\mu} \qquad (9)$$

and $|\,\vec{U}\,| = |\,\vec{u}-\vec{v}\,|$ is the average relative velocity between the particles and the fluid.

Within the DEM-software the solution is "Pseudo-3D". The Navier-Stokes equation is solved in two dimensions, but the porosity is calculated in terms of the volume of spheres. This is because the 2D area of the circles would underestimate the porosity and overestimate the particle forces compared to 3D observations.

For the purpose of the porosity calculation, the particles are assumed spherical, and the cells to have an out-of-plane length of the maximum particle diameter in the representative volume element of the macroscopic process chamber. The volume of particles with a diameter smaller than the maximum particle diameter is scaled by a factor of ($d_{max}/d_{particle}$) to maintain the Pseudo-3D geometry (for the porosity calculation only).[10]

3 EXPERIMENTAL

The sample used in the filtration experiments were micro glassspheres with an average particle size of 5.8 μm. The properties of the glass-spheres are shown in table 1. The suspensions were prepared with distilled water and 13% solid volume fraction. In order to improve the filtration behaviour, a chemical conditioning of the suspension is useful, which changes the interparticular repulsive and attractive forces by adding electrolytes or varying the pH-value.

Table 1 *Particle properties of glass spheres*

Filtration pressure p in kPa	200 - 800
Particle diameter d_{50} in μm (primary particles)	5.8
Solid density ρ_s in kg/m³	2520
Particle stiffness k_n/k_s in N/m	1350
Friction coefficient	0.5
Zeta potential ζ in mV	-40 (stabilized) ≈0 (flocculated)
Hamaker-Constant $C_{H,sls}$ in 10^{-20} J	0.43

To evaluate the model experimentally, measurement of material properties such as the packing density $\varepsilon_{s,0}$ and permeability k_0 at $p_s = 0$, compressibility index β, lateral pressure ratio λ_w and the filter medium resistance of pure medium $R_{FM,0}$ is necessary. These parameters were determined using the Press-Shear-Cell (Figure 1).[15] It is a combination of a laboratory filter, a compression-permeability cell and a ring shear cell in the medium pressure range. This test apparatus is most suitably applied in constant pressure filtration processes.

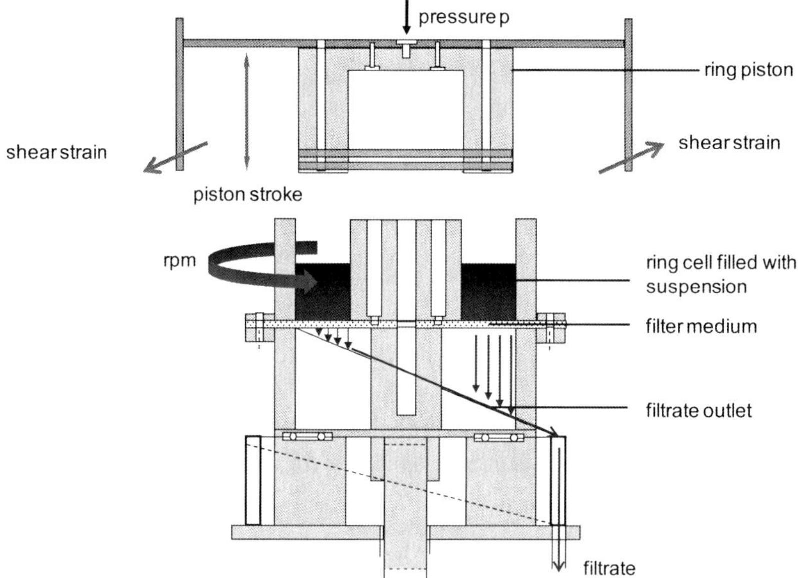

Figure 1 *Press-Shear-Cell*

4 RESULTS

In the present work, all particles are randomly located in the representative volume element having a starting velocity of v=0 without overlapping of particles for the initial setting. The filtration pressure during the simulation is *p=200 kPa*. Depending on the model size, the particle number is varied to obtain 13% solid volume fraction. The volume flow rate of the filtrate is calculated during each time step with the Darcy equation. Therewith the time-dependent filtrate volume can be calculated for each time step. The simulation ends, if all particles have been settled for the given pressure. The filter medium at the bottom is simulated with point walls, for which the maximum distance is 0.75 times the minimum particle diameter within the representative volume element to ensure that all particles are filtered. The initial state of the DEM simulation is shown in Figure 2.

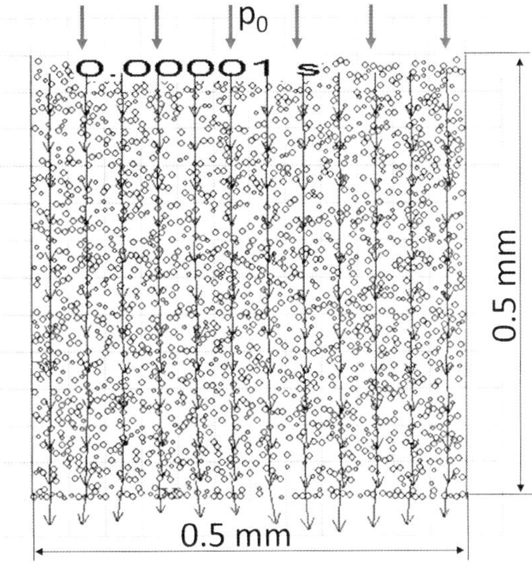

Figure 2 *The initial state of DEM simulations*

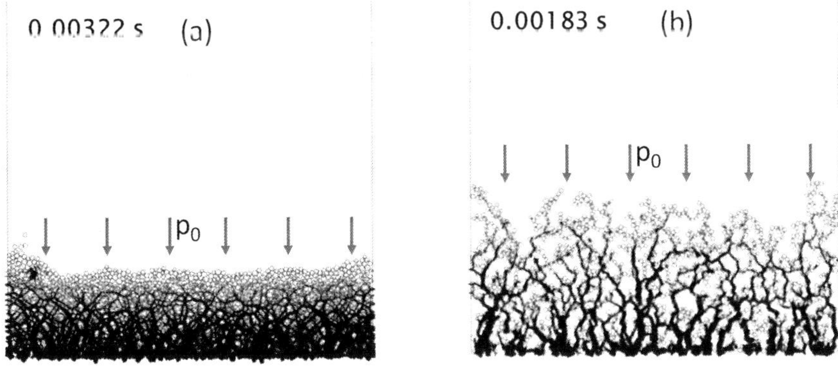

Figure 3 *Simulation result: (a) stable suspension and (b) flocculated suspension*

If large repulsive interactions are dominant (stable suspension) the particles are completely dispersed and deposit as single particles. Those suspensions produce a compact filter cake (Figure 3a) with large cake resistances and poor permeabilities.

Assuming that only attractive forces appear between the ultrafine particles within the suspension (comparable to the isoelectric point) the flocculation reaches its maximum value. Large flocs deposit very fast and produce a loose unstable structure (Figure 3b) which has a large compressibility. Due to the flocculation the filtration time can be decreased.

The time dependent curves of the porosity within the lowest fluid cell was recorded and compared for the two conditions. Figure 4 shows that the cake porosity decreased quickly at the beginning and then approached equilibrium values. The equilibrium state is reached faster in the case of a flocculated suspension, for which the porosity is much higher.

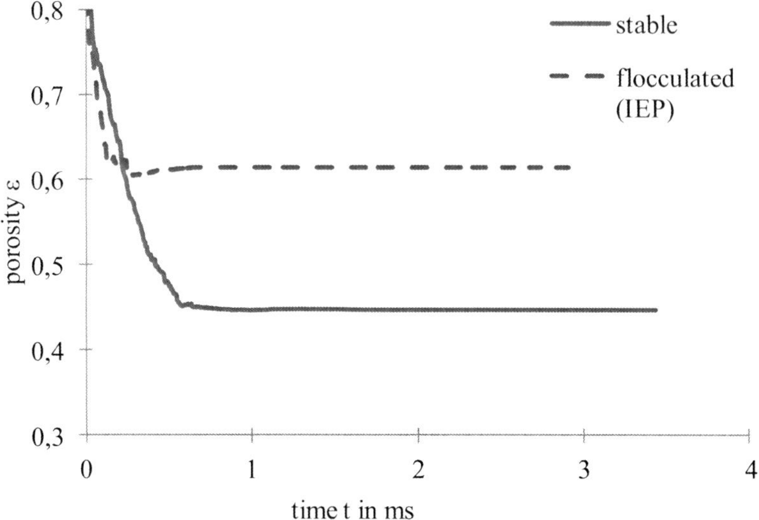

Figure 4 *Porosity of the lowest fluid cell versus time*

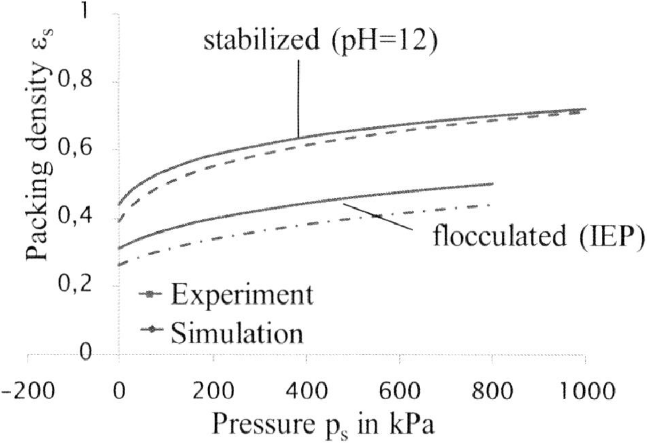

Figure 5 *Packing density ε_s versus filtration pressure p_s*

To predict the dependency of the filter cake structure on the applied pressure, the packing density ε_s for different filtration pressures p_s were adapted by the compression function

$$\varepsilon_s(p) = \varepsilon_{s,0}\left(1 + \frac{p_s}{p_0}\right)^{\beta} \tag{10}$$

where β is the compressibility index and p_0 the compression modulus.

The results in Figure 5 show that larger pressures lead to more compacted filter cakes. The compressibility of the flocculated cake is higher compared to the stable suspension and can be categorized as very compressible. The simulation results agree quantitatively with the experimental values.

5 CONCLUSIONS

In this work the filtration dynamics of ultrafine particle suspensions were determined and simulated using a coupled DEM and CFD. The numerical results were compared with the experimental data. A Press-Shear-Cell was used to investigate the compression behaviour of the consolidated particle packing and to measure the most important material properties. It was shown that the time dependency of specific filtrate volume and cake height for the flocculated and non-flocculated suspension can be predicted using this method.

It is observed that a flocculation causes an increase in porosity and permeability of the filter cake and a decrease of filtration time. A stable suspension with dominant repulsion produces a compact filter cake with small porosity, large filter cake resistance and poor permeability.

References

1   C. Alt, *Chem. Ing. Tech.,* 1976, **48**, 115.
2   B. Reichmann, *Modellierung der Filtrations- und Konsolidierungsdynamik beim Auspressen feindisperser Partikelsysteme*, Dissertation, Universität Magdeburg. 1999.
3   J.X. Bouillard, R.W. Lyczkowski and D. Gidaspow. *AIChE Journal.* 1989, **35**, 908.
4   S.V. Patankar, *Numerical Heat Transfer and Fluid Flow.* Hemisphere Publishing. 1980.
5   Y. Tsuji, T. Kawaguchi and T. Tanata. *Powder Tech.,* 1993, **77**, 79.
6   T. Kawaguchi, T. Tanata and Y. Tsuji. *JSME (B),* 1992, **58**, 79.
7   Y. Shimizu, R. D. Hart and P. Cundall. in *Proceedings of the 2nd International PFC Symposium*, Kyoto, Japan, 2004. 281.
8   P.A. Cundall, in *Proceedings of the Symposium of the International Society of Rock Mechnics,* Nancy, France, 1971, **1**, Paper No. II-8.
9   P.A. Cundall and O.D.L. Strack, *Géotechnique*, 1979, **29**, 47.
10   Itasca Consulting Group Inc. *PFC2D, Particle Flow Code in 2 Dimensions.* 3. Minneapolis, 2008.
11   B. Derjaguin and L. Landau *Acta Phys. Chem.* 1941, **14**, .633.
12   E.J.W. Verwey and J.T.G. Overbeek, *Theory of the stability of lyophilic colloids. The interaction of sol particles having an electrical double layer*, Elsevier. 1948.
13   S. Ergun, *Chemical Engineering Progress.* 1952, 89.

14   C.Y. Wen and Y. H. Yu, *Chemical Engineering Progress Symposium series 62*. 1966, 100.
15   B. Reichmann and J. Tomas, *Filtration & Separation*. 2000, **37**, 45.

Granular Flows

DEM MODELLING OF SUBSIDENCE OF A SOLID PARTICLE IN GRANULAR MEDIA

C.H. Goey, C. Pei and C.-Y. Wu

School of Chemical Engineering, University of Birmingham, Edgbaston, Birmingham B15 2TT, UK

1 INTRODUCTION

Granular materials possess complex mechanical behaviour and can manifest themselves as solid-like or liquid-like materials under different stress states.[1-3] On one hand, granular materials behave like solids when they are consolidated under high pressures, e.g. compacted tablets using dry powders and consolidated foundations made of soil materials. One the other hand, they can behave like liquids when they are fluidized or subjected to high shear stresses, such as in fluidized beds and in chute flows. Striving for a better understanding of the complex mechanical behaviour of granular materials, many researchers have investigated the associated mechanics and physics.[1-31] Nevertheless, due to the diversity and complexity of stress states of granular materials, their mechanical behaviour is still not well understood.

Taking the penetration of a spherical intruder into a granular bed as an example, experimental evidence reveals that the granular bed behaves like a solid material and resists the penetration of the intruder when the intruder impacts with the granular bed consisting of particles of similar density at a low initial velocity.[5] The intruder cannot penetrate into the granular bed, but sit on the surface of the granular bed or only produce a small indent on the bed surface. This is similar to the interaction of a sphere with a solid material. However, when the intruder collides with the granular bed at a high impact velocity[1] or even when the impact velocity is negligible but the density of the intruder is much higher than that of the particles in the granular bed,[4,5] the bed would behave like a liquid. In this case, an intruder can penetrate into the granular bed, which is similar to the penetration of the intruder into a viscous liquid.[4,5] Furthermore, a granular jet can be formed in some cases that mimic the formation of water jet when a pebble is thrown into the water.[3] It is clear that the behaviour of the granular material is very complicated even for such a simple case.

For the penetration of a spherical intruder into a granular bed, there has been an increasing interest to explore the penetration dynamics,[1-31] in particular, to examine how deep the intruder can penetrate into the granular bed; in other words, what the maximum penetration is? Ambroso *et al.*[5] experimentally investigated the penetration of an intruder into a granular bed with zero initial velocity but the density ratio of the intruder to the particles in the granular bed was varied from 0.1 to 6. Based on their experimental data, they proposed an empirical equation for shallow impact and found that the penetration depth, Z_{max}, can be given as

$$Z_{max} = \left(\frac{0.14}{\mu}\right)^{3/2}\left(\frac{\rho_i}{\rho_p}\right)^{3/4} D_i \tag{1}$$

where μ is the tangent of the angle of repose of the granular material in the bed, D_i is the diameter of the intruder, ρ_i and ρ_p are the densities of the intruder and the particles in the granular bed, respectively. Equation (1) indicates that Z_{max} is proportional to the density ratio ρ_i/ρ_p to the power of ¾ and is linearly proportional to the size of the intruder.

Similar penetration experiments were also performed by Lohse *et al.*[17] in which the density ratio of the intruder to the particles in the granular bed was varied between ~0.5 to 3. From their experimental observations, Lohse *et al.*[17] found that Z_{max} linearly depends on the mass of the intruder density m, i.e.

$$Z_{max} = \frac{2mg}{\kappa} \tag{2}$$

where g is the gravitational acceleration and κ is a constant that is related to the properties of the granular bed. In terms of the density and size of the intruder, Equation (2) can be rewritten as

$$Z_{max} = \frac{\pi\rho_i g D_i^3}{3\kappa} \tag{3}$$

Equation (3) implies that the Z_{max} depends linearly on the ρ_i and is proportional to the size of the intruder to the power of 3.

It is clear that Equations (1) and (3) give different predictions to the maximum penetration depth and this difference is primarily due to the limited experimental data available in developing the empirical equations. The investigations of Lohse *et al.*[17] and Ambroso *et al.*[5] were limited to penetration of less than 23 cm, i.e. relatively shallow to examine the fluidity of granular media; and the density ratio considered is limited to less than an order of magnitude. However, recent studies by Pacheco-Vazquez *et al.*[35] showed that the intruder can penetrate much deeper into the granular bed if the density ratio is high (say $\rho_i/\rho_p \sim 300$). Thus, it is interesting to examine if these empirical equations can be used to predict the penetration depth for the deep penetration with larger density ratios and to explore how accurate the predictions are. Therefore, the objective of this study is to explore the penetration of the intruder into a granular bed with a wide range of density ratio and to explore the penetration dynamics. For this purpose, an advanced discrete element method (DEM)[36] is employed to simulate the penetration of an intruder into the granular bed with zero initial velocity. The emphasis will be placed on the effect of boundary constraints and the effect of density ratio on maximum penetration depth.

2 DEM MODELS

Subsidence of a spherical intruder into a granular bed was analysed using DEM in 2D. All particles in the granular bed were treated as spheres. In the DEM used in this study, the classical contact mechanics was used to calculate the mechanical contact forces, in which the Hertz theory was used for normal contacts and the theory of Mindlin and Deresiewicz was used for tangential contacts. The motion of particles is calculated by integrating the equation of Newton's second law within each explicit time step using the central difference method.

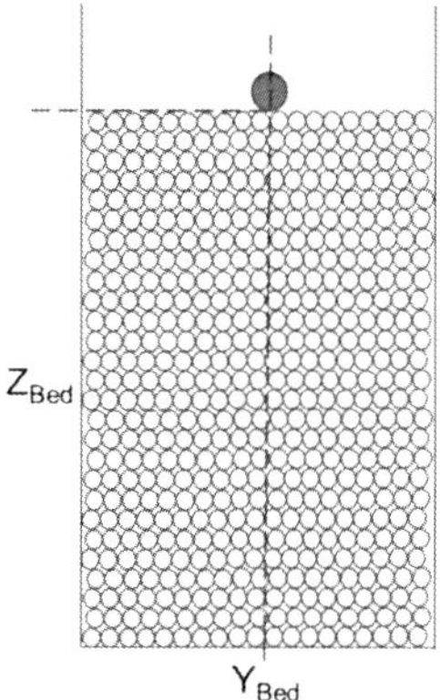

Figure 1 *Illustration of the DEM model set up*

Figure 1 illustrates the model set-up. An intruder was initially positioned on the top surface of the granular bed and released to move under gravity with zero initial velocity. In order to explore the influence of boundary constraints, two granular beds were considered, (i) a shallow bed and (ii) a deep bed. Both beds had the same width of $Y_{bed} = 0.4$ m that was wide enough (i.e. $Y_{bed} > 8D_i$) [12] to minimize any boundary effects from the side walls.

The shallow bed was generated using a mono-disperse system of 8,652 particles, which were randomly created in a container with three physical walls. These particles were then deposited under gravity until they settled to a steady state with negligible kinetic energy, i.e. when all the particles in the bed had settled with a negligible velocity of $<$ 1×10^{-5} m/s. The deep bed was generated by depositing additional layers of spheres on top of the shallow bed, and a total of 13,894 particles were used. The heights of the generated beds, Z_{bed}, were defined by the distance between the centre-point of the particle at the highest position on the granular surface and the bottom wall. Once the granular bed was settled, the intruder was generated at the middle point of Y_{Bed} on top of the granular surface (the initial position of intruder: $Z_0 = Z_{bed}$ + radius of intruder). The subsidence process was started with the release of the intruder from stationary, without any impact velocity.

To examine the effect of density ratio, ρ_i/ρ_p on the subsidence behaviour, the density of granules ρ_p was varied from 14 kg/m^3 to 4900 kg/m^3 while keeping the density of intruder ρ_i constant. To maintain the initial packing pattern of the granular bed, ρ_p was varied from the lowest value considered (14 kg/m^3) to a new value only after the particles were deposited completely. After changing the ρ_p value, a few more cycles of simulation were run to settle the new bed. The heights of the individual beds, Z_{bed}, vary slightly with ρ_p. The height of shallow bed is in the range of 0.5169 m$< Z_{bed} <$0.5304 m (with standard deviation $\sigma = 0.003139$) and the height of the deep bed in the range of 0.8269 m$< Z_{bed} <$0.8303 m (with $\sigma = 0.001112$).

The intruder and the particles in the granular bed were assumed to be of steel and polystyrene spheres, respectively, which are identical to those used in the experiments of Pacheco-Vázquez and Ruiz-Suárez,[37] and the corresponding material properties are given in Table 1. In the experiments of Pacheco-Vázquez and Ruiz-Suárez,[37] the material density for the intruder was altered from its original value by drilling holes of different sizes through the centre to produce particles with different masses but constant volume. In this study, a typical value of $\rho_i = 4900$ kg/m^3 was chosen but the range of ρ_i/ρ_p were selected to be wide enough to cover the ranges used in Ambroso *et al.* [5], Lohse *et al.*[17] and Pacheco-Vazquez *et al.*[35]

Table 1 *Material properties considered*

	Particles in the granular bed	Intruder	Container walls
Diameter *(m)*	0.005	0.01	-
Density *(kg/m^3)*	14 to 4900	4900	4900
Young's modulus *(GPa)*	3.37	210	210
Poisson's ratio	0.15	0.3	0.3

3 RESULTS AND DISCUSSION

Typical subsiding behaviour of the intruder into a granular bed is shown in Figures 2 - 4, in which the results for a density ratio of 200 are presented. Figure 2 shows the positions of the intruder at various time instants after it was released from the top surface of the deep granular bed (Figure 2a). It is clear that for this particular case once the intruder is released it sinks into the granular bed (Figures 2b & 2c) and penetrates to a finite distance (Figure 2d) until it stops. A close examination of the kinematics of the particles in the granular bed reveals that as the intruder subsides under gravity, the particles surrounding the intruder are pushed to the sides. A tail of cavity is formed (See Figure 2b & 2c) directly above the intruder, and the size of the cavity appears to depend on the position of the intruder and the density ratio. It is also observed that a loosely packed band is formed on the trail of the intruder as the particles near the trail re-arrange themselves once the intruder passes.

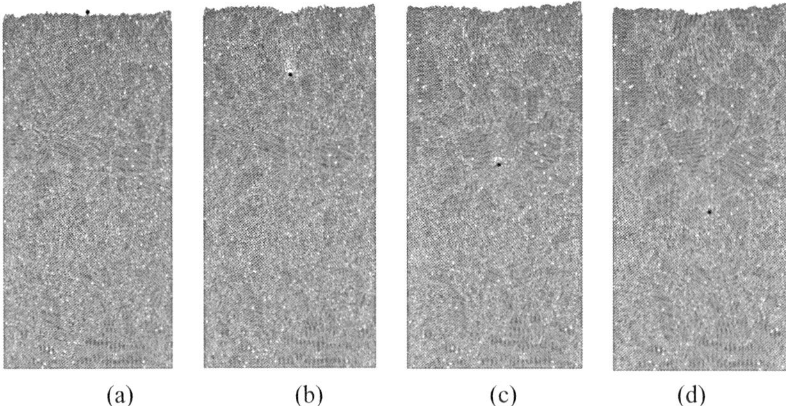

(a)   (b)   (c)   (d)

Figure 2 *Positions (Z) of the intruder at various time instants (t) during the subsidence into the deep granular bed with $\rho_i/\rho_p = 200$: (a) t = 0 s, z = 0 m; (b) t = 0.2 s, z = 0.165 m; (c) t = 0.4 s, z = 0.369 m; and (d) t = 0.6 s, z = 0.468 m*

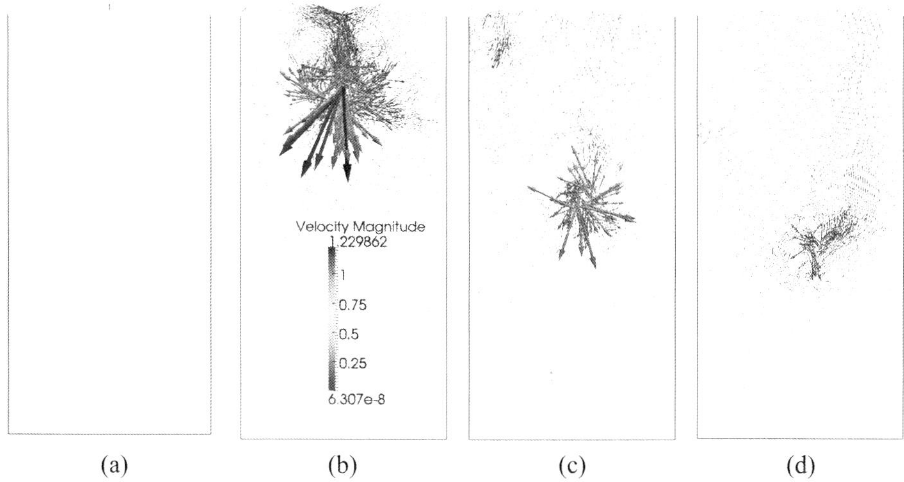

(a) (b) (c) (d)

Figure 3 *Velocity profiles at various time instances during the subsidence with*
$\rho_i/\rho_p = 200$: *(a) t = 0.0 s; (b) t = 0.2 s; (c) t = 0.4 s; and (d) t = 0.6 s*

The corresponding velocity profiles are shown in Figure 3. Initially, the particles in the bed were settled and no movement of particles was noticed before the intruder was released (Figure 3a). Once the intruder was released into the granular bed, a bulk granular motion was observed, as shown in Figure 3b, where the particles close to the trail of the intruder circulated on each side of the intruder. As the intruder travelled further down the granular bed, the velocity of the intruder and the particles decreased drastically within a very small time interval of 0.2 second (Figure 3c). At this moment, the particles no longer moved in bulk and the granular motion became intermittent. When approaching the end of the subsidence, re-arrangement of the particles in the upper region of the granular bed could be observed before they eventually settled down (Figure 3d).

Figure 4 shows the corresponding contact force network of the granular bed throughout the subsidence process. From Figure 4a, it was observed that the force chains were initially anisotropic and inhomogeneous, and they ramified throughout the tightly packed granules in the bed. It can be seen that the contact force network was most dilated at the top surface, but became denser, stronger and got connected in a more regular arrangement with depth. When the intruder started to subside, as shown in Figure 4b, the intruder broke the contact force network in the bed and left a clear region behind where no contact between the particles was observed. Furthermore, it was observed that the magnitude of the overall force network reduced and the bed appeared to be expanded during the penetration. As the intruder continued to subside, as shown in Figure 4c, a new contact network force was formed at the upper region of the bed when the particles rearranged and got in contact. This process of breaking the old contact force network and forming a new one continued until the intruder stopped at Z_{max}. From Figures 4b - 4d, it was clear that the region above the intruder with zero contact force decreased as the intruder approached the final penetration depth. At the end of subsidence, dense contact force network near the bottom of the granular bed was re-established (Figure 4d).

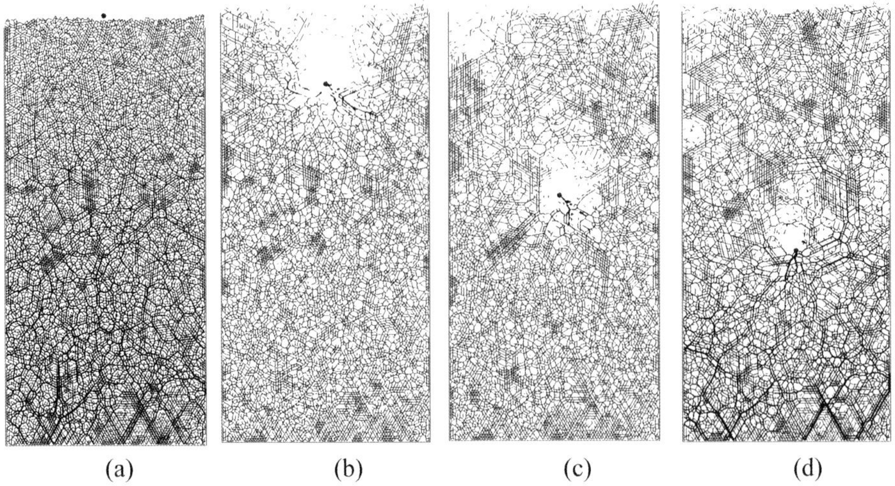

(a)   (b)   (c)   (d)

Figure 4 *Contact force network at various time instants during the subsidence with ρ_i/ρ_p = 200: (a) t = 0 s; (b) t = 0.2 s; (c) t = 0.4 s; and (d) t = 0.6 s*

Figures 5 - 8 compare the subsiding behaviour of the intruder into deep beds with density ratios ρ_i/ρ_p of 1, 50, 100, 200 and 300. Figures 5 & 6 show the subsiding velocity of intruder V, with respect to time and penetration distance. Except $\rho_i/\rho_p = 1$, a similar velocity pattern could be observed for all other ρ_i/ρ_p: the subsiding velocity V increases initially until it reached a maximum value V_{max}, followed by an intermittent decrease until the motion of the intruder was halted. The acceleration occurred only for a very short period compared to deceleration, and it is essentially constant regardless of ρ_i/ρ_p. On the other hand, the deceleration was observed to be intermittent, and was similar to the stick-slip condition commonly occurred in granular media.[4, 24]

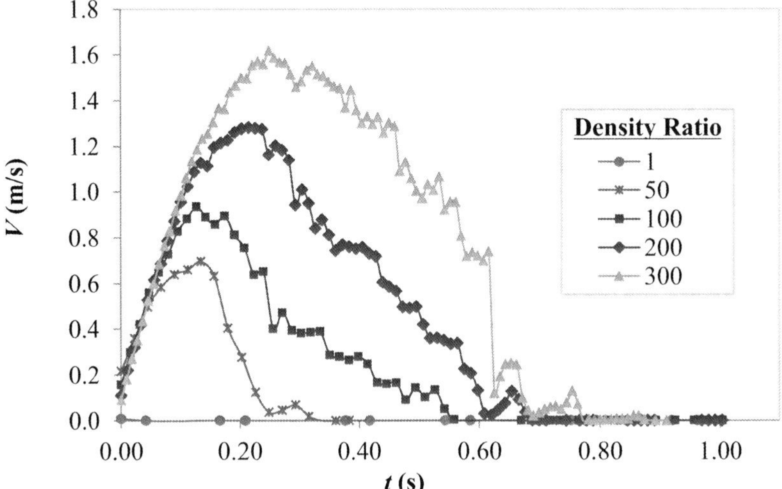

Figure 5 *Instantaneous subsiding velocity V of the intruder in the deep bed with various density ratio ρ_i/ρ_p*

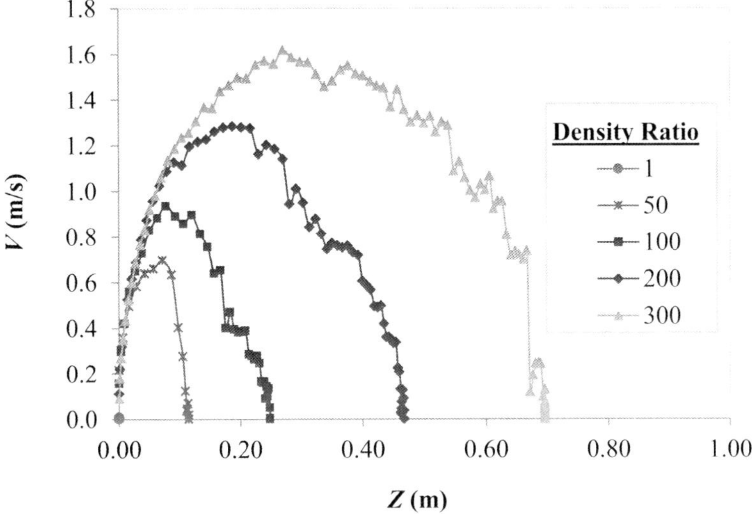

Figure 6 *Subsiding velocity V as a function of penetration distance Z in the deep bed for various density ratio ρ_i/ρ_p*

Several effects of ρ_i/ρ_p on the subsidence of intruder could likewise be observed. Firstly, as shown in both figures, the maximum subsiding velocity V_{max} increases with ρ_i/ρ_p. Secondly, the travelling time (Figure 5) and travelling distance, i.e. penetration depth Z_{max} (Figure 6) also increases with increasing ρ_i/ρ_p. Another interesting phenomenon was observed where the intruder did not come to rest with a smooth deceleration but went through a sudden halt. As shown in Figure 6, the degree of abruptness also increases with ρ_i/ρ_p, which was particularly prominent for $\rho_i/\rho_p = 300$.

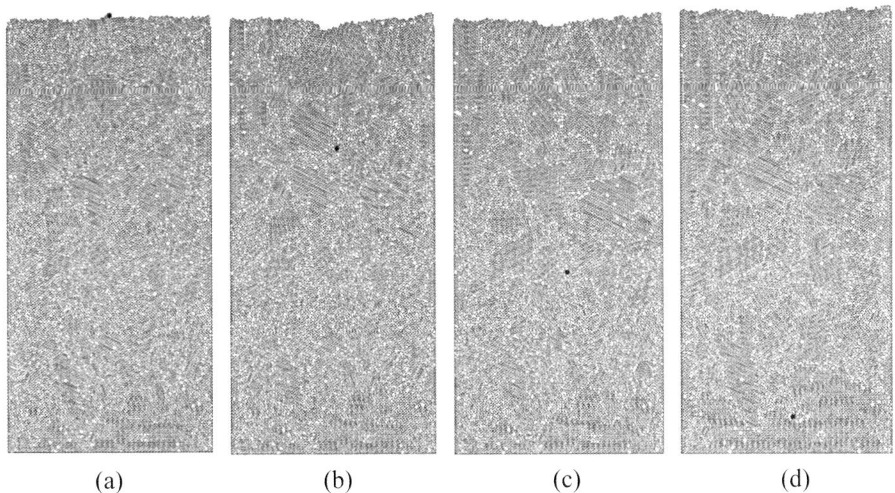

(a)   (b)   (c)   (d)

Figure 7 *Final positions of the intruder in the deep granular bed of (a) $\rho_i/\rho_p=1$; (b) $\rho_i/\rho_p=100$; (c) $\rho_i/\rho_p=200$ and (d) $\rho_i/\rho_p=300$*

Figure 7 shows the final position of intruder. At $\rho_i/\rho_p =1$, i.e. when the densities of intruder and particles are at unity, the intruder was unable to subside into the bed (Figure 7a) even though the diameter of intruder D_i was twice of the diameter of particles D_p (see Table 1). The same phenomenon was also observed in shallow beds (not shown). However, as ρ_i/ρ_p increases, the intruder subsided deeper into the bed as shown in Figures 7b & 7c. It is interesting to note that the granular bed is very fragile to allow the intruder to subside even at ρ_i/ρ_p as low as 25 (not shown). When the density ratio ρ_i/ρ_p was high enough (Figure 7d), the intruder could travel through the granular bed and reach the bottom of the bed.

Figure 8 shows the time evolutions of the penetration distance of the intruder Z in the granular beds with various density ratios. For $\rho_i/\rho_p = 1$, Z remains unchanged at zero throughout the simulation, indicates that no subsidence took place as observed in Figure 7a. For granular beds of $\rho_i/\rho_p > 1$, similar patterns were observed: the gradient of Z increased initially, followed by a gradual decrease until Z reached a plateau. It is also clear that the level of plateau increased with the value of ρ_i/ρ_p, and the time taken for Z to reach its respective plateau also increased with ρ_i/ρ_p.

Figure 9 shows the maximum penetration depth Z_{max} of the intruder in (i) shallow beds and (ii) deep beds with density ratios of $1 \le \rho_i/\rho_p \le 350$. The standard deviation of Z_{max} is very small for both shallow and deep beds (based on simulations on 3 shallow beds and their 3 corresponding deep beds generated). It can be seen that Z_{max} is essentially linear with ρ_i/ρ_p when a deep bed is used. For the shallow bed, Z_{max} is linear with ρ_i/ρ_p at low values of ρ_i/ρ_p (i.e. when $\rho_i/\rho_p \le 200$), and it has the same gradient as the deep bed. At high values of ρ_i/ρ_p, Z_{max} increases and get saturated at $Z \approx 0.5$ m without reaching the bottom of the bed, i.e. $Z_{max} < Z_{bed}$ (for shallow bed, 0.5169 m $\le Z_{bed} \le 0.5304$ m, referred to Section 2), which deviates from the results obtained with a deep bed. This indicates that the boundary effect only becomes significant when the density ratio ρ_i/ρ_p is very high (say >200). In other words, the effect of boundary constrain is negligible when $\rho_i/\rho_p \le 200$.

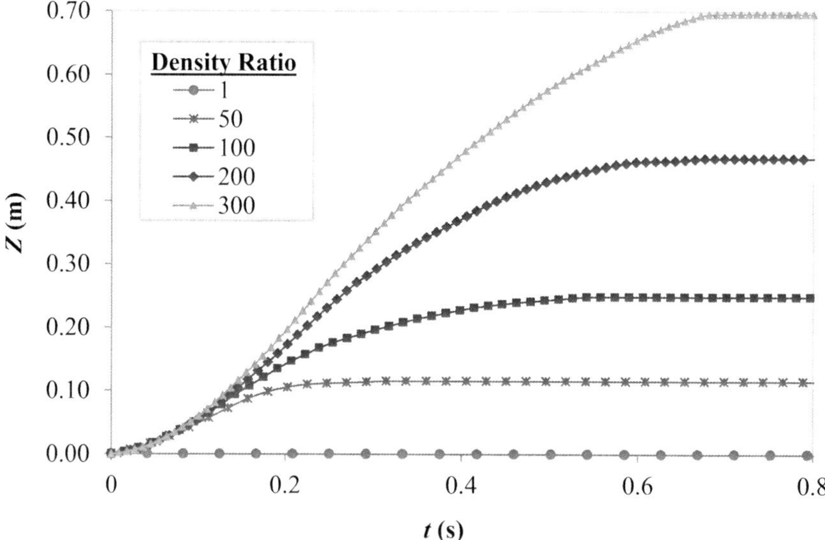

Figure 8 *Time evolutions of the penetration distance of the intruder Z in the deep bed with various density ratios ρ_i/ρ_p*

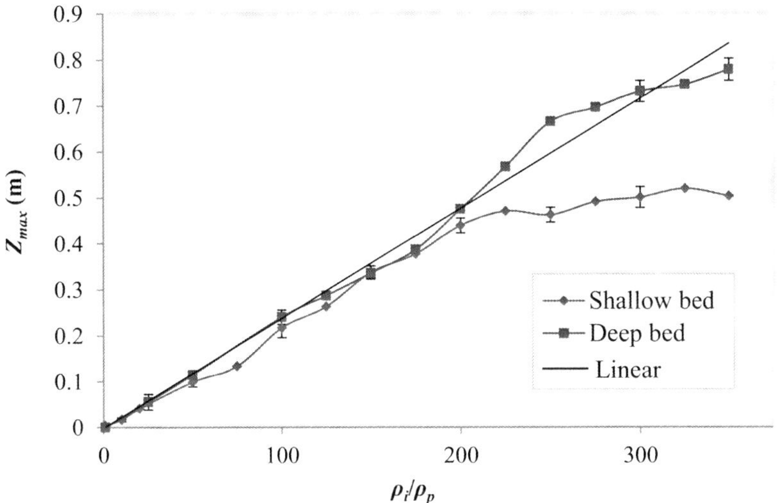

Figure 9 *Penetration depth Z_{max} as a function of density ratio ρ_i/ρ_p*

4 CONCLUSIONS

In this study, the penetration behaviour of an intruder into granular beds of various density ratios ρ_i/ρ_p was investigated. It was found that all granular beds with a density ratio higher than unity allowed subsidence of the intruder up to a certain depth. The subsidence process can be subdivided into two distinct stages: (1) acceleration and (ii) deceleration according to the interaction between the intruder and the granular bed. The subsidence depth was found to be linearly proportional to the density ratio. Besides, the degree of bulk motion of particles and the strength of the contact force network also increase with the density ratio. The effect of boundary constraint on the penetration behaviour and final penetration depth were also examined with the subsidence of the intruder in granular beds of different depths. It has been found that the boundary effect only becomes significant when the density ratio ρ_i/ρ_p is very high, while it is negligible at low and intermediate density ratios.

References

1   H. Jaeger, S. Nagel and R. Behringer, *Rev. Mod. Phys.*, 1996, **68**, 1259.
2   L. Kadanoff, *Rev. Mod. Phys.*, 1999, **71**, 435.
3   P. de Gennes, *Rev. Mod. Phys.*, 1999, **71**, 8374.
4   M. Ciamarra, A. Lara, A. Lee, D. Goldman, I. Vishik and H. Swinney, *Phys. Rev. Lett.*, 2004, **92**, 149902.
5   M.A. Ambroso, C. Santore, A. Abate and D. Durian, *Phys. Rev. E*, 2005, **71**, 051305.
6   M.A. Ambroso, R. Kamein and D. Durian, *Phys. Rev. E*, 2005, **72**, 041305.
7   M. Hou, Z. Peng, R. Liu, K. Lu and C. Chan, *Phys. Rev. E*, 2005, **72**, 062301.
8   H. Katsuragi and D. Durian, *Nat. Phys.*, 2007, **3**, 420.
9   D. Goldman and P. Umbanhowar, *Phys. Rev. E*, 2008, **77**, 021308.
10   F. Bourrier, F. Nicot and F. Darve, *Granular Matter*, 2008, **10**, 6.

11 J. de Bruyn and A. Walsh, *Can. J. Phys.*, 2004, **82**, 439.

12 A. Seguin, Y. Bertho and P. Gondret, *Phys. Rev. E*, 2008, **78**, 010301.

13 K. Newhall and D. Durian, *Phys. Rev. E*, 2003, **68**, 060301.

14 R. Albert, M. Pfeifer, A.-L. Barabasi and P. Schiffer, *Phys. Rev. Lett.*, 1999, **82**, 1.

15 P. Umbanhowar and D. Goldman, *Phys. Rev. E*, 2010, **82**, 010301.

16 J. Uehara, M. Ambroso, R. Ojha and D. Durian, *Phys. Rev. Lett.*, 2003, **90**, 194301.

17 D. Lohse, R. Rauhe, R. Bergmann and D. van der Meer, *Nature*, 2004, **432**, 689.

18 F. Pacheco-Vazquez, G. Caballero-Robledo, J. Solano-Altamirano, E. Altshuler, A. Batista-Leyva and J. Ruiz-Suarez, *Phys. Rev. Lett.*, 2011, **106**, 218001.

19` A. Tripathi and D.V. Khakhar, *Phys. Rev. Lett.*, 2011, **107**, 108001.

20 D. Wang, X. Ye and X. Zheng, *Eur. Phys. J. E*, 2012, **35**, 7.

21 P. Cixous, E. Kolb, N. Gaudouen and J.-C. Charmet, *AIP Conf. Proc.*, 2009, **1145**, 539.

22 M.E. Cates, J.P. Wittmer, J.-P. Bouchaud and P. Claudin, *Phys. Rev. Lett.*, 1998, **81**, 1841.

23 A. Liu and S. Nagel, *Nature (London)*, 1998, **21**, 396.

24 R. Candelier and O. Dauchot, *Phys. Rev. E*, 2010, **81**, 011304.

25 J. Royer, B. Conyers, E. Corwin, P. Eng and H. Jaeger, *EPL*, 2011, **93**, 28008.

26 L. Tsimring and D. Volfson, in *Powders and Grains*, A.A. Balkema, Rotterdam, 2005, Volume 2, p. 1215.

27 F. Pacheco- Vazquez, G. Caballero-Robledo, J. Solano-Altamirano and E. Altshuler, *Phys. Rev. Lett.*, 2011, **106**, 218001.

28 M. Forrestal and V. Luk, *Int. J. Impact Eng.*, 1992, **12**, 427.

29 Y. Boguslavskii, S. Drabkin and A. Salman, *J. Phys. D*, 1996, **72**, 905.

30 I. Albert, J.G. Sample, A.J. Morss, S. Rajagopalan, A.-L. Barabási and P. Schiffer, *Phys. Rev. E*, 2001, **64**, 061303.

31 I Albert, *Ph. D Thesis*, University of Notre Dame, 2001.

32 J. Zarnecki, M. Leese, B. Hathi, A. Ball, A. Hagermann, M. Towner, R. Lorenz, J. Macdannell, S. Green, M. Patel, T. Ringrose, P. Rosenberg, K. Atkinson, M. Paton, M. Banaszkiewicz, B. Clark, F. Ferri and M. Fulchignoni, *Nature (London)*, 2005, **438**, 792.

33 E. Nelson, H Katsuragi, P. Mayor and D. Durian, *Phys. Rev. Lett.*, 2008, **101**, 068001.

34 M.B. Stone, R. Barry, D.P. Bernstein, M.D. Pele, Y.K. Tsui, and P. Schiffer, *Phys. Rev. E*, 2004, **70**, 041301.

35 F. Pacheco-Vázquez, G.A. Caballero-Robledo, J.M. Solano-Altamirano, E. Altshuler, A.J. Batista-Leyva and J.C. Ruiz-Suárez, 2011, *Phys. Rev. Lett.*, **106**, 2018001.

36 Y. Guo, K.D. Kafui, C.-Y. Wu, C. Thornton and J.P.K. Seville, *AICHE Journal*, 2009, **55** (1), 49-62.

37 P. Pacheco-Vázquez and J.C. Ruiz-Suárez, *Nat. Comm.*, 2010, **1** (123), 1-7.

NUMERICAL SIMULATION Of THE COLLAPSE OF GRANULAR COLUMNS USING DEM

T. Zhao[1], G.T. Houlsby[1] and S. Utili[2]

[1] Department of Engineering Science, University of Oxford, Oxford, OX1 3PJ, UK
[2] School of Engineering, University of Warwick, Coventry CV4 7AL, UK; formerly at Department of Engineering Science, University of Oxford, Oxford, OX1 3PJ, UK

1 INTRODUCTION

The behaviour of flowing dense granular media is of interest for many diverse areas. This behaviour is particularly important to understand the propagation and the prediction of the run-out of geophysical mass flows (landslides, pyroclastic flows, rock and debris avalanches). For instance, rock and debris avalanches, can travel very long distances along flat or almost flat surfaces, and represent a class of exceptionally dangerous geomorphologic processes.

Among the various issues raised by these flows, the prediction of their run-out length is of primary importance, mainly due to their destructive power. Long run-out landslides travel distances several times larger than the height of the source topography sweeping populated areas located considerably far away from the original mountainside source.[1]

The collapse of a granular column is of great interest in this context and has been recognized as an important phenomenon useful for studying transient granular flow conditions and for modelling geophysical flows of granular materials. This type of process is similar to the dam break problem in fluid mechanics but involving a granular material.[2,3] A series of experiments are available in the literature.[4,5,6] They have been performed with different granular materials and under different boundary conditions. Great effort has been put into finding a governing equation to quantify the relationship between the initial model configuration and the geometry of the final granular deposit with the purpose of finding some general governing factors for real landslides in order to create a generalized model for the complex phenomenon of debris flow.[4,5,7]

2 LITERATURE REVIEW

2.1 Experimental Tests

Lube *et al.*[8] carried out experiments on granular flows in plane strain conditions. In these experiments, various granular materials (rice, sand, glass) were initially consolidated within a rectangular box resting on top of a channel with the same width as the granular column. After the confining box was quickly released, all the grains spread out along the channel. These experiments showed that the most important governing parameter is the initial column aspect ratio, defined as the ratio between the initial height, H_{ini}, and the initial length, L_{ini}. From this set of experiments, Lube *et al*[8]. inferred an empirical equation relating the initial aspect ratio to the final run-out distance of the granular flow:

$$\frac{L_{fin}-L_{ini}}{L_{ini}} \simeq \begin{cases} 1.2a, & a \leq 2.3 \\ 1.9a^{2/3}, & a > 2.3 \end{cases} \tag{1}$$

with L_{fin} being the final deposit length.

2.2 Numerical Simulations Using the Discrete Element Method

Recent research on this problem shows that the initial column aspect ratio controls the flow pattern and determines the final deposit geometry and the energy dissipation of the column.[7,10] Zenit *et al.*[10] ran 2D discrete element simulations to investigate the mechanical behaviour of granular columns with different initial aspect ratios. They identified a critical value of the aspect ratio, as the value corresponding to the transition between two final types of final profiles: one with a flat top and one without it. For initial aspect ratios smaller than the critical value, the final profile of the column tends to be a truncated cone with a flat top, whereas for aspect ratios larger than the critical, the final deposit lies along an inclined line (see Figure 1b).

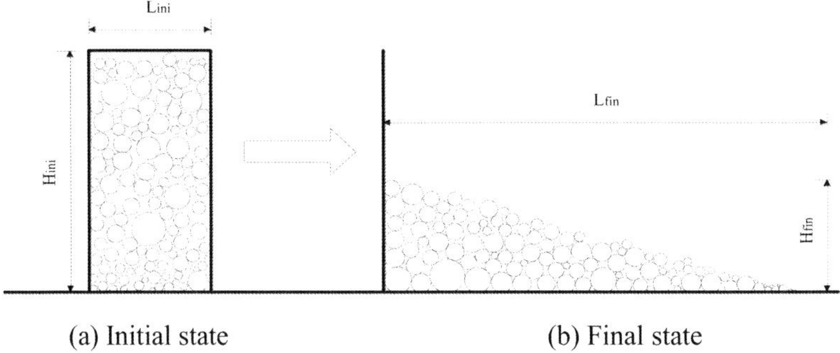

(a) Initial state   (b) Final state

Figure 1 *Geometry of initial and final deposits of granular material*

Also Staron[7] ran 2D DEM[12] analyses of the collapse of granular columns. He noticed that the collapse was mainly driven by particle free fall when the column aspect ratio is large, whereas there is no free fall process for small aspect ratios. These results indicate that the initial column aspect ratio governs the final shape of the granular deposit. Staron proposed the following relationship for the run-out distance:

$$\frac{L_{fin}-L_{ini}}{L_{ini}} \simeq \begin{cases} 2.5a, & a \leq 2 \\ 3.25a^{\alpha}, & a \geq 2 \end{cases} \quad (\alpha = 0.705 \pm 0.022) \tag{2}$$

Equation (2) summarizes the relationship between the run-out distance and the initial column aspect ratio, which quantitatively predicts the final deposit length of a landslide.

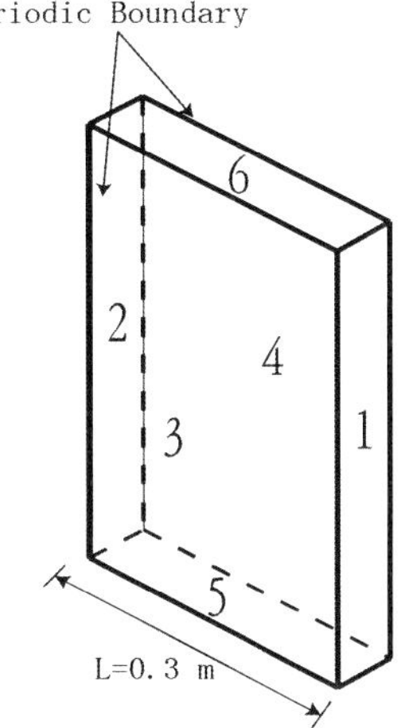

Figure 2 *The granular column boundary configuration: (1): right rigid wall; (2): left rigid wall; (3): front periodic boundary; (4): back periodic boundary; (5): bottom rigid wall; (6): top rigid wall*

3 NUMERICAL SIMULATIONS

The spheres were generated randomly within the boundary walls (see Figure 2). A uniform particle size distribution ranging from 3.2 mm to 4.8 mm was employed. The contact law adopted was the Hertz – Mindlin no-slip solution[11,14] together with a moment – relative rotation law which could be formally described as "linear elastic perfectly plastic" although plasticity of grains is not the phenomenon catered for by this law. In fact, the rolling law was employed with the only aim of taking into account the effect of non-sphericity of the sand grains on the angular moment equilibrium of the granular assembly. The adopted rolling law accounts for the rolling moments acting on sand particles arising from the fact that the line of action of the normal contact forces in case of non-spherical particles no longer passes through the centre of mass of the particles hence generating rotational moments. The moment – relative rotation contact law was implemented in the code in an incremental fashion as:

$$\Delta M = -k_r \Delta \theta_r \tag{3}$$

where k_r is the rolling stiffness and $\Delta\theta_r$ is the incremental relative rotation between two particles in contact. The limit moment M_{limit}^L, is given by the following equation:

$$M_{limit}^L = \eta \cdot r \cdot \|F_n\| \tag{4}$$

in which η is a coefficient to be calibrated, r is the average radius of the two spheres and $\|F_n\|$ is the norm of the normal force.

All the parameters used in the numerical simulations are listed in Table 1. During the simulation, the granular mass collapses downwards under gravity. Depending on the value of the initial aspect ratio a of the granular column, we can observe three distinct evolving morphologies for the deposit which are illustrated by four sequences of successive profiles displayed in Figure 3. (Note: In the following figures, "*a*" denotes the column aspect ratio. To get a clear observation of the profile evolution process, different scalars have been employed along the x and y axes.)

Table 1.   *Parameters used for granular column collapse simulations*

Parameter	Value	Parameter	Value
Solid Particle:		**Hertz-Mindlin contact law:**	
Solid Particle Number, N_p	8500	Normal Damping Coefficient, β_n	0.7
Shape	Sphere	Shear Damping Coefficient, β_s	0.7
Average Particle Radius, r	0.004 m	Coefficient of limited moment, η	1
Particle Size Distribution	[0.0032, 0.0048] m	Rolling Stiffness, k_r	1 Nm/rad
Density, ρ_s	2600 kg/m3	**Simulation Setting:**	
Young's Modulus, E	1.5E8 N/m2	Gravity, g	9.81 m/s2
Poisson Ratio	0.4	Time step, Δt	0.01455 s
Internal friciton Angle, θ_{int}	30 degree	Maximum strain rate (s^{-1})	(-1e-4, -1e-4, -1e-4)
Basal friciton Angle, Φ_{bed}	30 degree		

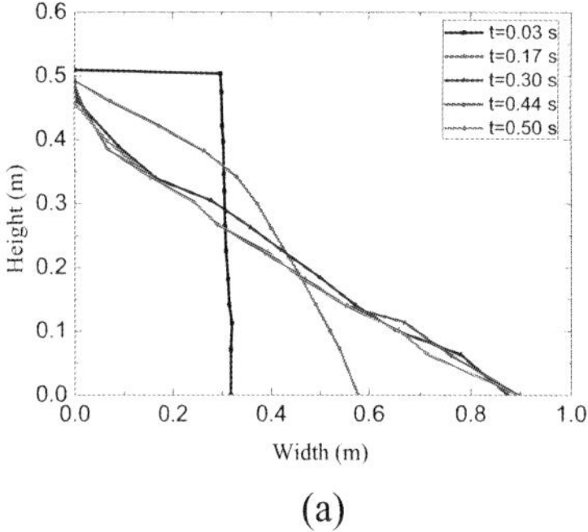

(a)

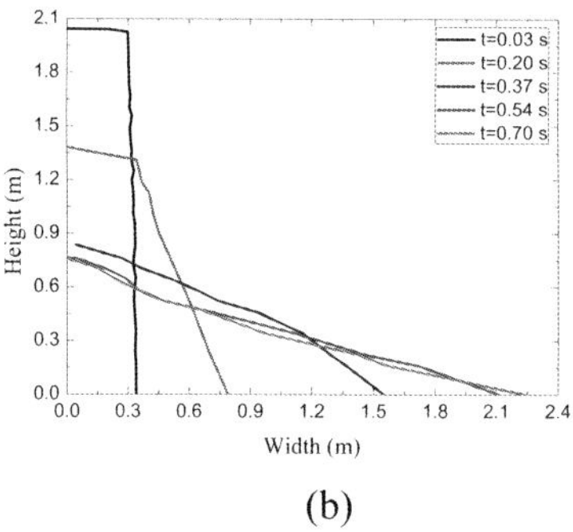

(b)

Figure 3 *Evolution of column profile over time observed in 3D DEM simulations for different initial aspect ratios: a) a=1.72 b) a=7.44.*

In Figure 3, we can see that the profile of the final deposit depends on the initial aspect ratio of the column. Upon release, the inner part remains static conserving its shape while the outer part of the column spreads out. As time elapses, the granular column continues to spread out, becoming progressively flatter and lower. For small values of the initial aspect ratio (see Figure 3a), the granular assembly aligns along an inclined line. For values of *a* larger than the critical aspect ratio (see Figure 3b), the whole column becomes involved in the downward movement.

In Figure 4, the final profiles of the granular assembly are plotted. It emerges that the normalized profiles have nearly the same shape regardless of the initial aspect ratio. As Lube *et al*,[8] we found that the profiles of the granular deposit are largely dependent on the value of initial aspect ratio. This suggests that the best way to plot results is making them dimensionless.

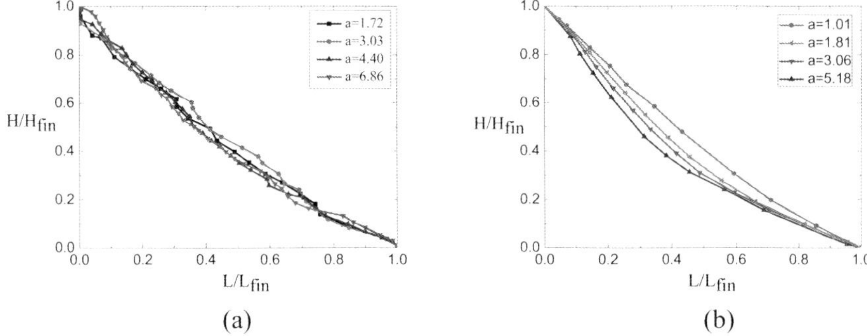

(a)   (b)

Figure 4 *Final normalised profiles of the granular column deposit: (a) 3D DEM results; (b) experiments after Lube et al.*[8]

3.1 Dimensional Analysis

In the literature, different column heights and widths are employed in experiments and numerical simulations. It is often difficult to compare results expressed in dimensional form. Hence, results should be given in dimensionless form. In this section the dimensional analysis of the problem is performed with the aim of expressing all the physical variables of importance by means of dimensionless groups. First, we need to find out the possible dimensionless parameter groups from the parameters in Table 1. In our problem, there are 14 independent variables and three fundamental dimensions: mass (M), length (L) and time (T). According to the Buckingham's Π theory, we can expect a relationship between 14-3=11 independent dimensionless groups. Since υ, θ_{int}, Φ_{bed}, η and β are already dimensionless parameters, six dimensionless groups are needed from the combination of nine dimensional variables: H_{fin}, L_{fin}, H_{ini}, L_{ini}, d, ρ, g, E, K_r. The six dimensionless groups were chosen as:

The initial column aspect ratio: $a = H_{ini}/L_{ini}$
The normalized final run-out distance: $k_L = (L_{fin} - L_{ini})/L_{ini}$
The normalized final height: $k_H = H_{fin}/L_{ini}$
Particle packing number: $N_c = L_{ini}/d$

Characteristic mobility of the assembly: $M_0 = d\sqrt[3]{\dfrac{E}{K_r}}$

The initial basal strain at the bottom of the granular column: $\varepsilon_L = \dfrac{\rho g L_{ini}}{E}$

Following dimensional analysis, the collapse of the granular column can be expressed by a dimensionless functional relationship between dimensionless input (on the right hand side) and output (on the left hand side) variables:

$$(k_L, k_H) = f(a, \Phi_{bed}, \theta_{int}, M_0, \eta, \varepsilon_L, \beta, \upsilon, N_c) \tag{5}$$

4 RESULTS

In this section, the collapse of the column is investigated by looking at the normalized run-out distance, k_L, only. The influence of the various input parameters on k_L will be illustrated.

4.1 Initial Aspect Ratio

Figure 5 shows the simulation results of granular column collapse for both cases of free and constrained rotation models. In the numerical simulations, we can see a large difference between the results obtained with particles free to roll and particles with the rolling resistance model. With particles free to roll, the mobility of the granular assembly is significantly higher. Best fitting the points from the rolling resistance model, plotted in Figure 6, to a power function, we obtained the following relationship between the normalized run-out distance and the initial column aspect ratio:

$$\frac{L_{fin}-L_{ini}}{L_{ini}} \cong \begin{cases} 1.21a, & a \leq 3 \\ 1.58a^{0.762}, & a > 3 \end{cases} \tag{6}$$

The obtained equation is very close to the equation (Eq. (1)) achieved by Lube *et al*[8] from experiments. The difference between the results from 2D DEM simulations (Eq.(2)) and ours can be explained in terms of the different kinematics of the particles in 2D and 3D. In the reality, particles are three dimensional and their 3D motion cannot be adequately replicated by a 2D model which effectively assumes particles as Schneebeli rods.

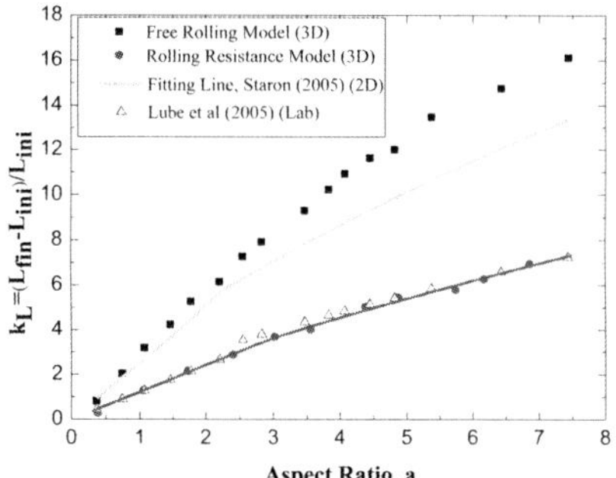

Figure 5 *Relationship between the initial aspect ratio and the final run-out length*

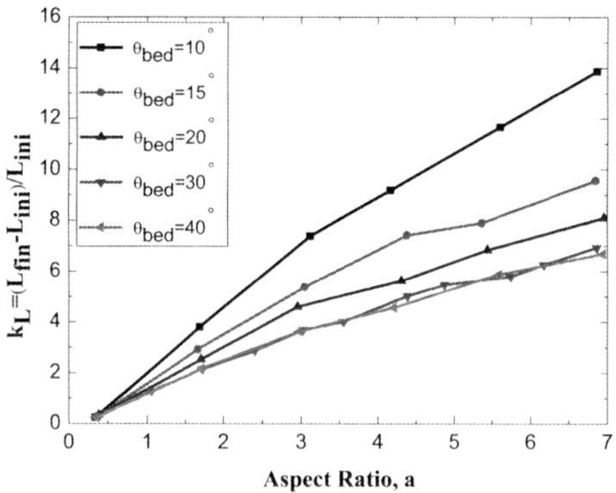

Figure 6 *The influence of basal friction angle*

4.2 Basal Friction Angle

The value of the basal friction is difficult to be measured and has proven to be an important parameter affecting the final run-out distance. Figure 6 shows that the run-out distance decreases monotonically with increasing basal friction angle. The rate of decrease becomes very small as Φ_{bed} approaches 30°. When the basal friction angle is greater than 30°, the granular run-out distance is no longer affected by the basal friction and all the curves remain unchanged. Thus, we can define $\Phi_{bed} = 30^\circ$ as the critical basal friction angle.

4.3 Internal Friction Angle

The internal friction angle plays an important role in the energy dissipation taking place during granular spreading. In Figure 7, we can see that the particles tend to run a long distance at low internal friction angles. The distance decreases as the internal friction angle increases. For values of the angle higher than 30°, the slope of the curve no longer changes.

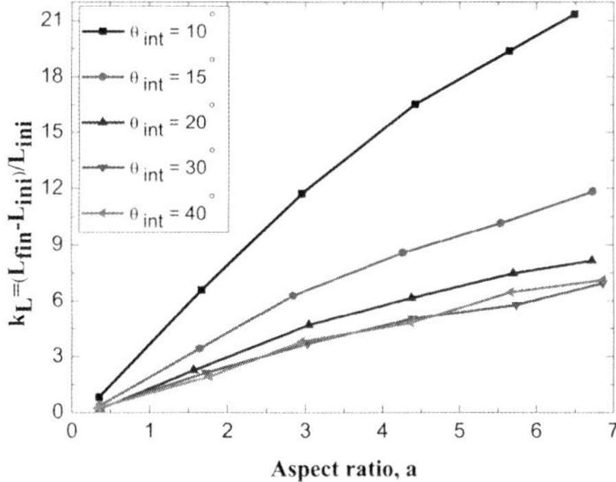

Figure 7 *The influence of internal friction angle*

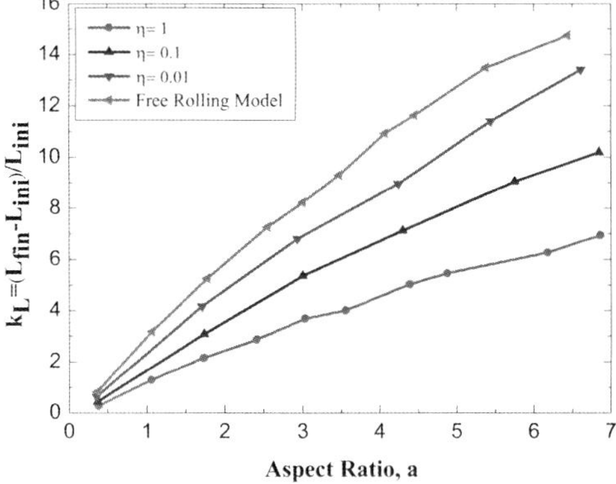

Figure 8 *The influence of limited moment*

4.4 Coefficient η

As discussed before, the microscopic parameters of the adopted contact laws in the DEM contact model are likely to influence the results significantly. Figure 8 illustrates the influence of the coefficient η of the rolling law on the aspect ratio – final run-out relationship. A simulation with particles free to roll, corresponding to the case of η=0, was also included. We can see that the final run-out length decreases for increasing values of η. Thus, the smaller the η coefficient is, the higher the mobility of the column is.

Other parameters such as the characteristic mobility, the initial column strain and the overall particle packing number can be analysed in the same way. From the analysis above, we can see that the major influence comes from the initial granular column aspect ratio, basal and internal friction and the microscopic rolling resistance contact model.

5   CONCLUSIONS

The adoption of a rolling resistance model to indirectly account for the effect of particle non-sphericity led to obtain quantitative agreement between the initial model configuration and the geometry of the final granular deposit. To deepen the understanding of the propagation mechanism of granular flow, dimensional analysis was used. The most important parameters governing the run-out distances of debris flow are the initial column aspect ratio and the parameters of the adopted rolling resistance model.

In this paper, only dry granular flows were considered. In reality, most of the times landslides take place during wet seasons, especially after short-term intensive rainfalls. The second stage of the research will focus on the introduction of fluid – soil coupling in the run-out simulations.

References

1   W. Dade and H. Huppert, *Geology*, 1998, **26**, 803.
2   A. Ritter, *Die Fortpflanzung der Wasserwellen. Z. Ver. Dtsch. Ing.*, 1892, **36**, 947.
3   A.J. Hogg, *Physics of Fluids*, 2007, **19**, 093301
4   E. Lajeunesse and J. B. Monnier, *Physics of Fluids*, 2005, **17**, 103302
5   G. Lube and H. Huppert, 2004, **508**, 175.
6   N. J. Balmforth and R. R. Kerswell, *J. Fluid Mech.* 2005, **538**, 399
7   L. Staron and E. J. Hinch, *J. Fluid Mech*, 2005, **545**, 1.
8   G. Lube, H.E. Huppert, R.S.J. Sparks. *Physical Review*, 2005, E **72**, 041301
9   G. Lube, H.E. Huppert, R.S.J. Sparks. *Physics of Fluids*, 2007, **19**, 043301.
10 R. Zenit. *Physics of Fluids*, 2005, **17**, 031703
11 N.J.P. Belheine, F.V. Plassiard, F. Donze and A. Darve. *Computers and Geotechnics*, 2008, **36**, 320.
12 P. A. Cundall and O. D. L. Strack. *Geotechnique*,1979, **29**, 47.
13 S. McDougall and O. Hungr. *A model for the analysis of rapid landslides motion across three-dimensional terrain*. NRC Research Press, 2004, 30 November
14 Yade Documentation, *https://www.yade-dem.org/sphinx/index-toctree.html*, 2009

DEM MODELLING OF THE DIGGING PROCESS OF GRAVEL: INFLUENCE OF PARTICLE ROUNDNESS

S. Miyai, T. Katsuo, T. Tsuji, T. Takayama and T. Tanaka

Department of Mechanical Engineering, Osaka University, 2-1 Yamada-oka Suita Osaka 565-0871 JAPAN

1 INTRODUCTION

To improve the energy efficiency of construction and mining machineries such as hydraulic excavators and bulldozers, it is important to optimize earthmoving processes and the shape of tools such as buckets and blades, which were used to handle ground materials directly. It is important to understand the interactions between mechanical tools and ground materials. However, due to its complexity, most designing has been conducted empirically so far. In the present study, a numerical model based on the discrete element method (DEM) is developed for the gravel excavation process using a hydraulic excavator bucket. It is generally known that gravel particles typically observed in mining sites are far from spherical and have angular shapes. It is expected that the non-sphericity of gravel particles will influence the shear strength and the flow pattern of particles during the excavation. A number of DEM studies considered the non-sphericity of particles. Generally, these can be classified into three models: multi-sphere,[1] super-quadric[2] and polyhedral[3] model. Granular materials in nature such as sand, gravel and rock include several geometrical factors over different scales. It is computationally expensive to model real particle shapes,[4-7] especially for practical engineering design problems. It is mandatory to develop a simplified model that only includes essential geometrical factors important for gravel-tool interactions.

It is generally known that the particle shape can be characterized by three shape indexes that are independent each other: sphericity, roundness and roughness.[8] Sphericity describes how spherical a particle is and is relevant to overall shape such as aspect ratio, elongation and flatness. Roundness describes the shape of the corners on a particle and roughness is much smaller surface features than a particle diameter. The effects of particle shape on mechanics of granules were reported experimentally and numerically. Frank and Cleary[9] presented computational studies of the collapse of granular columns using DEM. They showed that the geometry of final deposits was affected strongly by roundness than aspect ratio. Shamsi and Mirghasemi[6] conducted triaxial tests with different angular particle models and showed grains with high angularity have higher shear resistance and dilation. These studies were performed under quasi-static conditions and there are little studies under dynamic conditions that are important for the development of construction and mining machineries. According to Santamarina and Cho,[10] at large strains, roundness prevents the rotational motion of particles and roughness will hinder contact slippage between particles.

In the present study, we develop a new parallel DEM code based on the multi-sphere model and investigate the role of particle roundness on the dynamic behaviour of particles during excavation.

2 VERIFICATION TEST

A new parallel DEM code based on the multi-sphere model is developed in the present study. Non-sphericity of particle is expressed by combining multiple spherical elements. Different geometrical shapes can be represented by changing the combination of number, size and overlap of elemental particles. In the developed code, the motion of cluster particle is obtained by solving Newton's second law and Euler's equation of motion. The orientation of cluster particle is calculated using Quaternion and the contact detection between particles is conducted for each element.

A comparison study is performed to verify the accuracy of the code. A bench-scale experiment is performed in which a flat blade is translated in a container filled with particles. The angle of attack of the blade is 90 degee, which means that the working surface of the plate is normal to the translating direction, and the reaction force working on the blade is measured by a load cell. A blade which has a flat square surface of 100 mm high, 200 mm wide and 5.7 mm thick is used. The acrylic container has a length of 600 mm, width 200 mm and height 180 mm. There are small gaps between the blade and the side walls of container. The blade does not have a direct contact with side walls while the escape of particles from the sides of the blade is restricted.

In this section, a simplified particle geometry that enables direct comparisons between the multi-sphere model and the experiment is adopted. Namely, paired particles are prepared by brazing two steel spherical particles on their surfaces. Elemental particles used in the experiment have 3.97 mm diameter and the major axis length of the paired particle becomes 7.94 mm as a result. The particle used in DEM (i.e. DEM particle) is prepared to have the same length of major and minor axes while the geometry of the brazed neck is neglected. To compensate the loss of mass due to the simplification, the density of the DEM particle is determined to match the mass of paired particles used in the physical experiment.

In DEM, it is necessary to determine parameters such as spring constant k, coefficient of viscosity η and coefficient of friction μ. Table 1 shows DEM parameters used in the present study. It is usual that the coefficient of viscosity and friction are obtained through physical tests while a small spring constant within the range where particle behaviour is not affected so much is adopted. The coefficient of restitution was determined according to a drop test. It is required to define the coefficient of friction between particle-blade, particle-wall and particle-particle. In the case of particle-blade and particle-wall, it is obtained from the angle of friction obtained by sliding a set of spherical particles on inclined plates made of the same material as the blade and the walls. It is known that the coefficient of friction between particles is related to the angle of repose in bulk particles, we attempt to obtain the coefficient of friction between particles by comparing the results of angle of repose tests using spherical particle between simulations and experiments. In the range where the coefficient of friction is smaller than 0.5, the angle of repose increases monotonically as the coefficient of friction increases. Meanwhile, it almost converges in the range where the coefficient of friction is larger than 0.5 and a value of 23 degree is obtained as a result. The magnitude is similar to the experiment and, in the present study, the coefficient of friction between particles in DEM simulations is set to 0.5.

The bed is levelled without compaction and has a height of 90 mm in the experiment. In DEM simulation, to obtain a randomly-packed bed, after elevating particles to a proper height, random initial velocity is imposed, which enhances random packing in the container. The porosity after the packing is 0.42 in the experiment and 0.43 in the simulation, respectively. The digging depth that is the vertical length between the surface of particle layer and the blade tip is 30 mm. The travelling velocity of the blade is set to 100 mm/s.

Table 1 *Experimental and DEM parameters*

Container size		(m)	0.20×0.97×0.18
Density of particle	Experiment	(kg/m^3)	7.91
	DEM	(kg/m^3)	8.06
Mass of particle		(kg)	5.28×10^{-4}
Coefficient of friction	Particle-particle	(-)	0.50
	Particle-wall	(-)	0.42
	Particle-blade	(-)	0.30
Coefficient of restitution		(-)	0.55
Spring constant	Normal	(N/m)	2×10^4
	Tangential	(N/m)	5×10^3

Figure 1 shows the horizontal force working on the blade. It has a fairly good agreement between DEM and the experiment while a slight difference on its slope is also observed. This is believed to be due to that the coefficient of friction cannot be determined uniquely only by the angle of repose test performed. Figure 2 shows the instantaneous profiles of particle deposits formed in front of the blade at a displacement of 100 mm. A fairly good agreement is observed between DEM and the experiment.

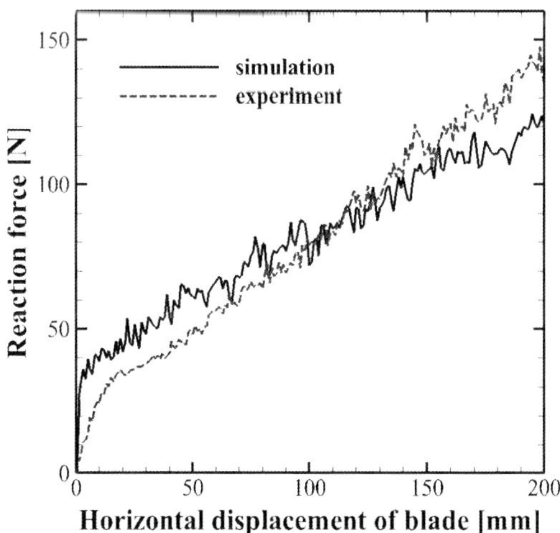

Figure 1 *Reaction force on the blade*

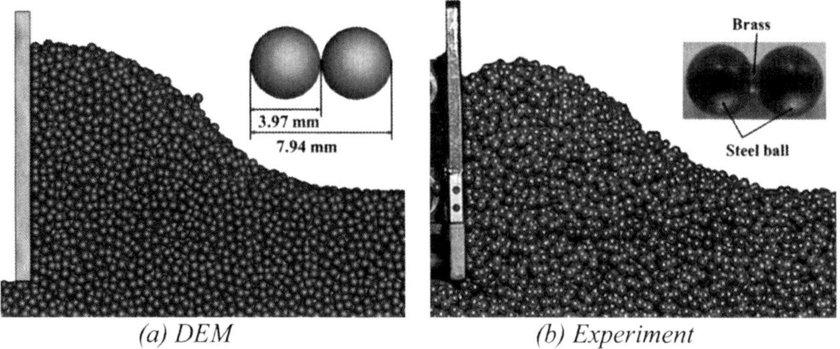

<table>
<tr><td align="center">*(a) DEM*</td><td align="center">*(b) Experiment*</td></tr>
</table>

Figure 2 *Profiles of particle deposits*

3 PARTICLE ROUNDNESS

Roundness and sphericity of particle are defined by the following equations:[11, 12]

$$R_w = \frac{1}{N}\sum_i^N \frac{d_i}{D} \tag{1}$$

$$\psi_P = \frac{\sqrt[3]{S^2}}{L \cdot I} \tag{2}$$

where N is the number of corners, d_i is the diameter of circles fitting corners, D is the diameter of inscribed circle in the maximum projection outline of particle, L is the longest diameter, I is intermediate diameter and S is the shortest diameter. In this study, in order to investigate the effect of particle roundness, non-spherical models with different roundness obtained by replacing the corners of a cube with different curvatures are developed. As shown in Table 2. in addition to the spherical model (R_w=1.0), five non-spherical models that have different R_w from 0.5 to 0.9 are prepared by changing the diameter and overlaps of elemental particles in the multi-sphere model. All non-spherical models are composed of mono-sized elemental particles arranged as a cluster particle with a centre of symmetry. The diameter of cluster particles D_p is defined as the side length of the circumscribed cube becomes 8 mm. Although a cubic shape can be described with a minimum of eight elemental particles, undesirable concavities become larger near the overlaps between elemental particles as R_w becomes smaller. For $R_w = 0.5$ and 0.6, twenty elements are used to minimize the effect of concavities. The density of cluster particles is 2500 kg/m³. Simulations are also performed for the translation problem of a flat blade discussed in the previous section.

In the deformation and fracture of particle assemblies, it is important to examine the development of shear bands characterized by the existence of particles with large rotation and large voids.[13, 14] Figure 3 shows the instantaneous angular velocity of particles normalized by the particle radius and the blade translating velocity at a travelling distance of 100 mm. These are the results in the vertical centre plane of the blade. It is confirmed that, inside the particle layer, particles having large angular velocity are restricted to a narrow zone between the blade tip and the free surface. The formation of shear band is expected in this zone. As the roundness decreases, the number of particles having higher angular velocity also decreases. Figure 4 is the reaction forces acting on the blade, only the results for R_w = 0.5, 0.8 and 1.0 (sphere) are shown. Comparing to the sphere case, the reaction forces finally are 80, 75, 60, 35 and 15 % higher for R_w = 0.5, 0.6, 0.7, 0.8 and 0.9, respectively. Particles with high roundness easily roll over each other due to the lower resistance on the rotation motion. This leads to the easy change in microscopic structures in the particle layer. Meanwhile, smaller roundness prevents the rotational motion of particles and strengthens the shear resistance of the particle layer. During the deformation of particle layer, the sliding between particles play the major role rather than rolling, which strengthens the shear resistance of the layer.

Table 2 *Multi-sphere model parameters*

Shape model						
Projective shape						
Roundness R_W	0.5	0.6	0.7	0.8	0.9	1.0
Number of elements	20	20	8	8	8	1
Diameter of elements (mm)	4.0	4.8	5.6	6.4	7.2	8
Bulk porosity	0.41	0.38	0.39	0.39	0.40	0.40
Bulk mass (kg)	16	17	17	17	16	16

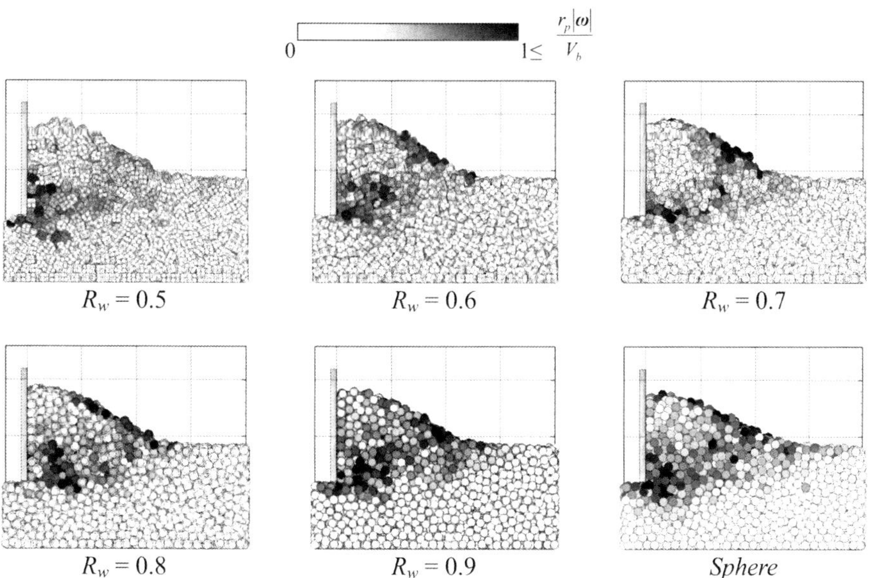

Figure 3 *Angular velocity distributions*

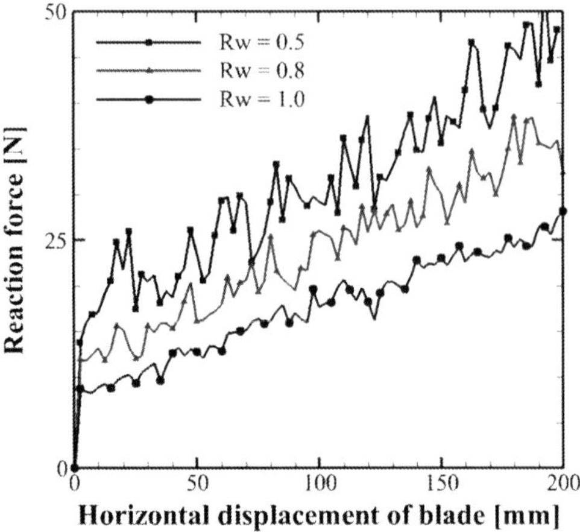

Figure 4 *Reaction forces on the blade for various R_w*

4 DIGGING PROCESS USING A HYDRAULIC EXCAVATOR BUCKET

In the previous section, the effects of particle roundness on particle behaviour and shear strength were investigated with a simple problem. In this section, as a more practical engineering application, the effect of particle roundness on the digging process using a hydraulic excavator bucket is investigated. Two cases using spherical (R_w = 1.0) and non-spherical (R_w = 0.6) particle models are examined and compared. The shape of bucket in the simulation is modelled according to a commercial hydraulic excavator bucket. The digging trajectory shown in Figure 5 is determined from experiments using a real hydraulic excavator. All computations in this section are performed in 1/10 scale. Figure 6 shows the forces acting on the bucket in horizontal (y) and vertical (z) directions during the digging process. The process can be divided into three stages according to the sign of vertical acceleration of the bucket's pivot point shown in Figure 5, namely, it becomes negative in the *penetration* stage ($0 \leq t < 0.8$ s), almost zero in the *translation* stage ($0.8 \leq t < 3.5$ s) and positive in the *scooping-up* stage ($3.5 \leq t \leq 4.35$ s). In the penetration stage, the horizontal force F_y increases rapidly and the positive value of force in the vertical direction F_z also is obtained due to the penetration resistance. Both horizontal and vertical forces of the non-spherical case are higher than the spherical case. In the translation stage, due to the deformation and fracture formation in front of the bucket tip, the horizontal force increases gradually. It becomes nearly 60% higher for the non-spherical case comparing to the spherical case. Meanwhile, vertical forces increase rapidly as particles flow into the bucket in both cases and no apparent difference is observed for different particle shapes considered. In the scooping-up stage, the bucket moves upward and the particles that cannot stay within the bucket fall out. As a result, horizontal and vertical forces decrease. Rapid increase of both forces around $t = 4.2$ s is due to the contact between the rear surface of the bucket and the particle layer.

Figure 7 show the normalised angular velocity of particles in the vertical centre plane of the bucket. It only includes the results at t =1.5 and 3.0s in the translation stage.

Similar to Figure 3, for the non-spherical case, the concentration of particles having higher angular velocity is observed in a narrow zone between the bucket tip and the free surface. Shear band play an important role in the digging process because it can be a starting point of material fracture. It is a great advantage of these simulations that the internal structures such as shear bands in the particle layers can be observed.

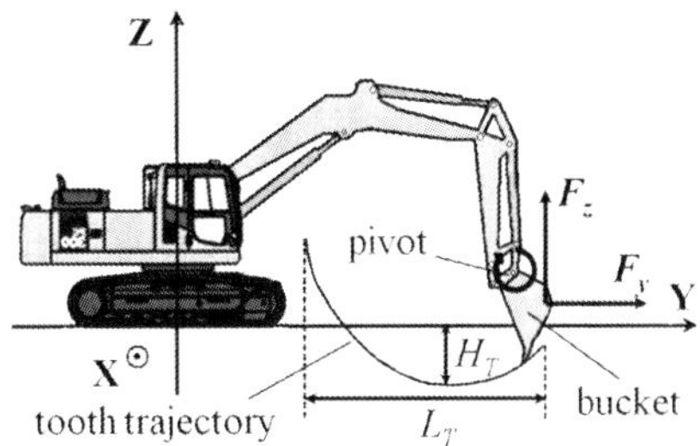

Figure 5 *Trajectory of bucket*

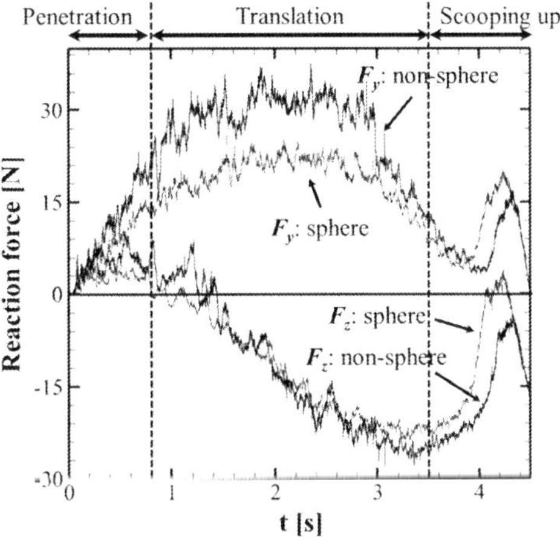

Figure 6 *Reaction force on the blade*

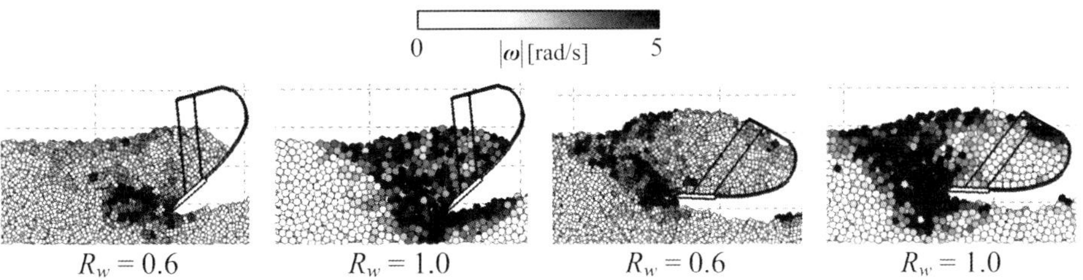

(a) $t = 1.5$ s (initial translation period)   (b) $t = 3.0$ s (last translation period)

Figure 7 *Angular velocity distributions*

5 CONCLUSIONS

This paper attempted to investigate the effects of particle roundness on the digging processes. Smaller roundness prevents the rotational motion of particles and strengthens the shear resistance of the layer. Particles with higher angular velocity were localized in the narrow zone where the formation of shear band is expected during digging processes. In order to predict the behaviour and shear strength of particle assembly precisely, it is necessary to take the effect of particle roundness into account when particles with higher angularity, such as gravels observed in mining sites, are involved.

Acknowledgements

This study is performed under the collaboration between Komatsu Ltd. and Osaka University. We would like to acknowledge Development Division, Komatsu Ltd. for their financial and technical supports.

References

1   X. Garcia, L. T. Akanji, M. J. Blunt, S. K. Matthai and J. P. Latham, *Phys. Rev. E*, 2009, **80**, 021304.
2   P. W. Clearly and M. L. Sawley, *Appl. Math. Model.*, 2002, **26**, 89.
3   P. A. Cundall, *Int. J. Rock Mech., Min. Sci. & Geomech. Abstr.*, 1988, **25**, 107.
4   T. Matsushima and H. Saomoto, *Proc. NUMGE2002*, 2002,239.
5   J. P. Latham, A. Munjiza, X. Garcia, J. Xiang and R. Guises, *Minerals Eng.*, 2008, 21, 797.
6   M. M. M. Shamsi and A. A. Mirghasemi, *Powder Technol.*, 2012, **221**, 431.
7   M. Price, V. Murariu and G. Morrison, *Proc. DEM Conf. 2007*, 2007, 27.
8   P. J. Barrett, *Sedim*, 1980, **27**, 291.
9   M. Frank and P. W. Cleary, http://www.mathematik.uni-kl.de/~frank/GranularCollapse SQ.pdf, 2006.
10 J. C. Santamarina and G. C. Cho, *Adv. in Geotech. Eng., Proc. Skempton Conf.*, 2004, 604.
11 H. Wadell, *J. Geol*, 1932, **40**, 443.
12 E. D. Sneed and R. L. Folk, *J. Geol*, 1958, **66**, 114.
13 K. Iwashita and M. Oda, *J. Eng. Mech, ASCE*, 1998, **124**, 285.
14 J. Maciejewski and A. Jarzebowski, *J. Terramech.*, 2002, **39**, 161.

DEM MODELLING OF HIGH SPEED DIE FILLING PROCESSES

C.-Y. Wu, F. Ogbuagu and C. Pei

School of Chemical Engineering, University of Birmingham, Birmingham, B15 2TT

1 INTRODUCTION

Die filling is a powder handling process in which particles/granules are deposited into a cavity (i.e. a die) using a feed shoe/hopper and is a critical processing stage in the manufacture of particulate products, such as pharmaceutical and detergent tablets, catalyst pellets, ceramic and powder metallurgical components. Two types of die filling systems have been employed in particulate product manufacture:[1] 1) active filling system, in which a die moves at a specified filling speed underneath a stationary feed shoe/hopper and powders flow into the die once an effective discharge area is created (i.e. when the die opening starts to transverse the orifice of the shoe/hopper). This is the typical filling system in rotary tabletting machines widely used in the pharmaceutical industries and elsewhere; ii) passive filling system, which employs a moving shoe with a specified filling speed to deliver powders into a stationary die and has been used in single station presses (such as compaction simulators) for manufacturing small batch of products for pharmaceutical research and development, and powder metallurgical and ceramic components. In most particulate product manufacturing processes, relatively low filling speeds (say < 1 m/s) are generally employed to ensure that the die cavity can be completely filled with specified passes/strokes.[1-8]

Nevertheless, active die filling systems operating at high filling speeds (up to 20 m/s) have recently been developed to dose granules of *ca* 500 μm in diameter. It is a challenging task to achieve consistent and precise dosing using these high-speed die filling systems due to the dynamic nature of the process and poor understanding of the mechanical behaviour of granules in this process. Therefore, in this study, for the first time, we explored the flow behaviour of granules during high-speed die filling processes using discrete element methods (DEM) and investigated how such high filling speeds affect the die filling behaviour.

2 DEM MODELS

In DEM, granular materials are modelled as assemblies of distinct particles that interact with each other using specified contact laws. The motion of particles is governed with Newton's second law and the kinematics is determined using numerical integration techniques, such as central difference methods, at specified time steps. Particles are generally modelled as disks in 2D and spheres in 3D, but irregular particles can also be modelled.[8] An in-house DEM code was employed in this study. In the DEM model, only elastic spherical particles were considered. The normal contact between those elastic particles was modelled using the classic Hertz theory[9] and the tangential interaction was modelled using Mindlin and Deresiewicz theory.[10] This DEM code has also been used for modelling die filling processes at low filling speeds and it has been shown that it is a robust tool for die filling simulations.[11-13] The following two die filling processes were then simulated using this DEM code in 2D:

1) Die filling with stationary shoe and die, in which both the shoe and the die are stationary and powders flow into the die once the orifice of the shoe is opened. This is similar to the typical hopper flow process;
2) High speed die filling, in which the shoe is stationary but the die moves at a speed in the range of 1 m/s and 20 m/s.

Case 1 was set up in order to further validate the DEM model, for which the numerical results will be compared with the Beverloo equation proposed for determining the mass flow rate in hopper flow. The DEM models for Cases 1 and 2 are shown in Figure 1.

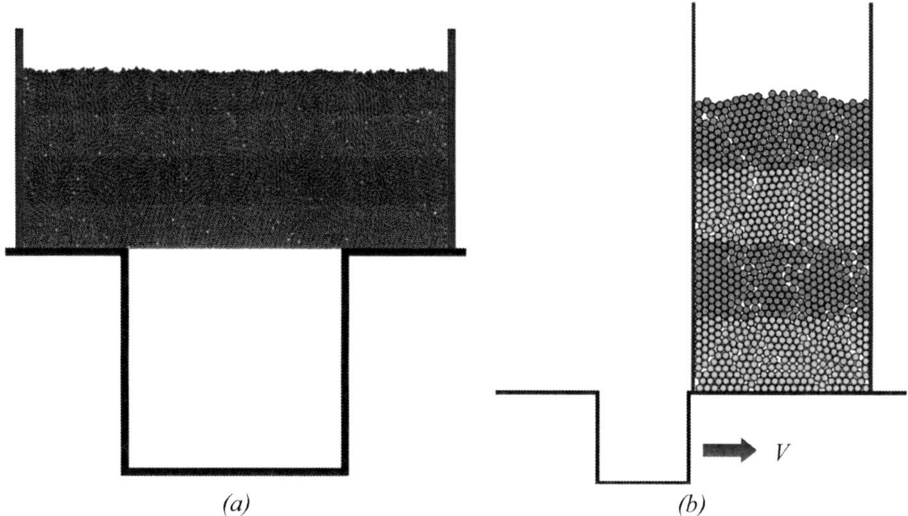

Figure 1 *DEM models for a) die filling with stationary die and shoe (i.e. Case 1) and b) high speed die filling (i.e. Case 2)*

For both models, the sizes of the die are identical of 6X6 mm and the shoes have the same width of 10 mm. A specified number of mono-sized particles are initially randomly generated in the shoe and then settle under gravity to a steady state with negligible overall kinetic energy. The particles are assumed to be of microcrystalline cellulose with a density of 1,500 kg/m^3, Young's modulus of 8.7 GPa and Poisson's ratio of 0.3. It is assumed that the die and the shoe are made of stainless steel with a density of 7,800 kg/m^3, Young's modulus of 210 GPa and Poisson's ratio of 0.3. The coefficients of friction between particles and between particle and walls are set to 0.3. For Case 1, particles with a diameter in the range of 62.5 μm to 500 μm are considered. The number of particles used in the simulations with different sized particles was varied between 250 (for 500 μm) and 16,000 (for 62.5 μm) in order to maintain the initial powder height in the shoe at 5 mm. Corresponding void fractions of the initial powder beds are in the range of 0.45 (for 62.5 μm particle) and 0.5 (for 250 μm particles). The particle starts to flow into the die once the shutter is removed. For Case 2, only the largest particle modelled in Case 1 (i.e. 500 μm) is considered as an attempt to model the high speed dosing of granules. For the simulations reported here, 1,000 particles were used. The die moves from left to right at a speed V, specified in the range of 1 m/s to 20 m/s. As the die moves past the shoe, the particles flow into the die under gravity. The filling process is stopped after the die has completely passed the shoe. The mass of deposited particles at various time instants is monitored so that the average mass flow rate can be determined.

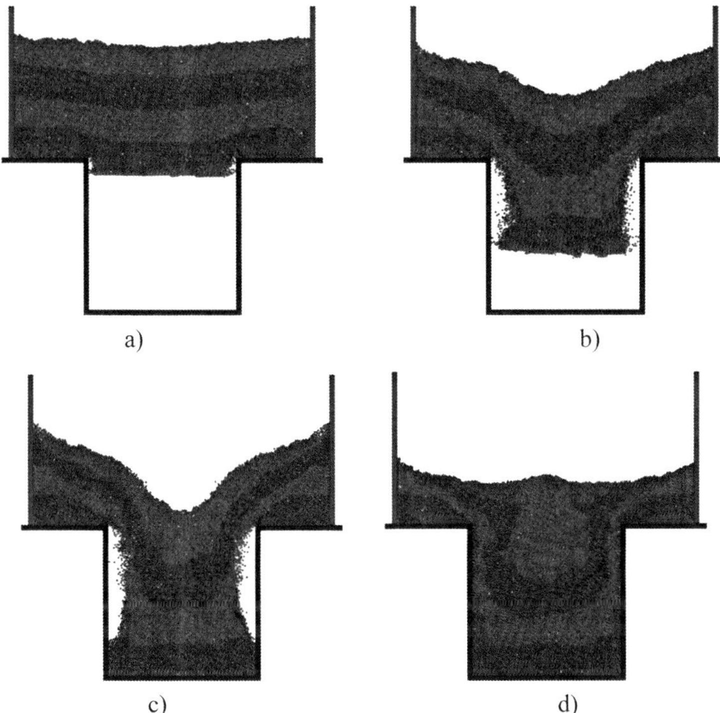

Figure centres a), b), c), d)

Figure 2 *Typical powder flow patterns during die filling with stationary die and shoe (particle size: 62.5 μm)*

3 RESULTS AND DISCUSSION

3.1 Die Filling with Stationary Die and Shoe

Typical powder flow patterns during die filling with stationary die and shoe are shown in Figure 2, in which the results for particles of diameter of *62.5* μm are presented. It can be seen that, once the shutter is removed, the particles fall vertically into the die in a smooth manner (Figure 2a). The previously flat top surface of the powder bed transforms into a concave shape (Figure 2b) as the powder flows into the die. Similar phenomena were also observed experimentally in Wu *et al.*[4] It is also clear that there are two narrow regions near the die walls, in which powder flow is retarded (Figures 2b & 2c). This feature is consistent with experimental observations during powder discharge from a hopper and a bin, and is known as the 'empty annulus' effect.[11,14] Although the powder flow behaviour is similar to that of hopper flow, it should be noted that there is subtle difference between die filling with stationary shoe and die considered here and hopper flow processes, as in the latter powders are generally discharged into an open space. While in the present study, the die of finite size sits below the orifice, discharged powder starts to pack in the die. And an enduring contact network will be developed (Figure 2c & 2d), which induce "back pressure" inhibiting the discharge of powders.

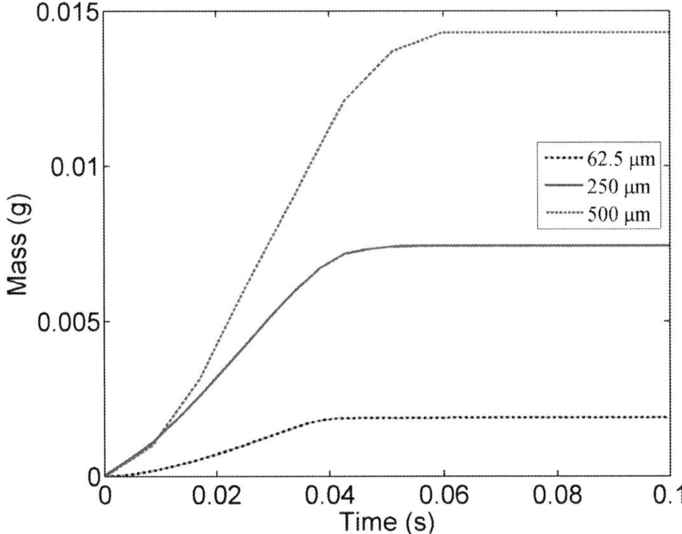

Figure 3 *Evolutions of the mass of particles deposited into the die during die filling with stationary die and shoe for different sized particles*

Figure 3 shows evolutions of the mass of particles deposited into the die during the filling with different sized particles. It is clear that as soon as the shutter is removed, discharge of the powder starts and the mass of deposited particles gradually increases until the die is completely filled, thereafter the mass of deposited particles remains constant. The pattern of the profiles shown in Figure 3 is consistent with those obtained by Guo *et al.*[11] The final mass of deposited particles appears to increase with the particle size, this is due to the fact that our DEM simulation was perform in Pseudo-2D, i.e., a slice of 3D model with a mono-layered particles was considered and the thickness of the powder bed is equal to the diameter of the particles. In other word, the volume of powders (hence the mass) increases with the diameter of the particles used. The filling time was defined as the duration in which powders are discharging, i.e. from the time instant when the shutter is removed to the time instant the die is just completely filled (i.e. the mass of the total deposited particles is just achieved). It can be seen that the filling time varies slightly with the particle size. Knowing the total mass of deposited particles, *M*, and the filling time, *t*, the average mass flow rate, $\overline{M}$, can hence be determined as

$$\overline{M} = M/t \tag{1}$$

The average mass flow rate for various particle sizes are shown in Figure 4, in which the lines show the predictions of the modified Beverloo equation for hoppers with a rectangular orifice, [11] *i.e.*

$$\overline{M} = C\rho_b \sqrt{g d_p} \left(b_0 - k d_p \right)^{3/2} \tag{2}$$

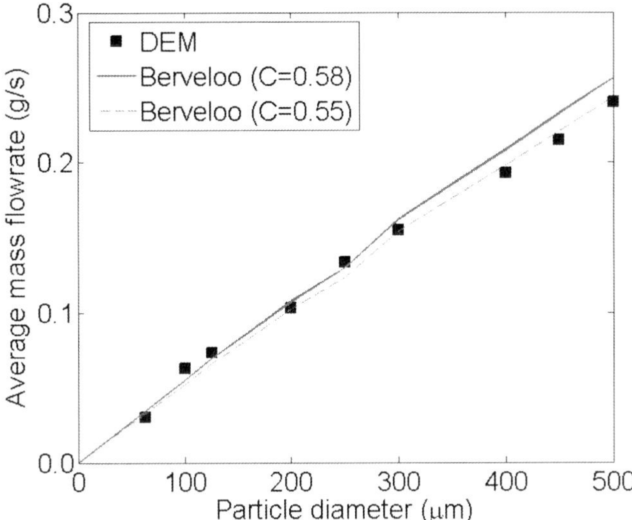

Figure 4 *Variation of average mass flowrate with particle diameters during die filling with stationary die and shoe*

where C is the Beverloo constant that has a value of $0.55\sim0.65$, and $C=0.58$ for discharging of spherical particles.[14] d_p is the particle diameter. g is the gravitational acceleration, b_0 is the width of the die opening (i.e. the width of the discharge area), k is a dimensionless parameter that is related to particle shape and has a value of ca 1.5 for spherical particles.[14] ρ_b is the bulk density of the powder in the hopper and can be determined as

$$\rho_b = (1 - \varepsilon)\rho_p \tag{3}$$

where ε is the void fraction of the powder bed and ρ_p is the particle density. Since we considered pseudo-3D hopper flow with a thickness of d_p, the nominal discharging area is hence $b_0 \times d_p$. The void fraction of powder bed, ε, is assumed to be the initial void fraction of the powder bed (i.e. before discharging) and can be readily determined from DEM simulations.

It can be seen from Figure 4 that the average mass flow rate increases as the particle size increases, and excellent agreements between DEM results and the predictions of Beverloo equation are obtained. DEM simulations only slightly underestimate the mass flow rate when the particles are sufficient large. This is attributed to the strong enduring contact network established at the later stage of the discharging process with large particles, which results in significant de-acceleration of particles, as clearly shown in the time evolutions of the mass of deposited particles of 200 and 500 μm (Figure 3). Nevertheless, this clearly demonstrates that present DEM models can be used to accurately predict the mass flow rate in die filling and hopper flow.

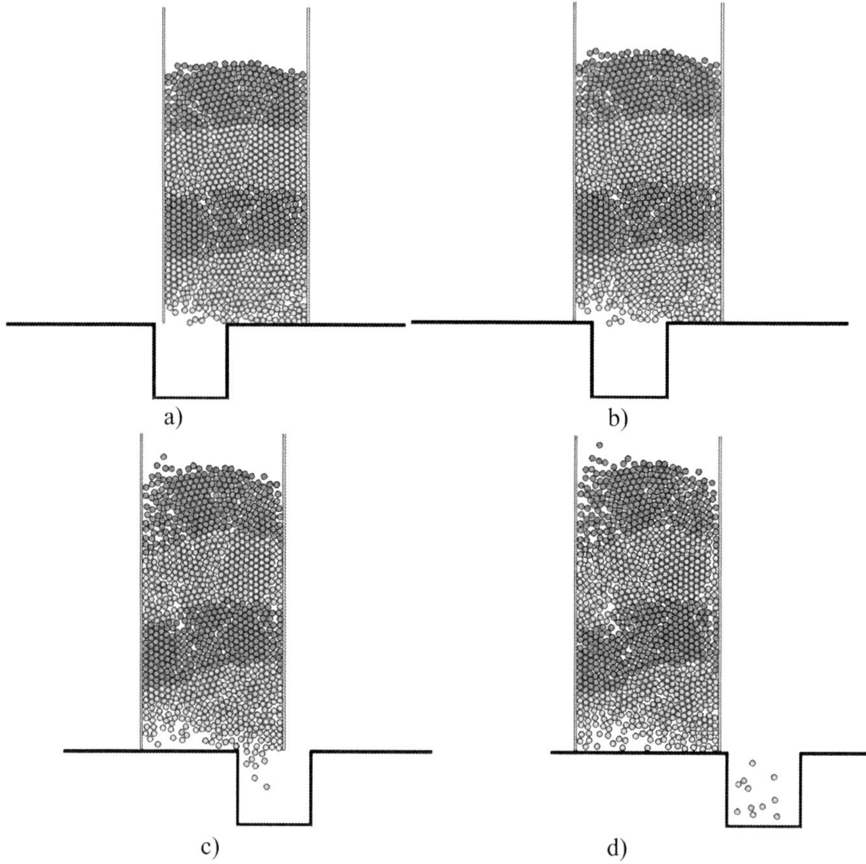

Figure 5 *Typical powder flow patterns during high speed die filling at a fill speed of 1 m/s*

3.2 High Speed Die Filling

As the die moves towards the shoe at a relative high speed, the friction between the particle bed in the shoe and the surface of the platform causes intensive shear at the interface. Consequently, significant dilation of the particle bed is induced as illustrated in Figure 5, which shows the typical powder flow patterns during high speed die filling with a filling speed of 1 m/s. Particles at the interface are pushed towards the leading side of the shoe (i.e. right hand side in Figure 5) and large voids are developed at the interface (Figures 5a and 5b). Some particles detach from the granule bed under gravity and flow into the die (Figure 5b). The tailing die wall that is rapidly moving collides with the falling particles and scrapes some particles into the die while it can also push some granules further into the particle bed at high speeds (Figure 5c). As a consequence of the intensive shear at the interface and the collisions between die wall and particles, the entire particle bed can be mobilised upwards causing a significant rise of granular temperature that characterises the fluctuating velocities of particles in rapid granular flow.

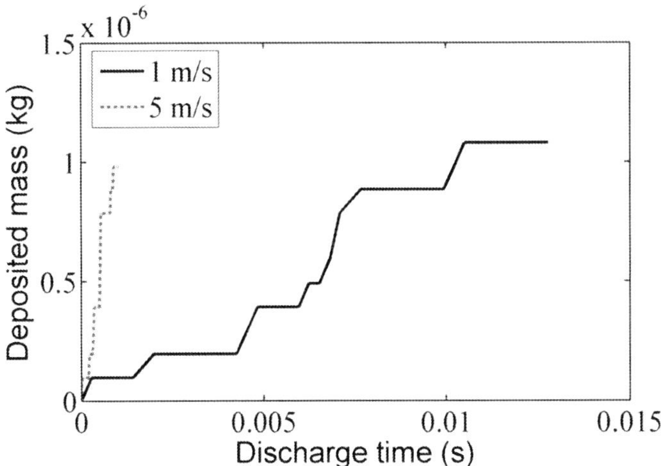

Figure 6 *Evolutions of the mass of particles deposited into the die during high speed die filling with filling speed of 1 & 5 m/s*

Evolutions of the mass of deposited particles during high speed filling at speeds of 1 and 5 m/s are shown in Figure 6. It is clear that the time evolution of the mass of deposited particles has a stepwise profile rather than smooth curves as shown in Figure 3. It is primarily attributed to the discontinuous discharging nature of the high speed die filling process, in which individual particles are deposited in a discrete manner. This is induced by the intensive shear and collision of die wall with falling particles. It was also interesting to note that although the total mass of deposited particles does not change much when the filling speed is increased from 1 m/s to 5 m/s, the effective filling time (defined as the time duration between particle starting to flow into the die and the total mass of deposited particles being first achieved) is reduced significantly. This indicates that the average mass flow rate defined by Equation (1) increases as the filling speed increases, as demonstrated in Figure 7, which show the variation of average mass flow rate with the filling speed. It can be seen that the average mass flow rate is almost linearly proportional to the filling speed, implying that the change in mass flow rate is mainly due to the reduction in effective filling time that decreases linearly as the filling speed increases. In Figure 7, the average mass flow rate for discharging the same particle system (*i.e.* particles of 250 μm) from stationary die and shoe discussed in Section 3.1, which is referred to as the static filling rate, is also superimposed. The average mass flow rate at high filling speeds (say > 5 m/s) is generally much higher than the static filling rate. However, when the filling speed is lower than 5 m/s, the average mass flow rate is lower than the static filling rate. This is believed to be the consequence of the reduction in effective discharging area during high speed die filling.

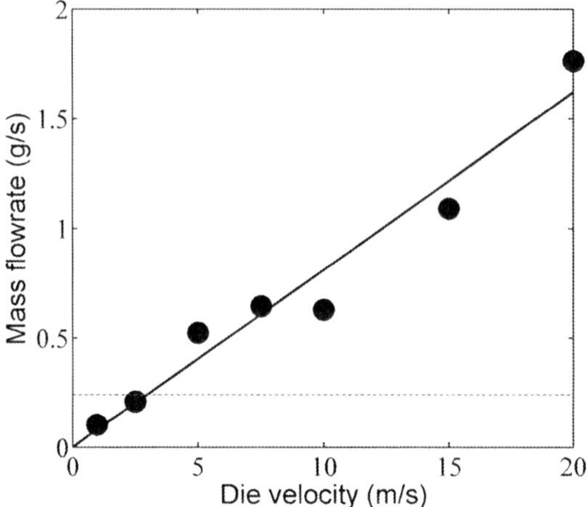

Figure 7 *Variation of average mass flow rate with filling speed during high speed die filling (Symbols: DEM results; dashed line: mass flow rate during hopper flow with the same particle system)*

4 CONCLUSIONS

High speed die filling has been developed for dosing individual granules. For this process, the mass flow rate is an important process parameter that controls the accuracy and consistency of granule dosing. In this study, high speed die filling process is modelled using a discrete element method (DEM). The DEM models is first validated with die filling with stationary die and shoe that is similar to hopper flow, for which mono-disperse systems of various particle sizes are considered and the mass flow rate is examined and compared with the prediction of the Beverloo equation. Excellent agreements are obtained, demonstrating the robustness and accuracy of the DEM model. DEM modelling of high speed die filling processes at various filling speeds are then performed and the average mass flow rate is determined. It is found that the mass flow rate in the high speed die filling process is essentially proportional to the filling speed. During high speed die filling, intensive shear at the interface between particle and platform surfaces and the collision between die wall and falling particles results in a discontinuous flow pattern.

Acknowledgements

The authors would like to thank V. Murthy for stimulating discussion.

References

1   C.-Y. Wu, *Particuology,* 2008, **6**, 412.
2   C.-Y. Wu and A.C.F. Cocks, *Powder Metallurgy,* 2004, **47**, 127.
3   C.-Y. Wu, A.C.F. Cocks, O.T. Gillia and D.A. Thompson, *Powder Technology,* 2003, **138**, 216.
4   C.-Y. Wu, L. Dihoru and A.C.F. Cocks, *Powder Technology,* 2003, **134**, 24.

5 C. Bierwisch, T. Kraft, H. Riedel and M. Moseler, *Powder Technology,* 2009, **196**, 169.

6 G.F. Bocchini, *Powder Metallurgy,* 1987, 30, 261.

7 I.C. Sinka, L.C.R. Schneider and A.C.F. Cocks, *International Journal of Pharmaceutics,* 2004, **280**, 27.

8 C.-Y. Wu and A.C.F. Cocks, *Mechanics of Materials,* 2006, **38**, 304.

9 H. Hertz, *Journal fur die reine und angewandte Mathematik, 1882, 92, 156.*

10 R.D. Mindlin and H. Deresiewicz, *Journal of Applied Mechanics*, 1953, **20**, 327.

11 Y. Guo, K.D. Kafui, C.-Y. Wu, C. Thornton and J.P.K. Seville, *AICHE Journal*, 2009, **44**, 49.

12 Y. Guo, C.-Y. Wu, K.D. Kafui and C. Thornton, *Powder Technology,* 2009, **197**, 111.

13 Y. Guo, C.-Y. Wu and C. Thornton, *Chemical Engineering Science,* 2011, **66**, 661.

14 J.P.K. Seville, U. Tuzun and R. Cliff, *Processing of Particulate Solids*, Blackie Academic & Professional, London, 1997

DEM ANALYSIS OF LOADS ON DISC INSERTS IMMERSED IN GRAIN DURING SILO FILLING AND DISCHARGE

R. Kobylka and M. Molenda

Institute of Agrophysics PAS, Doswiadczalna 4, 20-290 Lublin, Poland.

1 INTRODUCTION

Inserts are commonly used as construction components in storage silos during filling (to improve the mixing of material)[1] or discharge (e.g. to modify flow patterns or decrease dynamic loads shift).[2-5] Even though the first experiment with such objects was dated over 40 years ago,[6,7] procedures on the estimation of loads on objects embedded in granular materials are still not fully covered by standard design codes.[4]

Numerous attempts have been undertaken to measure or/and calculate insert loads based on continuum approach.[6,8-14] However the continuum approach, successfully used for calculation of the stress distribution inside grain silo, fails to describe the phenomena involving several particle dimensions, anisotropy of load or disturbed discharge. For such purposes, numerical methods, such as finite element methods (FEM) and discrete element methods (DEM) are considered as promising alternatives.[15]

In this study, 3D DEM simulations were performed to estimate vertical loads on disc inserts immersed in granular media filling cylindrical container. Results of numerical simulations have been compared with experimental results with wheat stored in a model silo,[16] in which five disc inserts (with ratios of disc diameter to silo diameter d/D of: 0.188, 0.253, 0.375, 0.55 and 0.675) were axially suspended at a height h of $h/D = 0.5$ in the model silo of 0.4 m in diameter, and vertical loads on discs during filling and discharge were estimated.

2 NUMERICAL SIMULATIONS

2.1 Contact Model

Interactions between particles or particle and wall lead to their deformation. Precise description of deformation in contact area usually cannot be directly treated because of its computing complexity. To enable efficient calculations simplified contact models are used. Among others the concept of overlap is introduced to calculate the contact force. The parameter is the deformation of the particle during the impact assuming only deformation at the contact area, and in the case of elastic spherical particles it can be calculated as follows:

$$h_{ij} = R_i + R_j - |x_i - x_j| \tag{1}$$

where indexes i and j stand for number of particles ($i \neq j$). Parameters R and x are used for radius and coordinate of sphere centre.

In case of collision between two spherical particles, the contact area increases with the contact force (depth of the deformation). With an increase of the contact area, force of interaction depends non-linearly on deformation (overlap). Value of the elasticity force (normal component in contact point) for two contacting spheres was evaluated in 1882 by H. Hertz.[17-19] According to the Hertz model, elastic interaction force can be calculated as:

$$\vec{F}_n = \frac{4}{3} \frac{E_i E_j}{E_i\left(1-v_j^2\right)+E_j\left(1-v_i^2\right)} \sqrt{\frac{R_i R_j}{R_i+R_j}}\, h_{ij}^{\frac{3}{2}} \vec{n}_{ij}, \tag{2}$$

where: E is Young modulus, v is Poisson ratio and $\vec{n}_{ij}$ is the normal vector representing direction of the force.

Numerous contact models that are currently used are based on elastic interactions as proposed by Hertz (e.g. Hertz-Mindlin, Kuwabara-Kono, Lee-Herman or Thornton contact models).[18] In this study, the Kuwabara-Kono contact model was chosen as it was proven useful in the analysis of impact between two colliding spheres[18] and impact of spherical particle against a flat surface.[20] The Kuwabara-Kono contact model had been proposed in 1987 as an extension of the Hertz model by including damping.[21] Damping component of the force was taken as:

$$\vec{F}_d = 2 \frac{B_i B_j}{B_i\left(1-\sigma_j^2\right)+B_j\left(1-\sigma_i^2\right)} \sqrt{\frac{R_i R_j}{R_i+R_j}}\, h_{ij}^{\frac{1}{2}} \vec{v}_{ij}^{\perp}, \tag{3}$$

where B and σ are coefficients representing energy loss during impact, $\vec{v}_{ij}^{\perp}$ is the normal component of the velocity.

Contact model was then expanded using a tangent force component to fully characterise the contact force:[22]

$$\vec{F}_t = \min\left(\frac{16}{3} \frac{G_i G_j |\delta \vec{s}|}{G_i\left(2-v_j\right)+G_j\left(1-v_i\right)} \sqrt{\frac{R_i R_j}{R_i+R_j}}\, h_{ij}^{\frac{1}{2}}, \mu_{ij}|\vec{F}_{ij}^{\perp}| \right) \frac{\delta \vec{s}}{|\delta \vec{s}|}, \tag{4}$$

where $\delta \vec{s}$ is the tangential deformation and the shear modulus $G_i = E_i/2(1 + v_i)$.

2.2 Model Setup

Numerical simulations to estimate the vertical load of grain on the disc were carried out using an open source code PAPA (Parallel Algorithm for PArticle flow).[22] Numerical experiments were performed with a model silo of 0.1 m in diameter and 0.33 m high (Figure 1). The silo was divided into two parts: a lower chamber of 0.25 m high, in which inserts were placed and for which essential computations were performed, and an upper chamber used as a preparatory chamber, where particles were randomly generated and settled under gravity. These chambers were separated by a conical hopper with an angle between walls and horizontal plane of approximately 8°. The loading orifice had a diameter $D_o=0.036$ m and a height $H_s=0.015$ m.

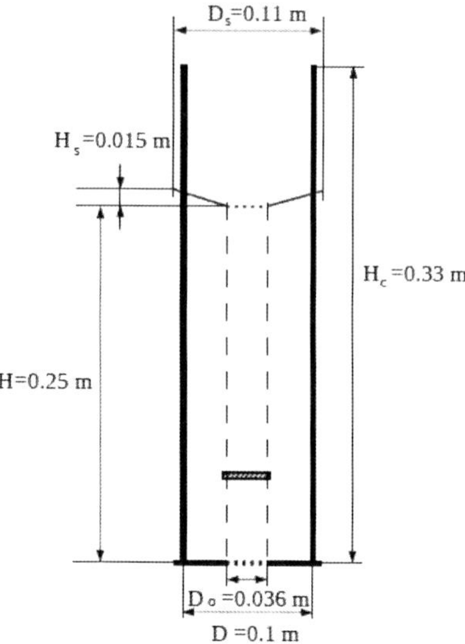

Figure 1 *Schematic diagram of the silo used in numerical experiments*

2.3 Material Parameters

In numerical simulations mono-sized spherical particles (diameter d_p=0.0038 m calculated as the equivalent volume diameter of the wheat kernel) with properties comparable to wheat kernels were used. Particle density ρ_p was 1290 kg/m^3 and silo wall density ρ_s of 7800 kg/m^3 (for steel), Young modulus for particles E_p=868 MPa and for walls E_s=2x10^5 MPa and Poisson ratio v_c=0.18 and v_s=0.3 for particles and walls, respectively, were taken after Wiacek.[23] Because of the difficulty in experimental determination of damping coefficients σ and B, these parameters were determined as follows: the rebound height during the impact of a steel particle against steel surface was measured, the damping parameters were systematically adjusted in DEM simulations until the same impact was reproduced numerically; a similar approach was adapted for the impact of wheat kernel against the flat steel surface (with damping parameters taken from previous step). The described method for determining σ and B did not give a single pair of values but rather a set of pairs, thus the assumption was adopted to fix σ=0.15 for both materials and adjust only parameter B. In case of steel, damping parameter B was chosen to have a value of 3x10^5 while for wheat kernels B=9x10^3. Dimensions of the silo and the insert sizes were taken to be the same as the experimental ones. The diameter of the discharge/loading orifice (equivalent to approximately 9.5d_p) is the minimum to ensure undisturbed flow. The silo diameter equivalent to approximately 26.3d_p was large enough to minimize the disturbance from the wall. The smallest, basic diameter of the disc was chosen to be larger than the discharge gate and the gap between insert and wall (approx. 8.2d_p) was wide enough to minimize flow disturbances.

Both, laboratory tests and numerical simulations were performed for five different disc diameters (with ratios of disc diameter to silo diameter d/D of 0.188, 0.253, 0.375, 0.55 and 0.675), which gave the surface area of discs in the range from 1 to 12.

3. RESULTS

In the Kuwabara-Kono contact model, damping of interactions between particles depends on the relative impact velocity and damping decreases with decreasing velocity. As a result increase of the falling height of the grain causes increasing load fluctuations inside the lower parts of the grain volume (clearly visible above 0.4 h/D).

Figures 2-6 present loads on inserts immersed in grain silo during discharge measured in laboratory experiment (black line, values given on left vertical axis) and obtained from DEM simulations (points, values given on right vertical axis). The ratio of mass of the grain above the insert from laboratory testing to the corresponding mass in numerical simulations was used to compare experimental and numerical results quantitatively.

For the smallest disc insert DEM results were in very good qualitative and quantitative agreement with the experimental data (see Figures 2 and 3). For the medium disc sizes, up to a height of $z_0 \approx 1.2$ h/D, DEM results agree fairly with the experimental data, but above that point they deviate from each other (Figure 4). With large disc sizes, the load determined from DEM analyses are much higher than the experimental measurement, and the gap between the insert and silo walls became too narrow to enable reliable flow of the material (Figures 5 and 6).

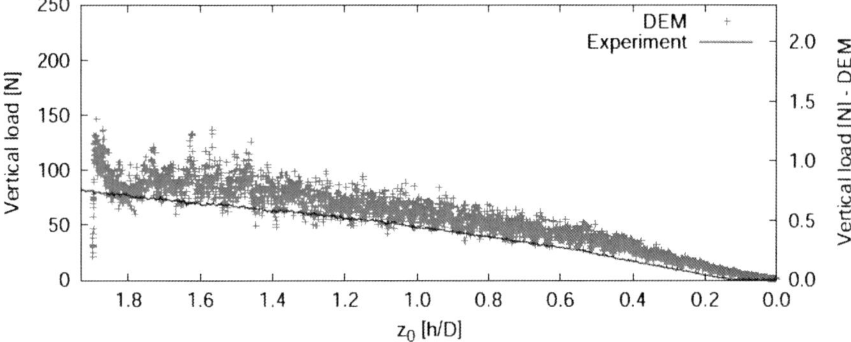

Figure 2 *Vertical load on the disc insert immersed in wheat grain during discharge phase (d = 0.1875D)*

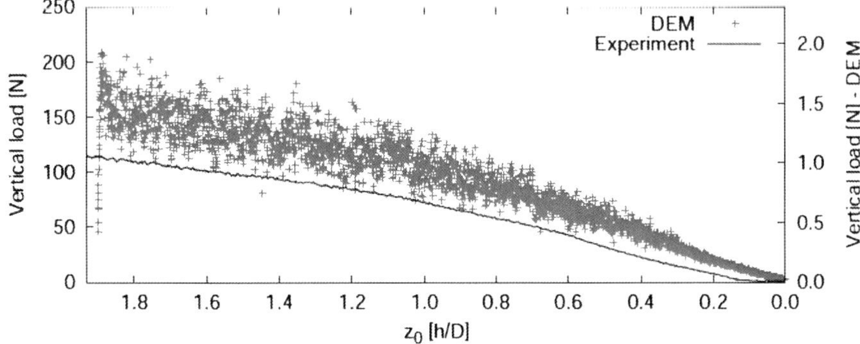

Figure 3 *Vertical load on the disc insert immersed in wheat grain during discharge phase (d = 0.252D)*

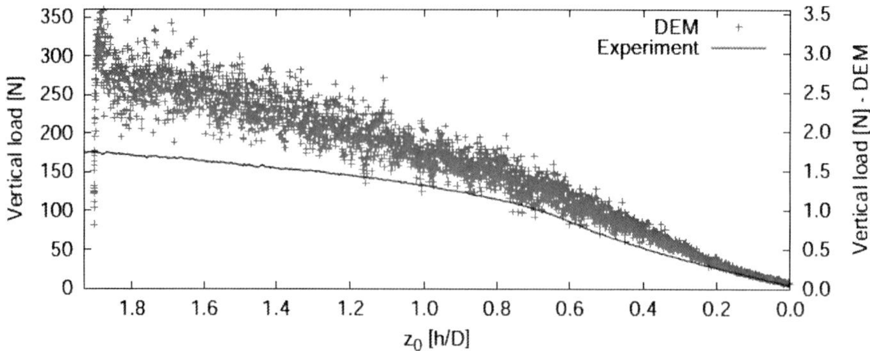

Figure 4 *Vertical load on the disc insert immersed in wheat grain during discharge phase (d = 0.375D)*

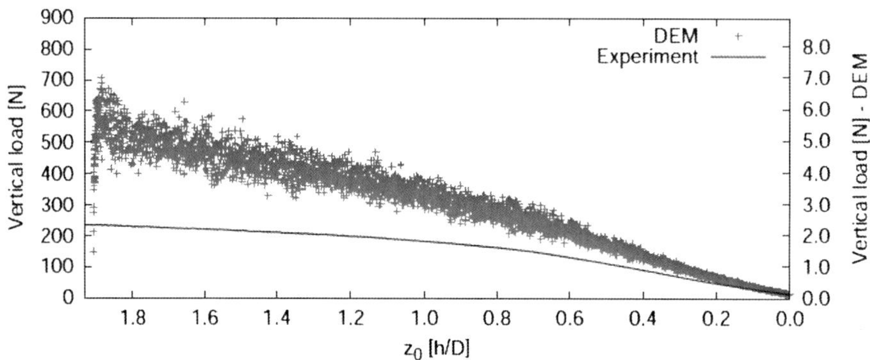

Figure 5 *Vertical load on the disc insert immersed in wheat grain during discharge phase (d = 0.55D)*

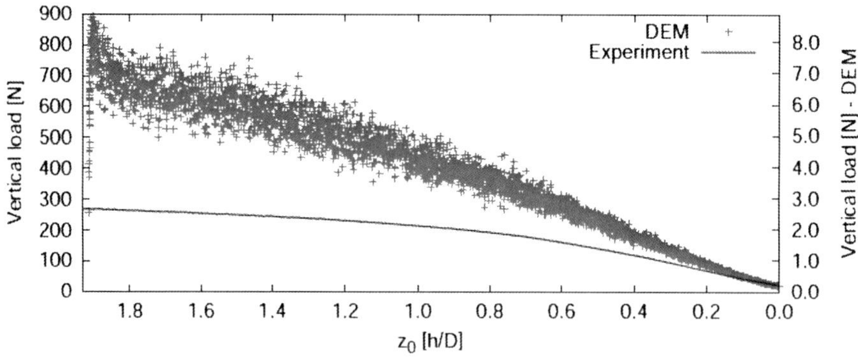

Figure 6 *Vertical load on the disc insert immersed in wheat grain during discharge phase (d = 0.675D)*

4. CONCLUSIONS

For numerical simulations, the Kuwabara-Kono contact model was chosen as it was proven working very well in case of two colliding spheres as well as in the case of sphere colliding with flat surface. Numerical results showed that the contact model did not cope with the increasing vibration at the lower level of the bed caused by the mass of the upper layers of the grain. Damping at low relative speed between particles appeared to be too weak.

Values of estimated vertical loads on discs during discharge corresponded fairly well to the experimental results. Ratio of the mass of material in laboratory and numerical experiments was found to be a good scaling factor.

Results of simulations and laboratory tests were in fairly good agreement in the case of two smallest discs, while for large discs (d/D=0.55 or higher) the loads obtained from numerical analyses were two times higher than the experimental measurements. This is probably due to arching as the gap between the disc and the wall is very small (less than ten particle diameters in the simulations), which precluded undisturbed flow.

Acknowledgement

This research was partially supported by the Ministry of Science and Higher Education, Poland (Grant No. N N310 305739).

References

1   D. Matuszek, M. Tukiendorf, *Inzynieria Rolnicza*, 2009, **9(118)**, 163.
2   J. R. Johanson, *Bulk Solids Handling*, 1982, **2(3)**, 495.
3   Q. Zhang, J. I. Bergen and M. G. Britton, *Can. Agric. Eng.*, 1997, **39(3)**, 215.
4   M. Wojcik and J. Tejchman. In *Proceedings of The 12th International Conference of International Association for Computer Methods and Advances in Geomechanics* (IACMAG), Goa, India 1-6 X, 2008.
5   J. Hartl, J. Y. Ooi, J. M. Rotter, M. Wojcik, S. Ding and G. G. Enstad, *Chemical Engineering Research and Design*, 2008, **86**, 370.
6   J. R. Johanson, *J. Engng Ind., Trans. ASME*, 1966, **88**, 224.
7   J. R. Johanson, *Powder Technol.*, 1967/68, **1**, 328.
8   D. M. Walker, *Chem. Engng Sci.* 1966, **21**, 975.
9   J. R. Johanson, *Powder Technology*, 1971/72, **5**, 93.
10 H. Tsunakawa and R. Aoki, *Powder Technology*, 1975, **11**, 237.
11 J. Strusch and J. Schwedes, *Bulk Solids Handling*, 1994, **14(3)**, 505.
12 S. Ding, M. Wojcik, M. Jecmenica and S. R. de Silva, *Task Quarterly* 2003, **7(4)**, 525.
13 C. Chou and J. Chang, *Advanced Powder Technol.*, 2006, **17(5)**, 505.
14 S. Ding, A. Dyroy, M. Karlsen, G. G. Enstad and M. Jecmenica. *Particulate Science and Technology*, 2011, **29(2)**, 127.
15 T. Schuricht, C. Furll and G. G. Enstad. In *Proceedings RELPOWFLOIV*, Tromso, Norway, 10-12 June 2008.
16 R. Kobylka, *Modelowanie obciazen obiektow zanurzonych w ziarnie pszenicy. (Modeling of loads on inserts immersed in wheat grain)*, Ph.D. thesis, Institute of Agrophysics PAS, Lublin, 2011.
17 B. K. Mishra, *Int. J. Miner. Process.* 2003, **71**, 73.
18 A. B. Stevens and C. M. Hrenya, *Powder Technology*, 2005, **154**, 99.
19 E. Dnitwa, E. Tijskens and H. Ramon, *Granular Matter* 2008, **10**, 209.

20 M. Wojtkowski, J. Pecen, J. Horabik and M. Molenda, *Powder Technology 2010*, **198**, 61.
21 G. Kuwabara and K. Kono. *Jpn. J. Appl. Phys.* 1987, **26**, 1230.
22 S. Schwarzer. PAPA - Parallel Algorithm for PArticle flow problems, Stuttgart, 2004.
23 J. Wiacek. *Discrete element modeling of quasi-static effects in grain assemblies*, Ph.D. thesis. Institute of Agrophysics PAS , Lublin, 2008.

THREE DIMENSIONAL DEM/CFD ANALYSIS OF SEGREGATION DURING SILO FILLING WITH BINARY MIXTURES OF DIFFERENT PARTICLE SIZES

C.-Y. Wu[1] and Y. Guo[1,2]

[1]School of Chemical Engineering, University of Birmingham, Birmingham, B15 2TT, UK
[2] Presently with Chemical Engineering Department, University of Florida, Gainesville, FL 32611, USA

1   INTRODUCTION

Silos are widely used in a number of industries to store bulk materials ranging from agricultural grains, coal, fine chemicals, pharmaceuticals, woodchips, food products and construction materials. These bulk materials generally consist of mixtures of various components that have different material properties, such as density, size, shape and morphology. It is a challenging task in processing and handling these bulk materials because the difference in materials properties is very likely to cause segregation, a common problem in processing and handling of bulk materials, that can create serious problems in product quality control.[1-4] It hence has attracted increasing attention over the last two decades.[2-9] However, segregation is a complicated phenomenon that can be attributed to many mechanisms, such as rolling, sieving, percolation, impact and air current.[2-4,10-20] Our understanding of segregation is far from complete.

As silos are generally filled with bulk materials in the presence of air, experimental studies showed that the presence of air had a significant impact on segregation behaviour in silo filling and segregation could be augmented by air currents induced by the down-flowing powder streams.[2,4,20] Silo filling in air involves complex two-phase flows and strong interaction between solid particles and air flow, especially when the particle size is small and/or the particle density is low.

To characterize the significance of air presence on the flow behaviour of solid particles, Guo $et\ al.$[21] proposed a new powder classification criterion in which powders can be classified as air sensitive or air inert materials using a dimensionless parameter ξ,[22]

$$\xi = A_r \Phi_\rho \tag{1}$$

where $\Phi_\rho\ (= \rho_s / \rho_a)$ is the ratio of particle density ρ_s to air density ρ_a and A_r is the Archimedes number for particle flowing in air and is given as

$$A_r = \frac{\rho_a(\rho_s - \rho_a)gd_p^3}{\eta^2} \tag{2}$$

in which d_p is the particle diameter, η is the air viscosity and g is the gravitational acceleration.

It was found that the presence of air had a negligible effect on powder flow behaviour if ξ is larger than a critical value $\xi_c = 9.56 \times 10^6$ and the powder is regarded as an air inert material. On the other hand, if $\xi < \xi_c$, the presence of the air can significantly affect the powder flow behaviour and the powder is air sensitive. Moreover, the smaller the value of ξ (i.e. the lower the particle density and the smaller the particle size), the more air-sensitive the powder is.

Although experimental studies showed that the presence of air could cause significant segregation, how the segregation tendency depends on the particle properties (size and density) is still not clear. Therefore, in this study, segregation behaviour during silo filling with particles of different sizes and densities in air is explored using a coupled Discrete Element Method and Computational Fluid Dynamics (DEM/CFD).

2   DEM/CFD MODELS

An in-house DEM/CFD code developed by Kafui *et al.*[23] was used in this study to simulate silo filling with binary mixtures in the presence of air. In this DEM/CFD the motion of solid particles is modelled using DEM, in which the translational and rotational motions of each particle are determined using Newton's equations of motion. Only spherical elastic particles were considered and the particle interactions are modelled using classical contact mechanics, for which the theory of Hertz is used to determine the normal force and the theory of Mindlin and Deresiewicz is used for the tangential force. The air is modelled as a continuum using CFD. A two way coupling between gas and solid (particle) phases is considered, i.e. how the motion of solid particle will affect the flow of air and vice versa are considered. A detailed description of this coupled DEM/CFD approach can be found in Kafui *et al.*[23] This DEM/CFD code has also been used to analyse the segregation behaviour during die filling.[24-26]

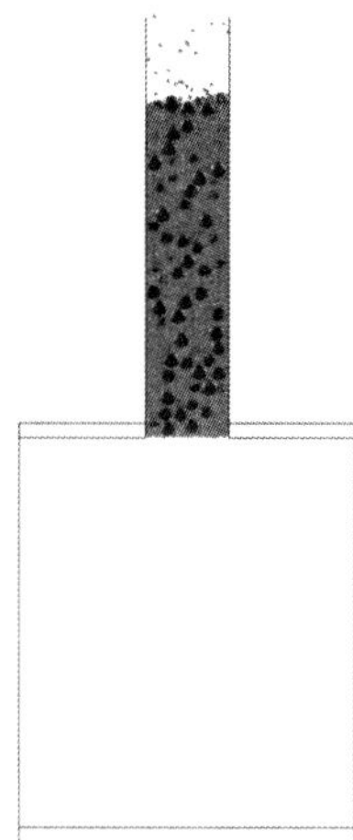

Figure 1 *The numerical model for silo filling*

The numerical model for simulating silo filling in air is shown in Figure 1. A slice of a cubic silo is modelled with a width of 16 mm and height of 20 mm, and two periodic boundaries are introduced in the thickness direction and separated by a distance of 4 mm. The side walls of the silo are physical walls that are assumed to be of aluminium with a Young's modulus of 100 GPa and Poisson ratio of 0.29. A tube with a square cross section (4×4 mm) is connected to the silo at the top centre and is used as the feeding channel, from which particles are deposited into the silo. The powder mixtures considered consist of two different construction material particles: sand particles (i.e., coarse particles) with a size of 760 μm and density of 2762 kg/m^3 and silicone glue particles (i.e., fine particles) with a size of 168 μm and density of 1210 kg/m^3. All particles are assumed to have a Young's modulus of 70 GPa and Poisson's ratio of 0.30. The particles are initially generated randomly in the tube with the bottom of the tube closed and deposited under gravity until the total kinetic energy becomes negligible. Thereafter, the bottom of the tube is opened and the particles are then filled into the silo. Two types of powder mixtures with a fines mass fraction of 25% and 50% respectively are considered. For the mixture with 25% fines, 221 coarse particles and 15,595 fine particles are used, while for that with 50% fines, 74 coarse particles and 15,595 fine particles are employed. In addition, silo fillings in a vacuum and in air are simulated in order to explore the effect of air presence. For silo filling in air, the side walls and the bottom of the silo are treated as no-slip wall boundaries for air flow. The top of the silo is modelled either as an impermeable and no-slip boundary (i.e. the silo is regarded as a closed container or an enclosed silo) or as an outflow boundary (i.e. an open silo). The latter is modelled as an attempt to explore how the air venting from the cover of the silo affects the flow and segregation behaviour in silo filling.

3   RESULTS

Numerical results for filling an enclosed silo with a powder mixture with a fine fraction of 25% are presented in Figures 2-4. Figure 2 shows the flow behaviour of the powder mixture during silo filling in a vacuum. It is clear that the powder mixture flows out of the discharge tube into the silo as one moving stream whose width is slightly larger than the tube width due to the dilation of the powder stream (Figures 2a & 2b). This is caused by the friction along the interface between the powder mixture and the tube surface and by the relaxation of the contact between compacted particles as they are discharged from the tube. The moving powder stream collides with the bottom of the silo and splashes towards the side walls (Figure 2c). During this process, the fine particles with the same initial kinetic energy rebound with a higher rebound velocity during the collision with the surface. Consequently, a dust of fine particles is formed in the silo that eventually settles down onto the top surface of the deposited powder bed (Figure 2c & 2d). The impact between the powder stream and the bottom of the silo results in a concaved packing profile as can be clearly seen from Figure 2d.

For the silo filling in air (Figure 3), the powder mixture initially flows out of the tube in the same manner as that in a vacuum, i.e., as one moving stream (Figure 3a). However, the width of the stream is much narrower than the tube width (Figure 3b) and the stream width observed in silo filling in a vacuum (Figure 2b). This is attributed to the build-up of the air pressure in the silo, which inhibits the discharge of the powder mixture from the tube. The discharged powder stream is then stretched and narrowed with a reduced width under gravity. The powder stream impacts with the silo bottom and spreads but with a much lower intensity when compared to that for silo filling in a vacuum (Figure 3c). It is also interesting to observe that the tail of the discharging powder stream primarily consists of fine particles (Figure 3c), which is in broad agreement with the experimental observations.[2] The powder mixture is packed in the silo with an essentially even top surface (Figure 3d).

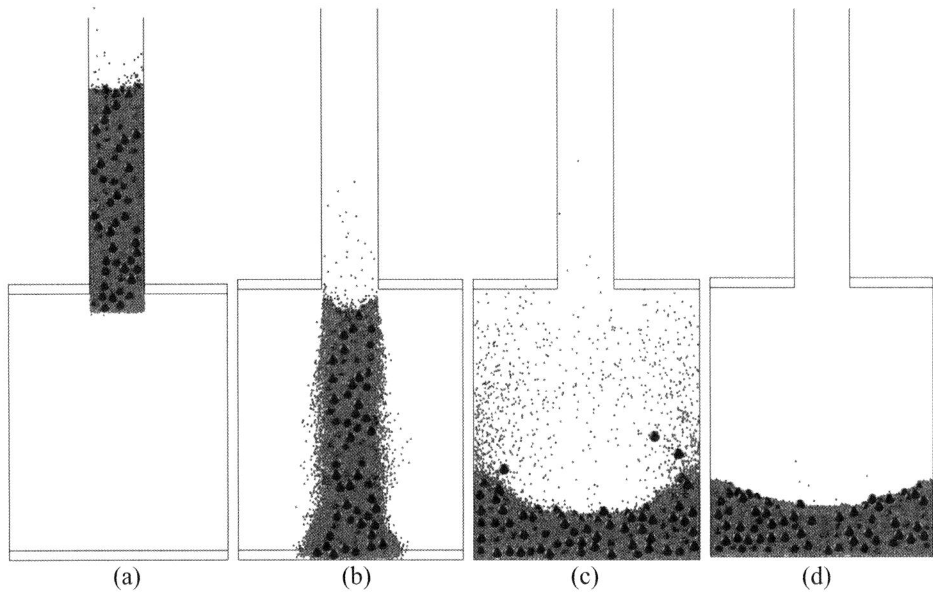

(a)   (b)   (c)   (d)

Figure 2 *Silo filling with 25% fines mass fraction in a vacuum*

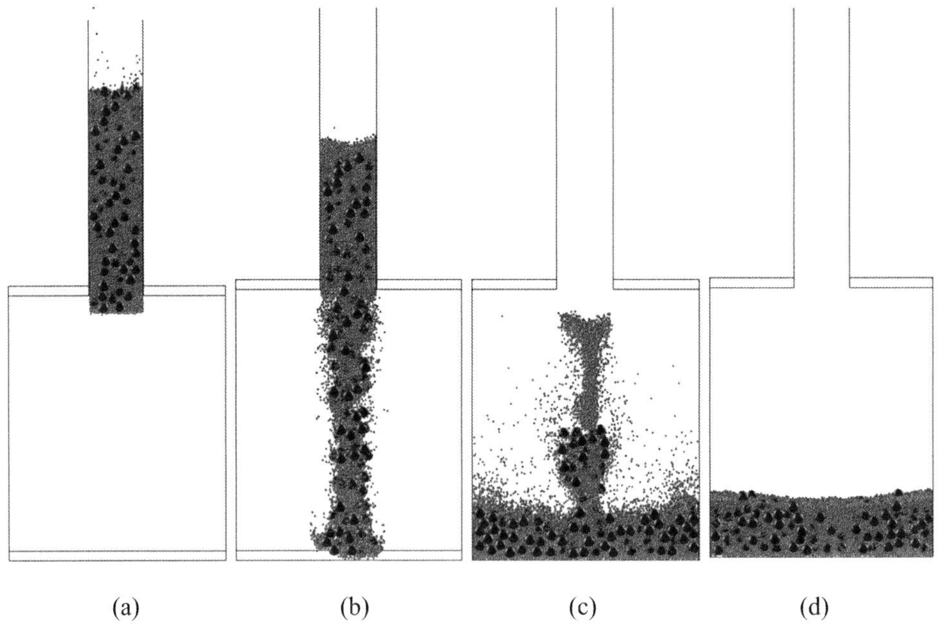

(a)   (b)   (c)   (d)

Figure 3 *Silo filling with 25% fines mass fraction in air*

To quantify the distribution of fine particles in the packed bed in the silo, the powder bed is divided into 5 equally spaced columns in the horizontal direction and 4 equally spaced layers in the vertical direction. The fine concentration in each column/layer f_i is determined as follows:

$$f_i = \frac{m_{fi}}{m_i} \tag{3}$$

where m_{fi} and m_i are the mass of fine particles and total mass in the region i, respectively.

Figure 4 shows the concentration profiles of the deposited beds in a vacuum and in air. It is clear that a uniform distribution of fine particles are obtained during silo filling in a vacuum as the fine concentrations in all regions fluctuate slightly around a value of 0.25 (i.e. 25%) that is the fines concentration in the mixture, indicating that there is no significant segregation induced in a vacuum. However, for silo filling in air, the fine concentrations in the middle and the top layer of the powder bed are much higher than the other regions, implying that significant segregation takes place in both horizontal and vertical directions with rich concentrations of fine in the middle and top surface of the powder bed.

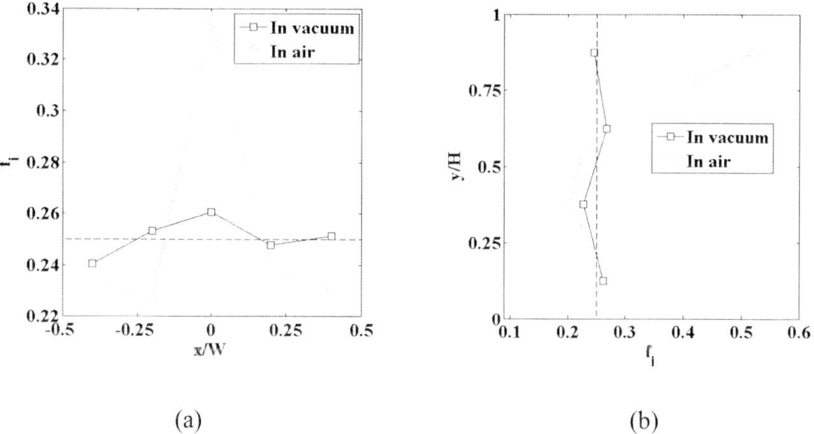

(a)   (b)

Figure 4 *(a) Horizontal concentration profiles and (b) vertical concentration profiles of fine particles in the deposited bed for the mixture with fines mass fraction of 25%*

The results for filling an enclosed silo with a powder mixture of 50% fine fractions are given in Figures 5-7. It can be seen that the overall flow patterns during silo filling in a vacuum (Figure 5) and in air (Figure 6) are similar to those for the mixture with 25% fine particles (Figures 2 & 3). Nevertheless, the impact and splashing intensity in a vacuum is much lower for the mixture with 50% fines mass fraction (Figure 5c) compared to that with 25% fines mass fraction (Figure 2c). For the silo filling in air, the down-flowing powder stream is distorted (Figure 6c) due to the significant influence of the entrapped air on the fine particles. Consequently, an uneven top surface of the packed bed is observed. When the fines mass fraction in the mixture is increased to 50%, significant segregation is induced during silo filling in a vacuum and in air, as can be clearly seen in Figure 7. The segregation pattern in air is similar to that with 25% fines mass fraction with more fine particles being deposited in the centre of the silo (Figure 7a) and on the top surface of the powder bed (Figure 7b). While a pronounced segregation in the vertical direction with more fine particles deposited in the top layers is induced during silo filling with 50% fines mass fraction in vacuum (Figure 7b), even though the horizontal distribution of fines concentration is similar to that for 25% fines mass fraction with the concentrations of fine particles fluctuating around the average concentration (Figure 7a).

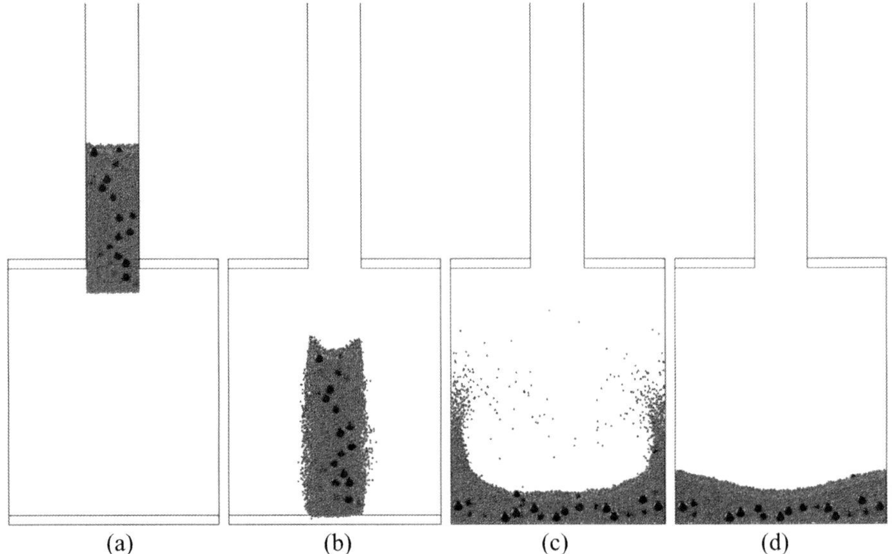

(a)   (b)   (c)   (d)

Figure 5 *Silo filling with 50% fine particles in a vacuum*

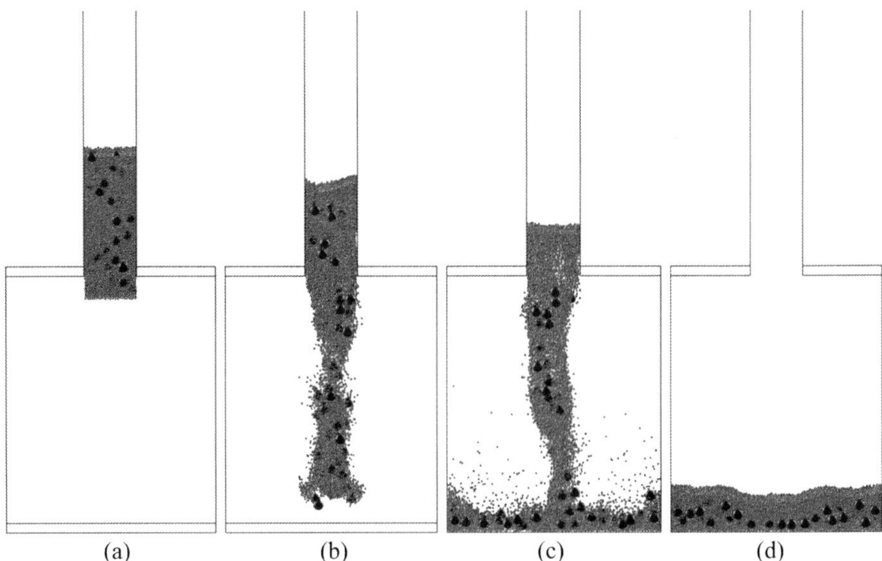

(a)   (b)   (c)   (d)

Figure 6 *Silo filling with 50% fine particles in air*

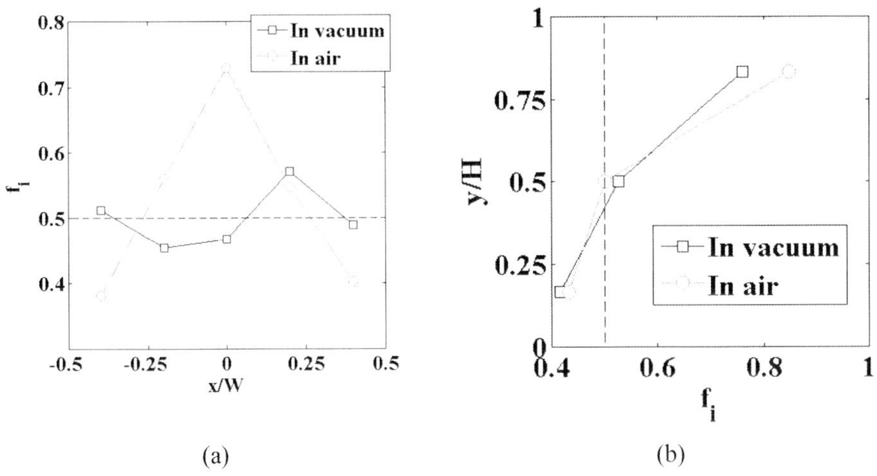

Figure 7 *(a) Horizontal concentration profiles and (b) vertical concentration profiles of fine particles in the deposited bed for the mixture with a fines mass fraction of 50%*

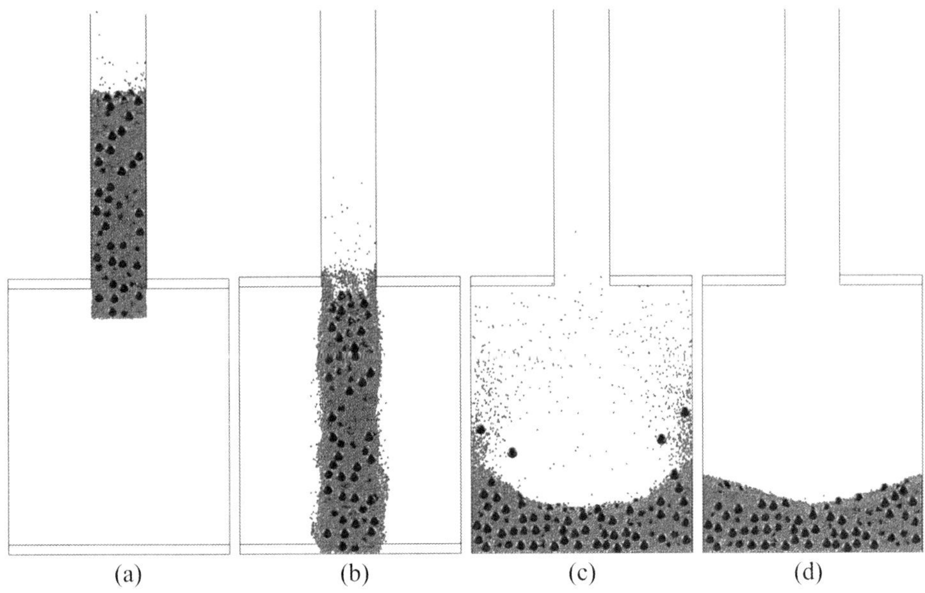

Figure 8 *Silo filling with 25% fine particles in air with venting*

Figure 8 shows the flow behaviour of the powder mixture with 25% fine particles during the filling of an open silo (i.e. with venting) in air, in which air inside the silo can escape from the top of the silo as the powder is discharged. The outflow of the air from the top can significantly reduce the air pressure in the silo and facilitate the filling process. It is clear that the flow behaviour appears to be similar to the silo filling in vacuum (Figure 2), in which the powder flows into the silo as a single stream with a lateral dilation (Figures 8a & 8b). Nevertheless, the tail of the powder stream is enriched with fine particles (Figure 8b) as observed in the filling of an enclosed silo in air (Figure 3b). As the powder stream collide with the silo bottom, it splashes and generates a dust of fine particles (Figure 8c) and a concaved powder bed is formed (Figure 8d), which is similar to the filling in vacuum (Figure 2d). The corresponding concentration profiles of fine particles are shown in Figure 9. It is clear that for the filling of an open silo in air, the concentration of fine particles near the silo walls are much higher than that in the centre (Figure 9a) and there is also a high concentration of fine particles in the top layer of the powder bed (Figure 9b). This pattern is different from that for silo filling in vacuum, illustrating that the flow of air (air current) affects the flow and segregation behaviour even though the top of the silo is open and air can be vented readily. It is worth mentioning that similar flow and segregation behaviour are also observed for the filling of an open silo using a mixture with a fines mass fraction of 50% (not shown).

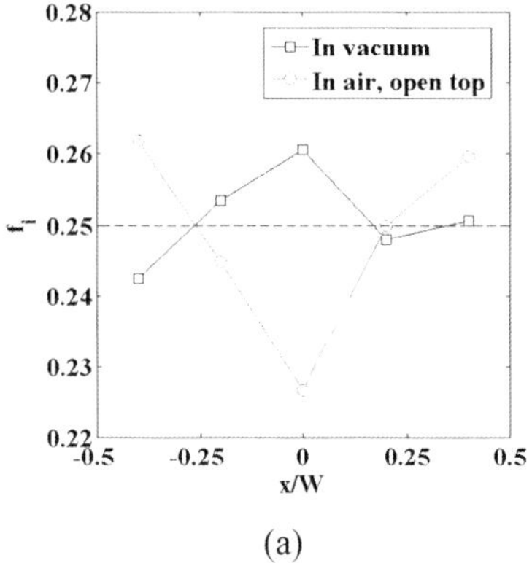

(a)

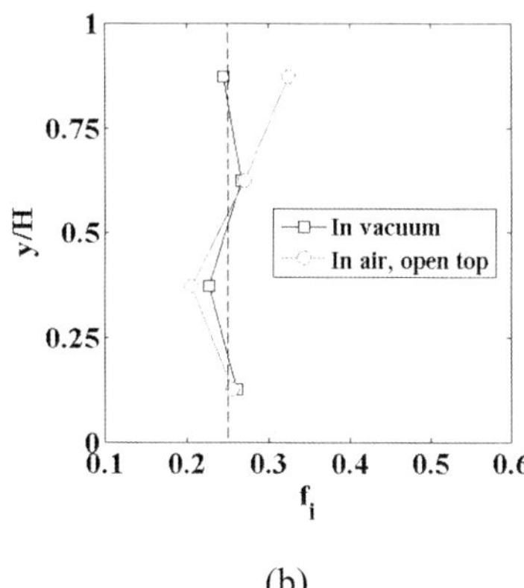

(b)

Figure 9 *(a) Horizontal concentration profiles and (b) vertical concentration profiles of fine particles in the deposited bed for the mixture with fines mass fraction of 25%*

4   DISCUSSION

For silo filling with a mixture of different sized particles, segregation may be induced due to the following mechanisms:

i) Air impediment - due to the strong interaction between air flow and solid particles, the motion of solid particles is retarded due to the presence of air. Guo *et al.*[21] showed that for the flow of particle into a confined space, the dimensionless mass flow rate can be given as

$$M^* = c\xi^{1/5} \tag{4}$$

where c is a constant related to the flow condition. From Equations (1), (2) and (4), it is clear that the smaller the particle is, the lower the flow rate. This indicates that the speeds of fine particles are reduced to a larger extent due to the interaction with air than that of larger particles. Air impediment mechanisms is similar to the 'trajectory' mechanism in the literature[20] and is responsible for the high concentration of fines particles observed in the tail of the down-flowing powder streams during the filling of silos in air as shown in Figures 3c, 6c and 8b, which subsequently results in higher concentration of fine particle in the top layer of the powder bed as shown in Figure 4b, 7b and 9b and in the centre of the silo as shown in Figure 4a and 7a.

ii) Air entrainment – fine particles can be entrained and carried away by the flowing air stream /current. This is the dominant mechanism of segregation in fluidisation beds and die filling, where fine particles are either suspended at the top of the fluidised bed or carried away in die filling. Air entrainment is responsible for the excessive fine concentration being induced near the walls during the filling of an open silo in air (see Figures 8c, 8d & 9a), because in this process the down-flowing powder stream induced circulating air currents from the centre to the walls that carry fine particles away from the centre. As a consequence, the concentration of fines in the centre is much lower that that near the walls (Figure 9a). Although air entrainment also plays a role in the filling of the enclosed silo (Figures 3 & 6), its contribution is overpowered by air impediment that becomes dominant when the silo is enclosed and the effect of entrapped air becomes significant. This is why different segregation patterns were observed during filling enclosed and open silos.

iii) Embedment – larger or heavier particles of higher inertia penetrate into the packed powder bed and become locked there. [20] This is the dominant mechanism of segregation during the silo filling in vacuum with a mixture of 50% fine fractions, in which the coarse particles embed in the powder layer initially occupied by fine particles during the collision between the flowing powder stream with the bottom of the silo. Consequently, there are more coarse particles at the centre and the bottom of the packed powder bed so that the fine concentration in these regions are much lower than the rest of the powder bed as shown in Figure 7.

iv) Percolation – fine particles fall through gaps between coarse ones. This mechanism becomes dominant when the mixture contains more coarse particles as in the case with 25% fine fraction. When the powder stream collides with the silo bottom during the filling in a vacuum, fine particles sift through the gap between coarse ones and become constrained (Figure 2). As a result, the fine concentration in the centre is slightly higher for the silo filling in a vacuum (Figure 4a).

It should be note that, in each of the silo filling processes considered in this study, all these mechanisms and other mechanisms of segregation identified in the literature may be incurred. However, the significance of their contributions depends heavily on the process itself that can be dominated by one or more mechanisms.

5 CONCLUSIONS

Segregation during the filling of silos with binary mixtures of particles of different sizes and densities has been investigated using a hybrid DEM/CFD approach. Three dimensional simulations have been performed to explore the effects of air current, fines mass fraction and silo venting design on the segregation behaviour. It has been found that the presence of air generally augments the segregation tendency. The segregation patterns (i.e. the distribution of fine concentrations) depend on the filling processes and four mechanisms have been identified as the dominant mechanisms of segregation in silo filling with particles of different sizes and densities.

References

1   K. E. Connally, *J. Met.,* 1982, **35**, A67.
2   N. Engblom, H. Saxen, R. Zevenhoven, H. Nylander and G. G. Enstad, *Powder Technol,* 2012, **215-16**, 104.
3   A. Kwade and O. Ziebell, *ZKG International,* 2001, **54**, 680.
4   S. Zigan, R. B. Thorpe, U. Tuzun, G. G. Enstad and F. Battistin, *Powder Technol,* 2008, **183**, 133.
5   A. Karolyi, J. Kertesz, S. Havlin, H. A. Makse and H. E. Stanley, *Europhys. Lett.,* 1998, **44**, 386.
6   P. Cizeau, H. A. Makse and H. E. Stanley, *Physical Review E,* 1999, **59**, 4408.
7   H. A. Makse, P. Cizeau, S. Havlin, P. R. King and H. E. Stanley, *Slow Dynamics in Complex Systems,* 1999, **469**, 53.
8   A. Samadani, A. Pradhan and A. Kudrolli, *Physical Review E,* 1999, **60**, 7203.
9   R. K. Goyal and M. S. Tomassone, *Physical Review E,* 2006, **74**, 051301.
10 P. E. Arratia, N. H. Duong, F. J. Muzzio, P. Godbole and S. Reynolds, *Powder Technol,* 2006, **164**, 50.
11   G. C. Barker and A. Mehta, *Nature,* 1993, **364**, 486.
12 G. Donsi, G. Ferrari and B. Formisani, *Powder Technol,* 1988, **55**, 153.
13 J. A. Drahun and J. Bridgwater, *Powder Technol,* 1983, **36**, 39.
14 D. A. Huerta and J. C. Ruiz-Suarez, *Phys. Rev. Lett.,* 2004, **93**, 069901.
15   D. A. Huerta and J. C. Ruiz-Suarez, *Phys. Rev. Lett.,* 2004, **92**, 114301.
16   K. Johanson, C. Eckert, D. Ghose, M. Djomlija and M. Hubert, *Powder Technol,* 2005, **159**, 1.
17 R. Jullien and P. Meakin, *Nature,* 1990, **344**, 425.
18   Shinohar.K, K. Shoji and T. Tanaka, *Industrial & Engineering Chemistry Process Design and Development,* 1972, **11**, 369.
19 D. J. Stephens and J. Bridgwater, *Powder Technol,* 1978, **21**, 29.
20 S.R. de Silva, A. Dyrøy and G.G. Enstad. in *IUTAM Symposium on Segregation in Granular Flows*, A.D. Rosato, D.L. Blackmore (Eds.), Kluwer Academic Publishers, Dordrecht, 2000, p. 11.
21 Y. Guo, K.D. Kafui, C.-Y. Wu, C. Thornton and J.P.K. Seville, *AICHE Journal*, 2009, **55**, 49.
22 C.-Y. Wu, Y. Guo, S. Nikoosaleh and C. Thornton, *Powders and Grains 2009,* **1145**, 985.
23 K. D. Kafui, C. Thornton and M. J. Adams, *Chemical Engineering Science*, 2002, **57**, 2395.
24 Y. Guo, C. -Y. Wu, K. D. Kafui and C. Thornton, *Powder Technol,* 2011, **206**, 177.
25 Y. Guo, C. -Y. Wu, K. D. Kafui and C. Thornton, *Powder Technol,* 2010, **197**, 111.
26 Y. Guo, C. -Y. Wu and C. Thornton, *Chemical Engineering Science,* 2011, **66**, 661.

MODELING PACKING OF SPHERICAL FUEL ELEMENTS IN PEBBLE BED REACTORS USING DEM

H. Suikkanen, J. Ritvanen, P. Jalali and R. Kyrki-Rajamäki

LUT Energy, Lappeenranta University of Technology, P.O. Box 20, FI-53851 Lappeenranta, Finland

1 INTRODUCTION

The discrete element method (DEM) is a useful tool for studying particle packing and dynamics of dense granular materials as it can provide highly realistic results and reveal details that are difficult or even impossible to measure from experiments. Although computationally more efficient algorithms for generating particle packing configurations exist,[1] they do not resolve the physics of the system the way that DEM does.

In this work, DEM is used to generate packed bed configurations representing the core of a pebble bed reactor (PBR). A PBR is a nuclear reactor of a very unique design. The fuel material is enclosed inside graphite spheres or "pebbles" that form a heat-producing column. The reactor is introduced in more detail in Section 2. Understanding the mechanical behaviour of the fuel pebbles inside the reactor core is important not only for assessing the loads on core structures but also because the packing characteristics affect the core neutronics and thermal-hydraulics. Although the pebbles flow through the core, the flow velocity is so small that a static pebble bed approximation is sufficient for most analyses. However, pebble flow and residence time information is needed in fuel burnup analyses, *i.e.* in determining the temporal changes in the isotopic composition of nuclear fuel.

An in-house DEM code is used to obtain three packed pebble beds with slightly different average packing densities $\bar{\phi}$. A simplified model of an actual PBR design with 450,000 pebbles is used. The packed pebble beds are analysed using Voronoi tessellation in three dimensions to provide local pebble-scale packing fraction data that is useful for investigating the packing structures in detail.[2] The data is used for statistical analysis, plotting packing density profiles in different parts of the reactor and for preparing three-dimensional visualisations. The usefulness of DEM for PBR analyses is also discussed.

2 PEBBLE BED REACTORS

A new generation of nuclear fission reactors with improved use of natural resources, economical competitiveness, proliferation resistance, safety and security is under development.[3] One of the next generation reactor concepts is the high temperature reactor (HTR), which is a gas-cooled reactor capable of producing both electricity and high temperature heat for various industrial processes. In addition, an HTR that has been designed in the right way has unique passive safety features that make a core meltdown practically impossible. The PBR is a specific HTR design, which at the moment is most intensively developed in the People's Republic of China.[4]

The core of a PBR contains a huge number of graphite pebbles (diameter ~60 mm). The fuel material is dispersed inside the pebbles in the form of coated particles (diameter ~1 mm). By using silicon carbide as the coating material, the coated particle can withstand temperatures up to 1600 °C without releasing significant amounts of radioactive materials. The fuel pebbles are fed from the top and removed from the bottom of the core so that they flow through the core slowly. Each pebble can be circulated through the core several times. Helium gas at high pressure enters the core from the top and flows through the pebble bed carrying the heat outside the core.

3 METHODS

An in-house DEM code is used for the packing simulations. Voronoi tessellation in three dimensions is constructed to calculate local packing fractions ϕ_{vor} in the investigated pebble beds. A brief description of the DEM implementation is given below along with a description of the used post-processing methods.

3.1 DEM Implementation

The DEM code implementation used in this work is based on the linear spring-dashpot model by Tsuji *et al.*[5] The normal and tangential forces for overlapping particles i and j (Figure 1) are calculated as

$$\mathbf{F}_{Cnij} = (-K_n \delta_{nij}^{3/2} - \eta_n \mathbf{v}_{rij} \cdot \mathbf{n}_{ij})\mathbf{n}_{ij}, \tag{1}$$

$$\mathbf{F}_{Ctij} = -k_t \delta_{tij} - \eta_t \mathbf{v}_{sij}, \tag{2}$$

where $\mathbf{n}_{ij}$ is the normal unit vector, $\mathbf{v}_{rij}$ and $\mathbf{v}_{sij}$ are relative velocity vectors and δ_{nij} and δ_{tij} are deformations in normal and tangential directions. The stiffness in normal direction is calculated from

$$K_n = \frac{4\sqrt{R_i R_j}}{3\left(\dfrac{1-\sigma_i^2}{E_i} + \dfrac{1-\sigma_j^2}{E_j}\right)\sqrt{R_i + R_j}}, \tag{3}$$

where R is the particle radius, σ is Poisson's ratio and E is Young's modulus. The stiffness in tangential direction is calculated from

$$k_t = \frac{8\sqrt{R_i R_j}}{3\left(\dfrac{2-\sigma_i}{G_i} + \dfrac{2-\sigma_j}{G_j}\right)\sqrt{R_i + R_j}}\delta_{nij}^{1/2}, \tag{4}$$

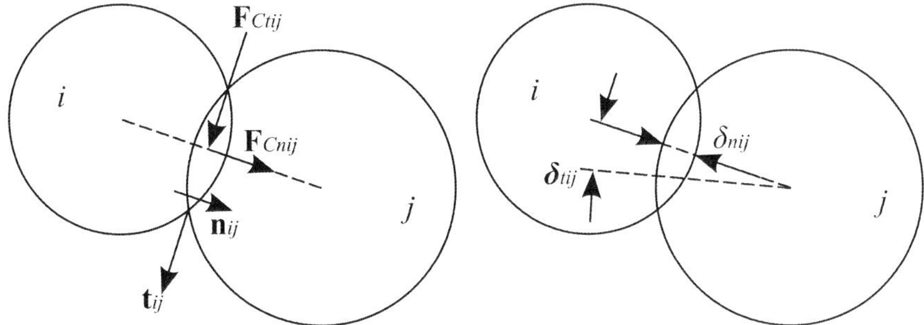

Figure 1 *Normal and tangential contact forces, unit vectors and deformations between two particles i and j.*

where G is the shear modulus related to σ and E as

$$G = \frac{E}{2(1+\sigma)}.$$ (5)

The normal and tangential components of the damping factor are assumed identical as

$$\eta_n = \eta_t = \alpha \left(\overline{m} K_n \right)^{1/2} \delta_{nij}^{1/4},$$ (6)

where the reduced mass is calculated based on the masses m of the two particles i and j as

$$\overline{m} = \left(\frac{1}{m_i} + \frac{1}{m_j} \right)^{-1}.$$ (7)

The coefficient α is related to the coefficient of restitution e by[6]

$$\alpha = \frac{-\sqrt{5}\ln e}{\sqrt{\pi^2 + \ln^2 e}}.$$ (8)

The individual forces and torques affecting a particle are solved and Newton's second law is used to give translational and rotational accelerations at each time step.

The method is implemented as a C++ code. A neighbour search routine similar to the link cell algorithm is used for efficient detection of contacts.[7] A cylindrical coordinate system is used in the neighbour search routine as an optimisation for cylindrical geometries. In addition, shared memory parallelisation is used to speed-up simulations.

3.2 Post-Processing Methods

Local packing fraction data is extracted from a packed pebble bed by first forming a three-dimensional Voronoi decomposition (Figure 2) and then calculating the volumes of the Voronoi cells. A C++ software library Voro++ is used,[8] which forms the Voronoi cells and calculates their volumes. Cylindrical convex walls that are formed of multiple planes can be used to cut the Voronoi volumes at the outer cylindrical wall boundary. Some additional effort is needed to cut the Voronoi volumes at the inner non-convex wall. A specific technique is used, where the near-wall pebbles are mirrored outside the boundary to resolve the inner wall (Figure 2). After calculating the Voronoi volumes these mirror pebbles, more specifically their Voronoi cells, are simply excluded from the analysis as their only purpose is to in the accurate calculation of the near-wall Voronoi volumes.

Voronoi data is used to calculate pebble-scale packing fractions ϕ_{vor} and to form packing fraction profiles for different parts of the pebble column. 11 axial and 14 radial bins are defined for which the Voronoi volume based average packing fractions $\bar{\phi}_{vor}$ are calculated. Profiles are then plotted for inner and outer wall regions and for middle region in the axial direction and for bottom, middle and top regions in radial direction. In addition, the Voronoi data is directly used to prepare three-dimensional visualisations of the columns for visual inspection.

4 CALCULATION MODELS

The packing simulations are done for a geometry representing a real PBR design, namely the 400 MW$_{th}$ Pebble Bed Modular Reactor (PBMR-400),[9] which was under development in South Africa. The reactor design has an annular cylindrical core where the fuel pebbles are packed between solid graphite reflectors. For the packing simulations, the actual core geometry is simplified so that the model does not include defueling chutes or any other geometrical details that the real reactor design has. A flat plane is used as the bottom boundary. The annulus inner and outer diameters, however, correspond to the real design. A realistic number of fuel pebbles with realistic material properties is used. The parameters used in the simulations are given in Table 1.

An initial dilute configuration is generated inside the reactor geometry using a random number generation algorithm. The randomly generated pebbles are then packed under gravity. Several cases are simulated varying the friction factor μ_f, the coefficient of restitution e, and the packing density of the initial configuration. Three packed pebble beds with different $\bar{\phi}$ are selected for further analyses.

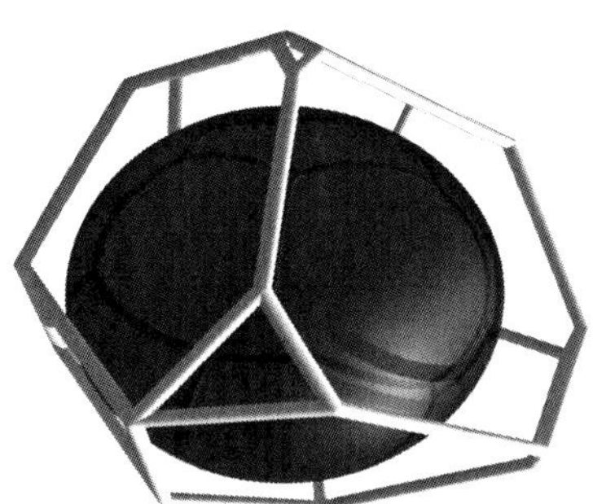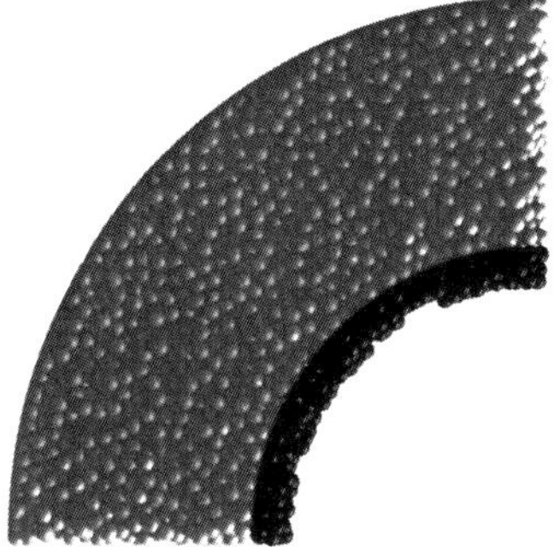

Figure 2   *Voronoi cell surrounding a pebble (left). Pebbles mirrored outside the wall boundary for accurate Voronoi tessellation of the pebbles located near the inner wall (right)*

Table 1   *Simulation parameters.*

Parameter	Symbol	Value
Annulus inner radius	R_{inner}	1.00 m
Annulus outer radius	R_{outer}	1.85 m
Number of pebbles	N	450 000
Pebble radius	R	0.03 m
Pebble mass	m	0.21 kg
Graphite Young's modulus	E	10^{10} Pa
Graphite Poisson's ratio	σ	0.13

5 RESULTS AND DISCUSSION

Pebble beds with average packing densities of 0.620, 0.626 and 0.632 are selected from the DEM simulation results for detailed analyses. Local pebble-scale packing fractions are calculated using Voro++.

Probability density functions are plotted to examine the distributions of ϕ_{vor} in the packings (Figure 3). It can be seen that the mean increases and the standard deviation decreases as the average packing density of the pebble bed increases.

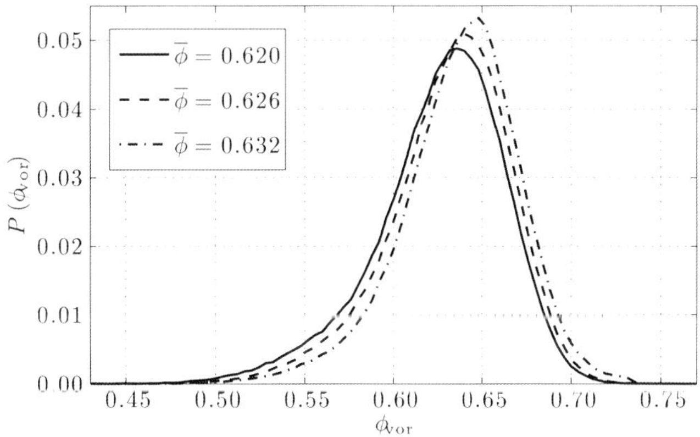

Figure 3   *Probability density functions of pebble-scale packing fractions for the three selected pebble beds*

Packing fraction profiles in different regions are plotted in axial and radial (z and r) directions (Figure 4). The axial profiles are formed using the average packing fraction data calculated for the bins located next to the inner wall, next to the outer wall and in the middle of the column. As can be expected, the profiles show denser packing in the middle region where the packing is not disrupted by the walls. The wall effect is also seen at the bottom bin as slightly lower values of $\overline{\phi}_{vor}$. Otherwise, pebbles are more tightly packed in the bottom region, especially for $\overline{\phi} = 0.620$. Packing appears to be more uniform in axial direction in the high $\overline{\phi}$ case. The sudden decline in $\overline{\phi}_{vor}$ at the top of the column is due to inaccuracy in determining the cutting plane for the topmost Voronoi cells, which leads to very low packing fraction values for the top pebbles. Similarly as the axial profiles, the radial profiles show denser packing at the bottom of the pebble bed in the $\overline{\phi}=0.620$ case. The wall effect is seen as smaller $\overline{\phi}_{\mathrm{vor}}$ in the $\overline{\phi}=0.620$ and $\overline{\phi}=0.626$ cases but there are peaks near the walls in the radial profile of the densest case. This suggests highly ordered structures in the wall vicinity.

Visualizations of the pebble beds confirm the increase of ordered packing structures at the walls as the average packing density increases (Figure 5). Locally dense regions are formed inside the pebble bed in connection to these crystallized regions.

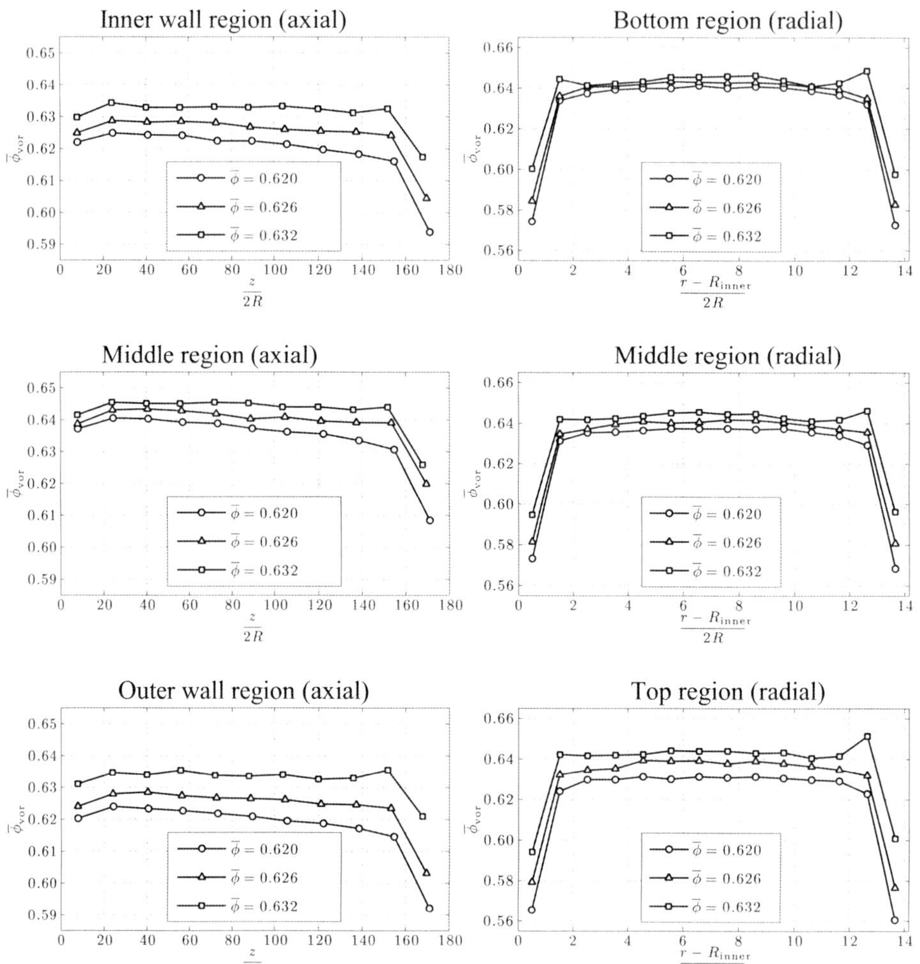

Figure 4 *Packing fraction profiles in axial and radial directions in different regions of the pebble beds. The plot points represent the average packing fraction in a bin calculated as the mean of the Voronoi cell based packing fractions of pebbles residing in the bin.*

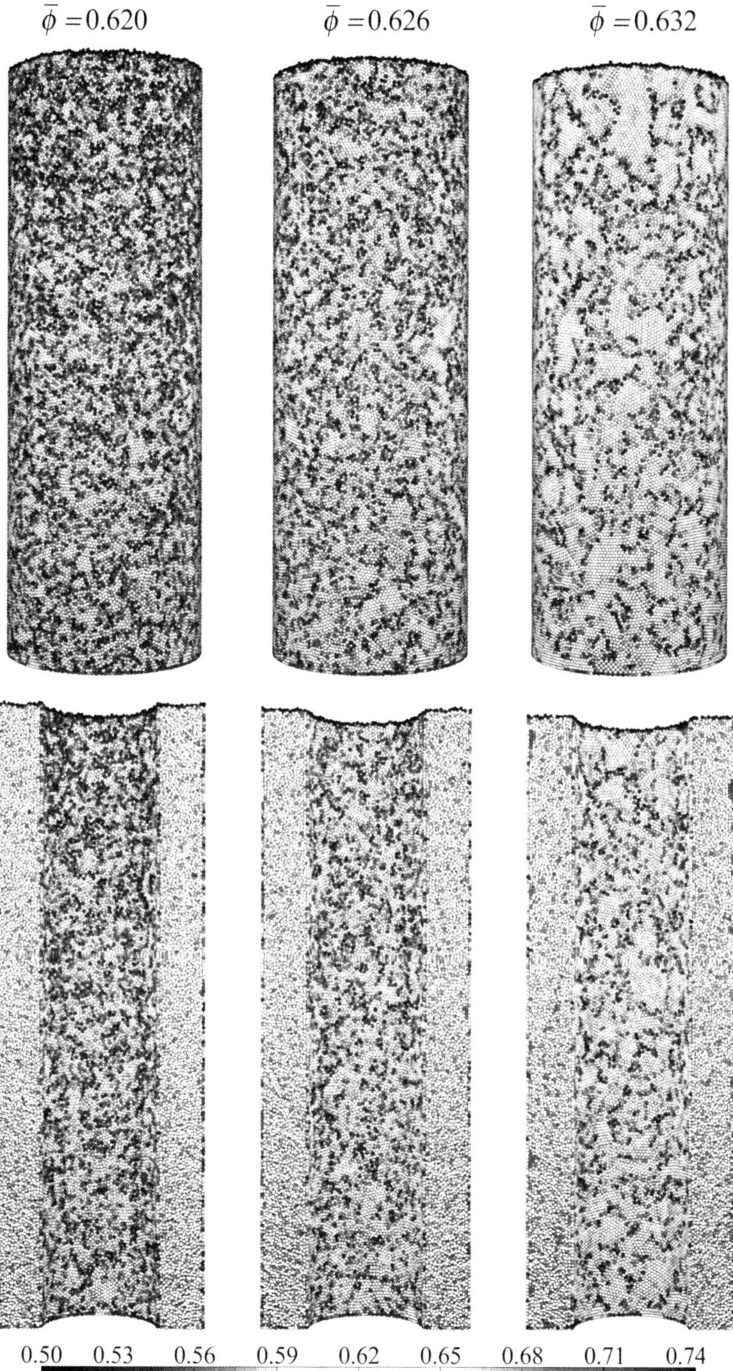

Figure 5   *Pebble-scale packing fractions for the outer (top three figures) and inner (bottom three figures) annulus surfaces of half pebble beds.*

6 REACTOR ANALYSIS CONSIDERATIONS

Detailed packing data that DEM simulations provide can be used in further reactor physics and thermal-hydraulic analyses. Accurate pebble positions can be used to build detailed geometrical models for stochastic reactor physics codes using the Monte Carlo method.[10] Small regions inside the pebble bed could be selected to study the coolant flow in detail with computational fluid dynamics methods. These analyses might give insight to the formation of local hotspots in the core.

Full core thermal-hydraulic calculations of pebble bed reactors are usually performed using the porous medium approach, where the pebble bed is considered a packed bed and a suitable porosity profile is used to account for the wall regions. Based on the results of this work, it is suggested that also the axial variation in porosity should be taken into account. Average porosities could be calculated for individual control volumes using pebble beds packed with DEM. Accuracy of the full core analyses should improve when the data of the same packing is used in the reactor physics calculations to determine the power profile of the core and the coupled reactor physics and thermal-hydraulics are solved.

DEM could also be used for investigating the slow flow of the fuel pebbles through the core, which would give useful data to be utilized in the fuel burnup calculations. Behaviour of the pebble bed during an earthquake is also an interesting topic that DEM can be used. Together with tribological models, DEM could be used to estimate the production of graphite dust during operation as the pebbles grind against each other and walls.

7 CONCLUSIONS

DEM was used to simulate packed beds inside an annular cylinder representing the core of a pebble bed nuclear reactor. Local packing fractions were calculated using Voronoi decomposition and the data was used to provide insight to the packing structure of the pebble beds. An increase in ordered regions was observed when the average packing density of the pebble bed increased. The use of DEM as a part of reactor analysis was discussed and it was found out to be a useful tool that can provide input data for reactor physics and thermal-hydraulics analyses.

Acknowledgements

This work was funded by the Academy of Finland (Grants No. 124264, 124368 and 123938) and Fortum Oyj.

References

1   A.M. Ouqouag, J.J. Cogliati and J.L. Kloosterman, in *International Topical Meeting on Mathematics and Computation, Supercomputing, Reactor Physics and Nuclear and Biological Applications*, Avignon, France, September 12-15, 2005.
2   G. Voronoi, *J. Reine Angew. Math.*, 1907, **133**, 97.
3   U.S. DOE Nuclear Research Advisory Committee and the Generation IV International Forum, *A Technology Roadmap for Generation IV Nuclear Energy Systems*, Report GIF-002-00, 2002.
4   Z. Zhang, Z. Wu, D. Wang, Y. Xu, Y. Sun, F. Li and Y. Dong, *Nucl. Eng. Des.*, 2009, **239**, 1212.
5   Y. Tsuji, T. Tanaka and T. Ishida, *Powder Technol.*, 1992, **71**, 239.

6   R. Mindlin, *J. Appl. Mech.*, 1949, **16**, 259.
7   T. Pöschel and T. Schwager, *Computational Granular Dynamics Models and Algorithms*, Springer Berlin Heidelberg, New York, 2005, ch. 2, p. 61.
8   C.H. Rycroft, *Chaos*, 2009, **19**, 041111.
9   P.J. Venter and M.N. Mitchell, *Nucl. Eng. Des.*, 2007, **237**, 1341.
10 H. Suikkanen, V. Rintala and R. Kyrki-Rajamäki, in *Proceedings of the 5th International Conference on High Temperature Reactor Technology*, Prague, Czech Republic, October 18-20, 2010.

Quasi-Static Deformation

A NUMERICAL INVESTIGATION OF QUASI-STATIC CONDITIONS FOR GRANULAR MEDIA

C. Modenese[1], S. Utili[2] and G.T. Houlsby[1]

[1]Department of Engineering Science, University of Oxford, Parks Road, OX1 3PJ, UK
[2]School of Engineering, University of Warwick, Coventry, CV4 7AL, UK

1 INTRODUCTION

The Discrete Element Method (DEM) is now widely used as a numerical tool to investigate the physics of granular media. The material is envisioned as a collection of bodies that are allowed to interact according to some physical relationships at the microscopic level. The attraction of the DEM for researchers is due to its simplicity as well as its ability to provide insight into complex physical phenomena, which are difficult or impossible to observe in laboratory experiments. However, due to computational limits, several assumptions are generally required. Different scaling techniques are often applied to artificially increase the minimum time step required to assure the stability of the numerical explicit scheme adopted in the analysis.[1] These techniques are particularly useful whenever quasi-static (QS) conditions are sought. The latter are a very common assumptions in large part of DEM experiments, although little evidence has been provided so far on the validity of such assumption, given the impossibility to run numerical experiments for realistic material properties (e.g. particle density and stiffness) and physical parameters (e.g. strain rate), which nowadays would necessitate unaffordable computing time, especially if parametric studies are carried out.

This paper presents a numerical investigation by the DEM method of the effects of density scaling on the macroscopic behaviour of an ideal material made up of spherical particles under very dense conditions. The open-source code YADE[2] was employed in the analysis. Triaxial compression tests are run for two extreme cell pressures, 1 kPa and 1 MPa. In particular, the focus is on the capability of the DEM to reproduce QS conditions when scaling techniques are employed. A sample can be considered under quasi-static conditions, when the response of the material is rate-independent. In traditional geotechnical tests, it is of particular importance to ensure that numerical experiments are effectively run under QS conditions in order to draw a meaningful comparison between numerical and experimental results. In the current literature, the so-called inertial number is used as the reference parameter to assess whether the system is under quasi-static conditions.[3] Results from tests under different inertial numbers will be presented in this paper. The numerical findings will focus on the physical meaning of dimensionless parameters defined by dimensional analysis. In particular, macroscopic results will be illustrated for different combinations of stiffness and inertial parameters. The mechanical behaviour at small and large strains will be investigated into detail. Particular emphasis will be given to macroscopic observations at the critical state.

2 THE NUMERICAL MODEL

2.1 The Contact Model and the Particle Size Distribution

The contact model used in this study follows the classical Hertz-Mindlin no-slip solution.[4] The Mohr-Coulomb criterion is employed to limit the maximum shear contact force. Non-linear contact viscous damping is only employed during the generation process to help decrease the maximum unbalanced force in the sample and reach a stable solution prior to the triaxial phase. However, no viscous damping was employed during the shearing phase.

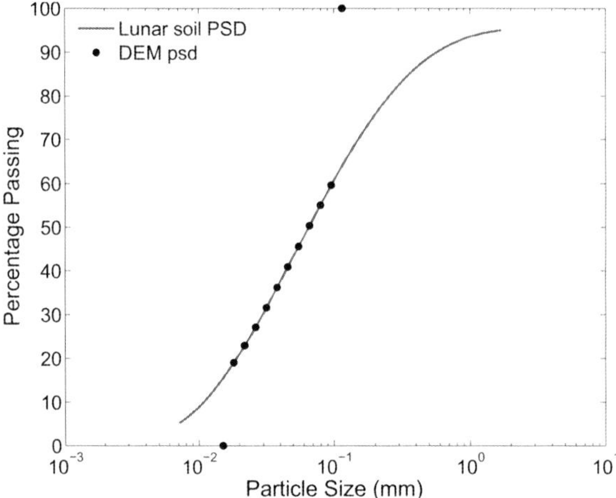

Figure 1 *Particle size distribution of the DEM model*

Table 1 *Input DEM parameters for the tests*

Number of particles	N	10,000	
Young's modulus	E	70	GPa
Poisson's ratio	ν	0.25	
Average particle size	R	72	μm
Particle density	ρ	2,650	kg/m^3
Strain rate employed	$\dot{\varepsilon}$	1.0e-1	1/s
Interparticle friction coefficient	μ	0.6	

A realistic discrete particle size distribution (PSD) was employed in the analysis, as shown in Figure 1. A very wide PSD was selected for this study ($R_{max}/R_{min} = 8$) to represent the PSD of Lunar Soil.[5] The research presented in this paper, in fact, originates from a research project aiming to characterise the mechanical behaviour of Lunar regolith (a 2-5 m thick layer of granular soil covering the Moon surface). The particle properties were taken from typical Lunar regolith (see Table 1).

2.2 Sample Preparation

Samples made of 10,000 particles were prepared under very dense conditions ($D_R = 100\%$), according to a dynamic procedure which can be summarized as follows:

- Isotropic compression under zero friction angle until the pressure of 1 kPa is reached and quasi-static conditions are assured;
- Gradual increase of friction angle until the realistic value is achieved and subsequent stabilisation of the sample is achieved through a sufficient number of cycles;
- Isotropic compression until the desired cell pressure is achieved.

3 DIMENSIONAL ANALYSIS

Dimensional analysis is an important tool in numerical experiments as the total number of parameters to be investigated can be substantially reduced down to only a few dimensionless numbers. In this paper we focus on the analysis of two main parameters, the so-called stiffness and the inertial numbers. These numbers govern the physics of a particulate system made of spherical particles.[3] Such parameters are related to both the properties of the grains as well as the macroscopic deformation of the granular assembly. They can be expressed using the following relationships, respectively:

$$k = \left(\frac{E}{P(1 - \nu^2)} \right)^{2/3} \tag{1}$$

$$I = \dot{\epsilon} d \sqrt{\frac{\rho}{P}} \tag{2}$$

where $\dot{\epsilon}$ is loading rate, d a characteristic length of the sample, generally identified with the average particles size, ν the Poisson's ratio, ρ and E the density and Young's modulus of the particles, respectively, and P the confining pressure. The inertial number, I, is meant to characterize the inertia of the system whilst the stiffness number, k, provides a parameter to describe the elastic stiffness of the granular assembly. The definition of k accounts for the Hertzian non linear contact model.[6]

Experimental values of I for quartz sand tested under triaxial test conditions are found approximately around 10^{-8}. As will be later discussed, it is extremely demanding to perform numerical tests at such a low inertia (e.g. low loading rate), particularly if large strains are sought.

This paper will show that a different definition of the inertial parameter is more appropriate when the combined effect of elastic and inertial forces cannot be neglected. The couple of parameters (I, k) would appear to be the most appropriate choice in this case. On the other hand, if it is generally true that the choice of dimensionless numbers is arbitrary, such a choice can be non-trivial depending on the response of the numerical model, i.e. depending on the physics of the problem to deal with. In fact, this paper will show that for deformable materials the use of the canonical definition of I, expressed by Equation (2), does not lead to results consistent with the critical state theory of soil mechanics.[7]

4 RESULTS

4.1 The Effect of the Inertial Number

Triaxial compression tests in a periodic cell were run at constant confining pressures of 1 kPa and 1 MPa. Stresses were measured in the usual manner, i.e. by considering the contribution of all the contact forces within the periodic cell and taking the average of those over the total cell volume. The average stress tensor for a granular assembly is calculated as the sum of two terms, the contact and kinetic stresses, which reads as follows:[8]

$$\sigma_{ij} = \frac{1}{V}\Sigma_p \Sigma_c l_i^c f_j^c + \frac{1}{V}\Sigma_p m u_i'^p u_j'^p \tag{3}$$

where l_i is the branch vector at the contact, f_j is the total contact force, V is the volume of the periodic cell and u' the fluctuating velocity of particle p with respect to the mean flow velocity. However, in all the simulations carried out the kinetic contribution to the stress tensor turned out to be negligible so that the tensor simplifies into the following:[9]

$$\sigma_{ij} = \frac{1}{V}\sum_p \sum_c l_i^c f_j^c \tag{4}$$

The two stress invariants, output from the analysis, are the deviatoric stresss, computed as $q = \sigma_1 - 0.5(\sigma_2 + \sigma_3)$, i.e. the difference between the axial principal stress and the mean lateral pressure, and the mean cell pressure equal to $p = (\sigma_1 + \sigma_2 + \sigma_3)/3$. Finally, strains were computed directly from the deformations of the periodic cell.

To start with, the inertial number was varied between 10^{-2} and 10^{-5} whilst the stiffness number was kept constant. The goal was to quantify the effect of inertia at both small and large strains. The inertial number was changed by progressively increasing the loading rate or the particle density. The results of the calculation are shown in Figures 2–4. Several remarks can be made:

- At small strain levels, the quasi-static solution was achieved for $I \leq 10^{-3}$;
- At large strain levels, the solution was still dependent on I even for very small values of I, such as 10^{-5}. This is an important finding since the peak stress is often evaluated against the rate of loading to prove that the condition is quasi-static whereas the current data show that the quasi-static conditions should be assessed at very large strains, too;
- The critical state porosity decreased non-linearly with decreasing I.

A first conclusion is that two different limits exist for the quasi-static condition at small and large strains. The difference could be related to the fact that different mechanisms prevail at different strains. For instance, at small strains the behaviour is mainly elastic as confirmed by some energy calculations (results not shown here). On the other hand, at larger strains, the friction dissipation becomes more significant compared to other forms of energies.

A second conclusion is that for some reasons the quasi-static limit proposed by Roux and Combe,[10] i.e. *I = 10⁻³*, did not apply to the tests illustrated herein. Roux and Combe[10] simulated 4,000 spherical with particle mechanical properties similar to the ones used in this work. However, Roux and Combe[10] analysed monodisperse samples whereas in this study, a very broad PSD is employed. In fact, it emerges that the sensitivity to the rate of loading increases with the broadness of the particle size distribution so that lower inertial numbers, are required to attain quasi-static conditions. Further studies on samples generated with various PSDs would be necessary to confirm this finding.

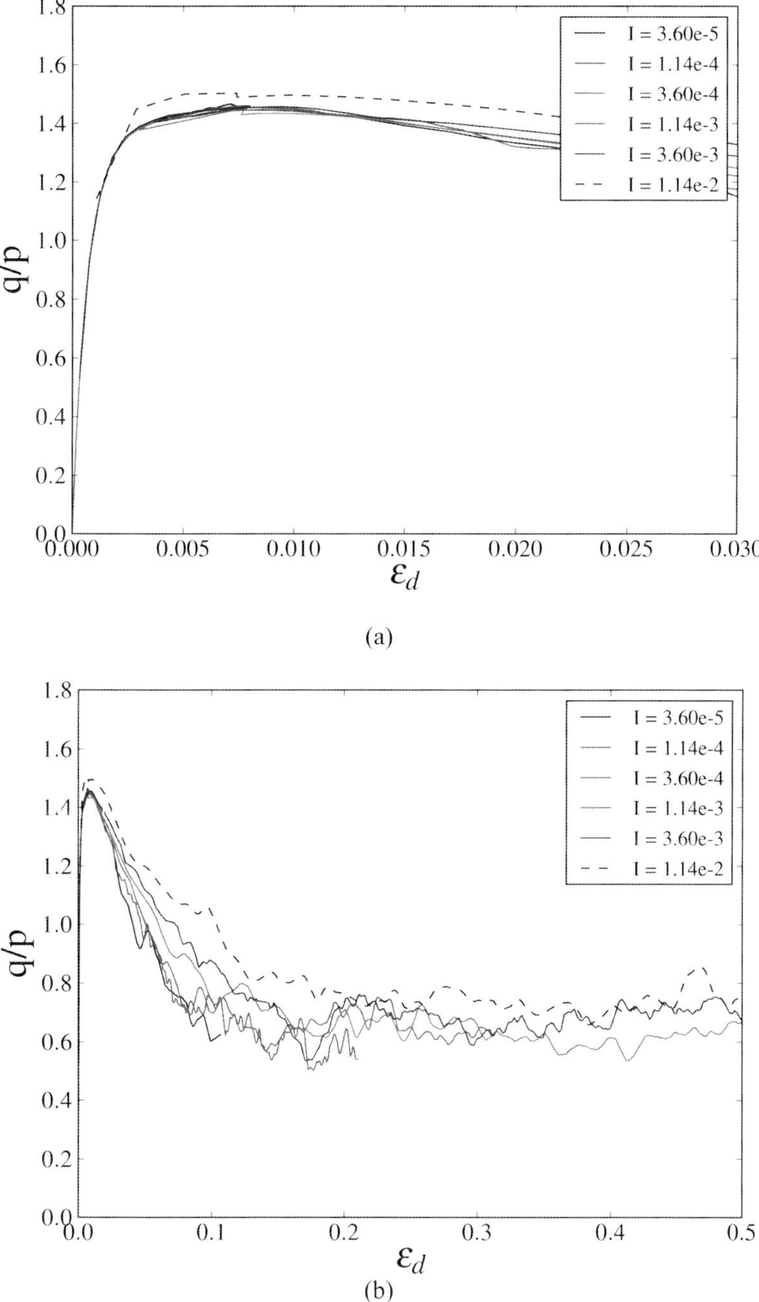

(a)

(b)

Figure 2   *Stress-strain behaviour at (a) small and(b) large strains*

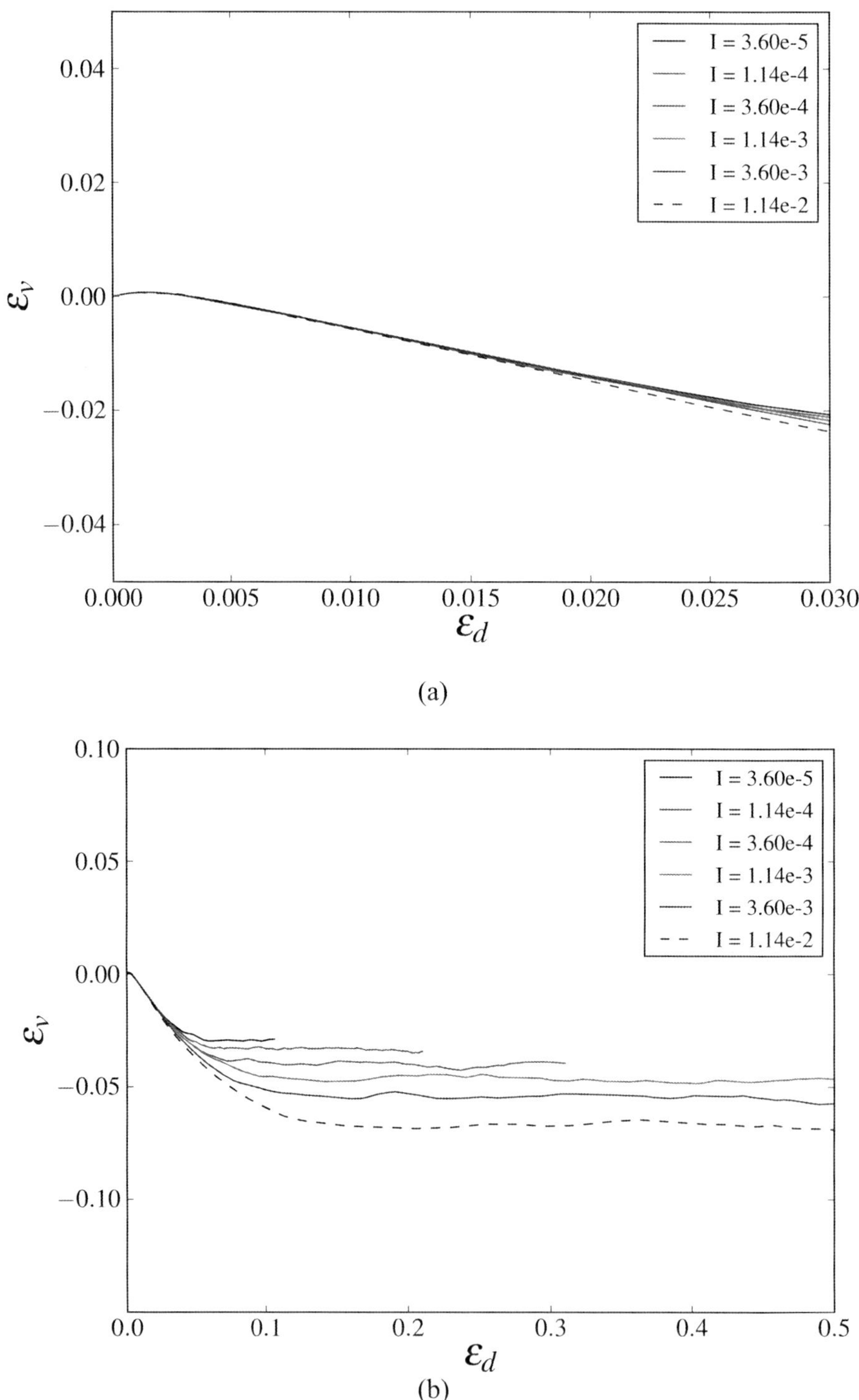

(a)

(b)

Figure 3 *Volumetric behaviour at (a) small and (b) large strains*

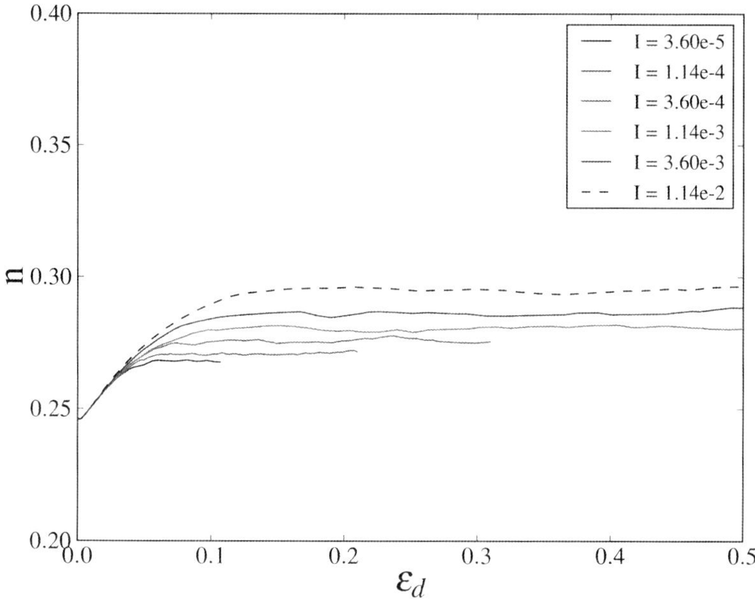

Figure 4 *Behaviour of macroscopic porosity for different inertial numbers*

4.2 The Combined Effect of the Inertial and Stiffness Numbers on the Critical State Porosity

This section analyses results obtained for different inertial and stiffness numbers. The goal is to show that pressure (or stiffness) has a non-negligible effect at the critical state for tests run under the same I. This is an important issue since it is related to the ability of the DEM method to reproduce the critical state theory of the soil.

At first, the confining pressure is varied but at constant I. The behaviour of the critical state porosity versus k is reported for two series of tests in Figure 5a for two very low values of I. It can be observed that the porosity at the critical state decreases with increasing k, i.e. increases with increasing pressure. This result is in contradiction with the principles of the critical state theory.[7] This is also in contradiction with what stated by Da Cruz et al.[8] who found that no influence of k exist on the critical state, for the same range of k adopted in this study. However, the results were based on 2D simulations and data were not shown in the paper.

The contradiction emerging from the results shown in Figure 5a lead us to consider a different definition of the inertial number. The modified definition reads as follows:

$$I^* = \dot{\varepsilon}d\sqrt{\frac{\rho}{E}} \tag{5}$$

A new series of tests is now considered in light of the new couple of parameters (k, I^*). Tests at different k were carried out under the same I^* value. The results are shown in Figure 5b. It can be observed that this time the critical state porosity tends to decrease with increasing pressure, at least for the higher value of I^* employed here. On the other hand, almost no change in critical porosity is observed as I^* decreases. There results are no longer in contradiction with the classical critical state theory and also show that as the condition approaches the quasi-static equilibrium, i.e. as I^* decreases, the solution no longer depends on k in the range of k considered in this study, i.e. $10^5 \leq k \leq 10^3$ (approximately 1 kPa $\leq P \leq$ 1 MPa, the inter-particle Young's modulus being equal to 70 GPa).

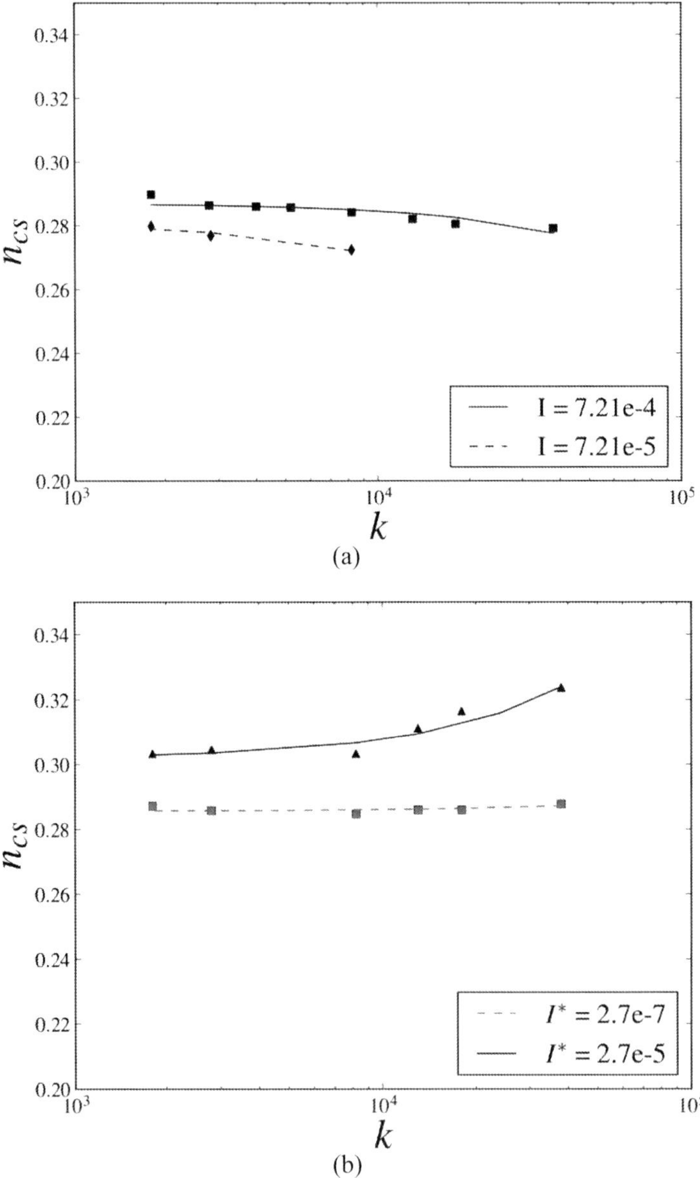

Figure 5 *Porosity at the critical state for different stiffness numbers at (a) constant I and (b) constant I**

5 CONCLUSIONS

This paper investigates the attainment of quasi-static conditions of a synthetic soil sample by means of DEM triaxial tests. A realistic non-linear contact model and a wide particle size distribution have been employed in the analysis.

The influence of two dimensionless numbers, the inertial number I and the stiffness number k, on the mechanical response of Lunar soil, has been studied in detail. It was found that at small strains the quasi-static limit is achieved for smaller I than at very large strains. Moreover, the quasi-static limit suggested in the current literature, $I = 10^{-3}$, was not sufficient at very large strains. This shows that the sensitivity of the material to the inertial number might depend on the simulated system, e.g. on the particle size distribution adopted.

Results from DEM triaxial tests have also been analysed at the critical state in the light of two definitions of inertial numbers, both equally valid but with slightly different physical meaning. The main outcome of the comparison is that the traditional definition of I does not lead to results in agreement with the principles of the critical state theory. Therefore, the second definition of inertial number, I^*, proportional to the speed of sound, is proposed as a better criterion to evaluate the inertia of the system particularly under different confining pressures.

References

1   C. Thornton and S. Antony, *Phil. Trans R. Soc. A*, 1998, **356**, 2763.
2   V. Šmilauer, E. Catalano, B. Chareyre, S. Dorofeenko, J. Duriez, A. Gladky, J. Kozicki, C. Modenese, L. Scholtès, L. Sibille, J. Stránský and K. Thoeni, *Yade Reference Documentation*, ed. V. Šmilauer, 2010.
3   M. Babic, H.H. Shen and H.T. Shen, *J. Fluid Mech.*, 1990, **219**, 81.
4   R.D. Mindlin and H. Deresiewicz, *J. Appl. Mech.*, 1953, **16**, 259.
5   C. Modenese, S. Utili, and G.T. Houlsby, *Proc. Earth and Space ASCE Conf.*, 2012.
6   F. Radjai and F. Dubois, *Discrete-element Modelling of Granular Materials*, ed. ISTE and Wiley, 2011.
7   A. Schofield and P. Wroth, *Critical State Soil Mechanics*, McGraw-Hill, 1968.
8   F. Da Cruz, S. Emam, M. Prochnow, J.-N. Roux and F. Chevoir, *Phys. Rev. E*, 2005, **72**.
9   C. Thornton, *Géotechnique*, 2000, **50**, 43.
10 J.N. Roux and G. Combe, *AIP Conf. Proc.*, 2010, **1227**, 260.

EXPLORING THE CONTROLLING PARAMETERS AFFECTING SPECIMENS GENERATED IN A PLUVIATOR USING DEM

L. Cui

Division of Civil, Chemical and Environmental Engineering, University of Surrey, GU2 7XH, UK

1 INTRODUCTION

Pluviation, tamping and vibration are three main methods to generate reconstituted granular soil specimens in laboratory. The specimens generated by different methods may exhibit different engineering properties and stress-strain behaviour. It has been found that pluviation, compared to vibration, caused negligible particle crushing, less segregation, but produced more accurate density measurement and repeatable specimen density.[1] Some other studies compared the pluviation and tamping method. It has been shown that specimens generated by pluviation have more uniform distribution of void ratio.[2, 3] To investigate the in-situ behaviour of granular soils, pluviation was considered as the preferred method to prepare specimen with a controlled density and fabric.

A number of factors may influence the specimen density generated in pluviation, e.g. falling height, depositional intensity (or flow rate), and pluviation condition. A model was proposed to predict the influence of falling height using a free-falling sphere of equivalent diameter D_{50} in a fluid.[4] A falling particle eventually reaches maximum (critical) velocity thus highest impact energy. However, as found by others, this model neglect the interaction between particles and the dynamic movement of air turbulence.[5] As agreed by many researchers, specimen density normally reduces as depositional intensity increases. Interestingly, it has been found that there is an optimum depositional intensity where a maximum specimen density is achieved.[6] Density reduces at lower depositional intensity, as only at the optimum depositional intensity, an "energetic" layer develops continuously and compacts the specimen most effectively. Comparing the specimens generated in air, water and vacuum pluviation, it has been found that water pluviation generated loosest specimens and vacuum pluviation generated densest specimesn.[5] This agreed with the free-falling sphere model.[4] It is also found that the vacuum pluviation achieved more uniform specimens.

Using the discrete element method, a number of other controlling parameters for the specimen density and fabric could be investigated. In this paper, a series of specimens were generated to explore the influence of inter-particle friction, damping coefficient as well as falling height. The resultant specimen density, coordination number, "at-rest" stress ratio (K_0), and fabric anisotropy were investigated. The coordination number (N) is the average contact per particle, i.e. $N=2N_c/N_p$ (N_c - total number of contacts; N_p- total number of particles). The "at-rest" stress ratio, K_0, is the ratio of the effective horizontal stress to the effective vertical stress (σ'_h/σ'_v). The fabric of specimen can be quantified using the fabric tensor. The 2nd ranked fabric tensor is defined as

$$\Phi_{ij} = \frac{1}{N_c} \sum_{N_c} n_i n_j \quad (i,j = x,y,z)$$

(1)

where n_i is the component of the vector, which connects the centroids of two contacting particles, in the i direction. The fabric tensor measures the spatial distribution of contact directions. The degree of fabric anisotropy can be quantified by the difference between the largest and smallest Eigen values of fabric tensor, i.e. Φ_1-Φ_3. The degree of fabric anisotropy has been found to be directly related to soil strength.[7]

2 VIRTUAL PLUVIATOR SET-UP

In the DEM simulations for pluviation, a loose specimen comprising 1000 spheres was firstly generated in a cubic box of size 100 mm × 100 mm × 100 mm. The density of the sphere is 2650 kg/m^3; the shear modulus is 28.7 MPa; the Poisson's ratio is 0.28. Five equally distributed sphere radii were used: 6.60 mm, 5.78 mm, 4.95 mm, 4.13 mm, 3.30 mm. Non-uniform spheres were used to prevent regular packing occurring. No sphere contacting with its neighbours initially. The cubic box was raised to various heights, then the gravity was turned on and spheres settled down. The Hertz-Mindlin contact model was used. Inter-particle friction and damping coefficient were varied systematically for each falling height. Boundaries were set to be frictionless to prevent the undesired frictional resistance and uneven specimen surface. The coefficient of inter-particle friction was varied between 0.0 and 1.0. The falling height attempted were 0, 10 mm, 50 mm and 100 mm. A system of global damping, which is proportional to its mass, was used in the present DEM simulations.[8] This mass damping was implemented in the DEM code using the formula as:[9]

$$Ma^t + Cv^t = F^t$$

(2)

and

$$C = \alpha M$$

(3)

where M is the mass matrix, a^t is the acceleration vector at time t, C is the damping matrix, α is the damping coefficient, v^t is the velocity vector at time t, and F^t is the force vector. Using the central difference time integration scheme, the velocity is updated as:

$$v^{t+\Delta t/2} = \frac{1-\alpha\Delta t/2}{1+\alpha\Delta t/2} v^{t-\Delta t/2} + \frac{\Delta t}{1+\alpha\Delta t/2} M^{-1}F^t$$

(4)

where Δt is the time step in a DEM simulation. The mass damping coefficient, α, used in this paper was 0.1, 0.6 and 6.0. Higher damping coefficient can significantly shorten the simulation time required to reach the equilibrium stage.

3 RESULTS AND DISCUSSIONS

3.1 Void Ratio

The void ratios were measured in three zones (top, middle, and bottom zones) to show the spatial distribution. The average value of these three void ratios is also calculated as the void ratio for the whole specimen. As there are large voids between the particles and boundaries, these three zones are defined 5 mm apart from boundaries. The initial average void ratio of the whole specimen before pluviation is 0.572. The void ratio (or specimen density) is found to be sensitive to the inter-particle friction. It is shown that the average void ratio increases at a reducing rate as the coefficient of inter-particle friction increases from 0.0 to 0.5. The increasing trend is negligible as coefficient of friction increases from 0.5 to 1.0. Typical plots for two falling heights are illustrated in Figure 1. This matches what is expected as that, with higher coefficient of friction, more frictional resistance were applied to particles while they settled down, and the potential energies and kinetic energies of particles were dissipated via frictional work. Consequently, particles would come to stop at higher positions resulting in a lower specimen density and higher void ratio.

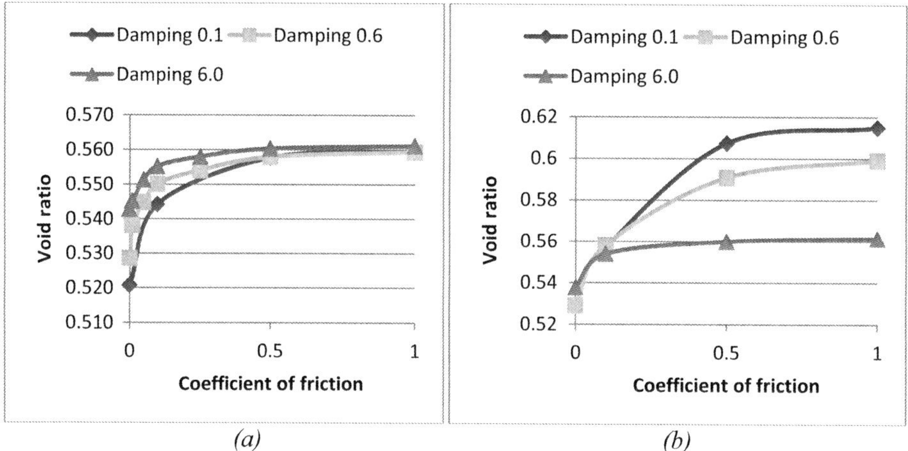

Figure 1 *Void ratio versus coefficient of friction with various damping coefficent: (a) Falling height =10 mm; (b) Falling height =100 mm*

The void ratio increases slightly with increasing damping coefficient for the falling height of 10mm, as illustrated in Figure 1a. This may be because that the velocities and hence kinetic energies reduced slightly at higher damping coefficient, and hence particles stopped at higher positions. However, for a higher falling height, the effect of damping coefficient is opposite. As illustrated in Figure 1b, the void ratio increases significantly with decreasing damping coefficient. It seems that, at higher falling height and lower damping coefficient, the impact velocities of particles before colliding with the bottom of the box are much higher. The higher impact velocities will cause wild collisions between particles, which result in looser specimens. This trend is illustrated more clearly in Figure 2, which shows the variation of void ratio against falling height. Severe particle collisions would result in loose specimens at higher falling height; however, the differences in the maximum falling velocities for higher falling height are eliminated with higher damping. Future study can be focused on the effect of damping on the maximum falling velocity.

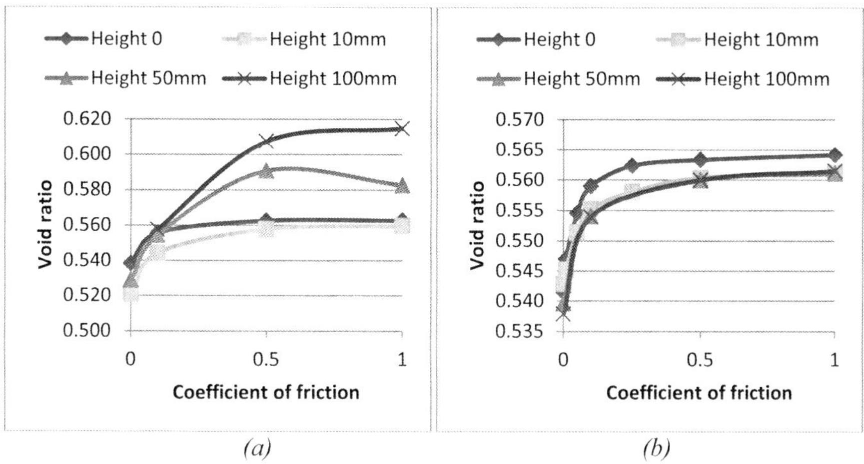

(a) *(b)*

Figure 2 *Void ratio versus coefficient of friction with various falling height: (a) Damping coefficient = 0.1; (b) Damping coefficient = 6.0*

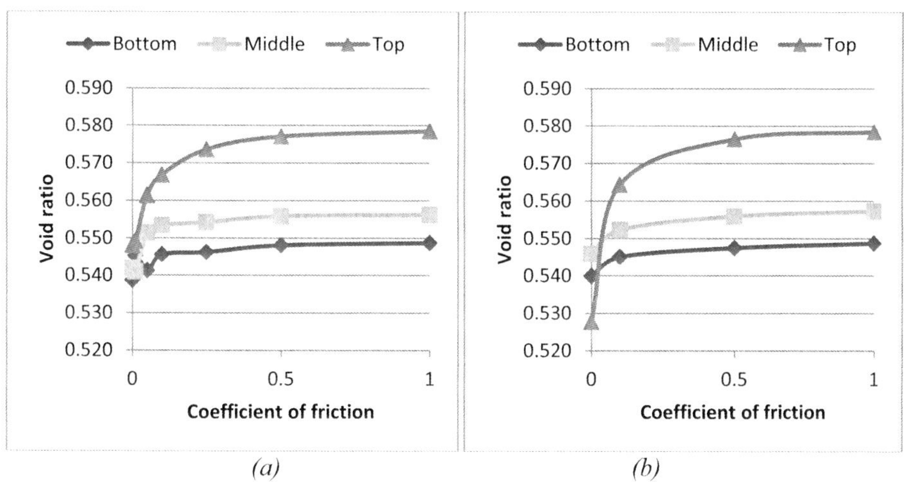

(a) *(b)*

Figure 3 *Void ratio versus coefficient of friction in different measurement zones (damping = 6.0): (a) Falling height =10 mm; (b) Falling height =100 mm*

The spatial distribution of void ratio can be visualised by plotting the void ratios in the top, middle, and bottom zones. With very high damping coefficient (6.0), there is a consistent trend of distribution of void ratio for all falling heights: void ratio is lowest in the bottom zone and highest in the top zone, as seen from Figure 3. It is a result of the action of gravity, i.e. particles in the bottom zone support more weights of particles above them, therefore packed denser than those in the top zone. However, with lower damping coefficient (0.1 or 0.6), the distribution is different, as shown in Figure 4. At falling height of 10 mm, the distribution of void ratio still matches what is expected; however, at falling height of 100 mm, the middle zone appears to be the loosest zone. It shows that the particle collisions had more significant influence on the final packing density than the gravitational force with lower damping.

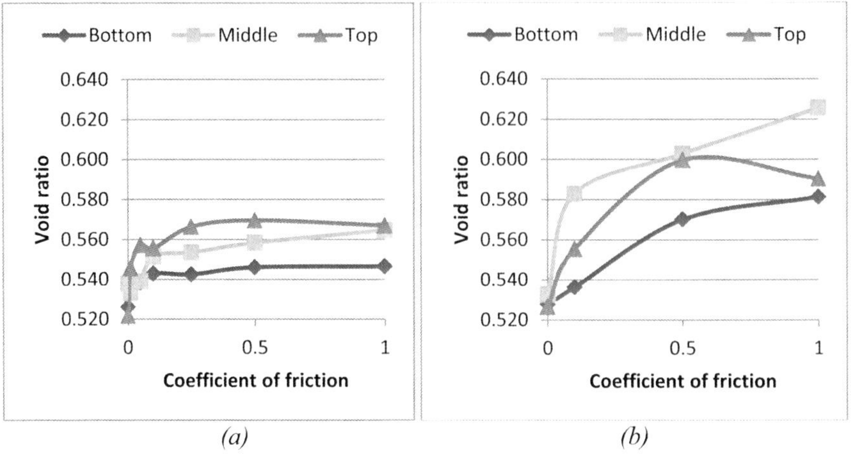

(a)   (b)

Figure 4 *Void ratio versus coefficient of friction in different measurement zones (damping = 0.6): (a) Falling height =10 mm; (b) Falling height =100 mm*

3.2 Coordination Number

The coordination number is also sensitive to the coefficient of friction, i.e. it decreases at a reducing rate as coefficient of friction increases from 0.0 to 1.0 (Figure 5). However, the coordination number does not show great sensitivity to the falling height or damping coefficient. The coordination number only decreases slightly with increasing damping coefficient at all falling heights, as seen from Figure 5a, although there are large differences in the void ratios at higher falling height (comparing with Figure 1b). Similar fact is found for damping coefficient. As seen from Figure 5b, the coordination number increases slightly with falling height, which is the same for all damping coefficients, although the void ratios for various falling height at low damping coefficient are quite different (Figure 2a). It is interesting to find that a higher void ratio (looser specimen) is not always related to a lower coordination number.

3.3 Anisotropy

The anisotropy properties of the specimen can be measured using both "at rest" stress ratio K_0, and fabric tensor. Fabric tensor only considers the directions of contacts between particles, while K_0 considers both the direction and the magnitude of the contacts. Representative variation of K_0 as a function of friction is illustrated in Figure 6. As the falling height is zero, K_0 reduces with increasing coefficient of friction. There is no big difference between various damping coefficients. As the falling height increases, the variation of K_0 shows a different trend. For very high damping coefficient, K_0 still reduces with increasing friction; while for lower damping coefficient, K_0 is lowest at zero friction and highest at friction coefficient of 0.1. The variation of K_0 at higher falling height is smaller than that at lower falling height.

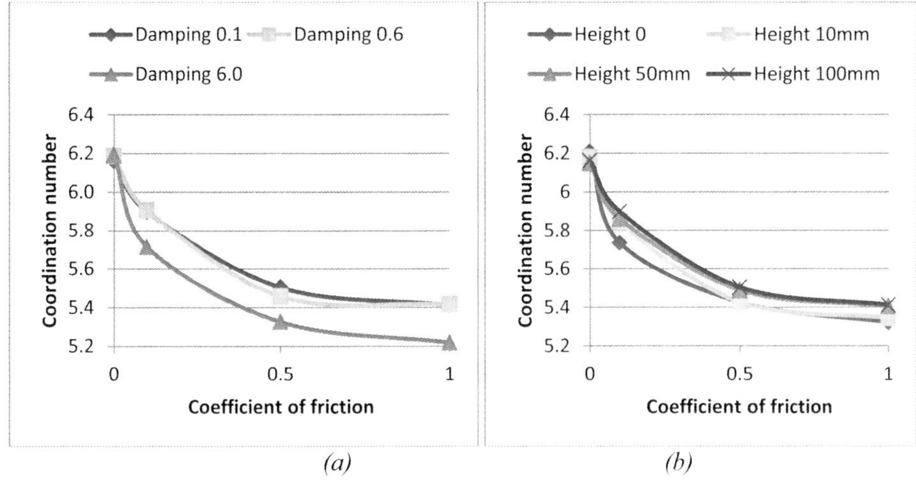

(a)   *(b)*

Figure 5 *Coordination number versus coefficient of friction: (a) Various damping coefficient (falling height 100 mm); (b) Various falling height (damping 0.1)*

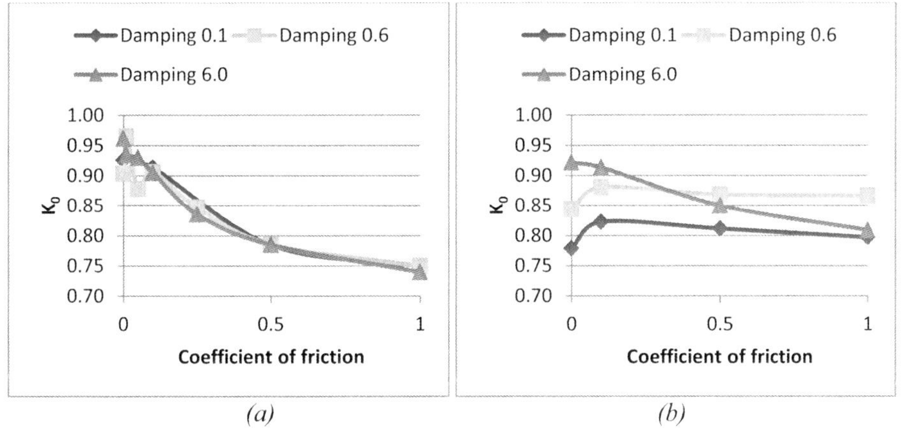

(a)   *(b)*

Figure 6 K_0 *versus coefficient of friction: (a) Falling height = 0; (b) Falling height = 100 mm*

The degree of fabric anisotropy, Φ_1-Φ_3, does not show a clear relationship with friction. But it can be found that, at higher falling height, the degree of fabric anisotropy increases with decreasing damping coefficient (Figure 7b). The degree of fabric anisotropy also increases with increasing falling height by comparing Figure 7a and Figure 7b.

4 CONCLUSIONS

Using the virtual DEM pluviator, the controlling parameters for pluviation specimen characteristics were investigated. The key findings are summaried as follows:

- The void ratio increases at a reducing rate as coefficient of inter-particle friction increases from 0.0 to 0.5, while the increasing trend is negligible as coefficient of friction increases from 0.5 to 1.0;

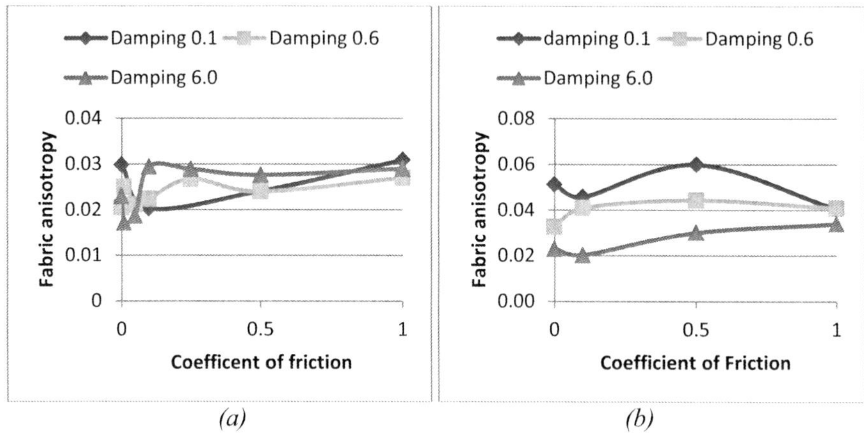

Figure 7 *Degree of fabric anisotropy versus coefficient of friction: (a) Falling height = 0;
(b) Falling height = 100 mm*

- The void ratio increases slightly with increasing damping coefficient at lower falling height, while it shows opposite trend (decreases remarkably) at higher falling height, which is a result of different intensity of particle collision with various damping;
- The void ratio increases significantly with increasing falling height with lower damping, while the difference vanished with higher damping;
- The void ratio is lowest in the bottom zone and highest in the top zone at higher damping, while it does not show this trend at lower damping, as particle collisions had more significant influence than the gravitational forces;
- The coordination number decreases at a reducing rate as friction increases, while it does not show significant sensitivity to falling height or damping coefficient;
- For very high damping or low falling height, K_0 drops with increasing friction, while for low damping or high falling height, the highest K_0 was obtained at friction coefficient of 0.1;
- The degree of fabric anisotropy does not show a clear relationship with friction, while it increases with decreasing damping coefficient at higher falling height.

References

1.   D. C. F. Lo Presti, S. Pedroni and V. Crippa, *geotech. Test. J.*, 1992, **15**, 180.
2.   T.-T. Ng, C. Hu and S. Altobelli, *Geotechnical Special Publication*, 2006, **149**, 104.
3.   S. Miura and S. Toki, *Soils Found.*, 1982, **22**, 61.
4.   Y. P. Vaid and D. Negussey, *Soils Found.*, 1984, **24**, 101.
5.   R. Lagioia, A. Sanzeni and F. Colleselli, *Soils Found.*, 2006, **46**, 61.
6.   A. Cresswell, M. E. Barton and R. Brown, *geotech. Test. J.*, 1999, **22**, 324.
7.   L. Cui and C. O'Sullivan, *Geotechnique*, 2006, **56**, 455.
8.   P. A. Cundall and O. D. L. Strack, *geotechnique*, 1979, **29**, 47.
9.   J. Bardet, in *Behaviour of Granular Materials*, ed. B. Cambou, Springer-Verlag, Editon edn., 1998, p. 99.

DEM TRIAXIAL TESTS OF A SEABED SAND

G. Macaro[1] and S. Utili[2]

[1]Department of Engineering Science, University of Oxford, Oxford, UK
[2]School of Engineering, University of Warwick, UK, formerly at Department of Engineering Science, University of Oxford, Oxford, UK

1 INTRODUCTION

The open source code YADE, which is based on the so-called soft-particle approach, was used to run numerical triaxial tests on an assembly of spheres replicating the particle size distribution (PSD) of a typical seabed sand in the North Sea, i.e. Leighton Buzzard (see section 2.2). These tests were carried out as part of a larger research program on the soil – structure interaction for pipelines lying on sandy seabeds. The described numerical model is currently employed in 3D DEM tests involving a plane strain section of an unburied pipeline (Figure 1) to investigate the lateral pipe-soil interaction for on-bottom pipelines subject to lateral buckling.

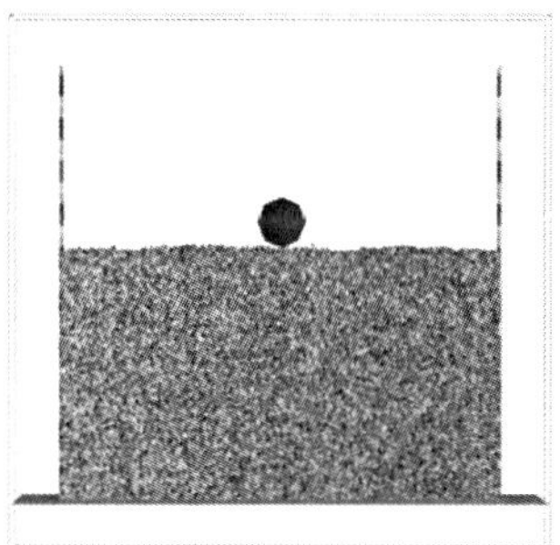

Figure 1 *Plane strain 3D DEM simulation of the lateral soil-pipeline interaction*

2 NUMERICAL MODEL

2.1 Contact Model

The contact law adopted is the Hertz – Mindlin no-slip solution[1] together with moment – relative rotation law which could be formally described as "linear elastic perfectly plastic" although plasticity of grains is not the phenomenon catered for by this law. In fact, the rolling law was employed with the only aim of taking into account the effect of non-sphericity of the sand grains on the angular moment equilibrium of the granular assembly. It is well known that an assembly of free rolling spherical particles is featured by values of peak and critical state friction angles which are significantly lower than the angles observed in real sands. The adoption of a rolling law makes it possible to achieve realistic values of peak and critical state friction angles for the granular assembly. Indeed, this is only one way of achieving realistic friction angles for the assembly and several other methods have been proposed in the DEM literature. However, the appeal of a rolling law lies in its computation efficiency. The adopted rolling law intends to account for the rolling moments acting on sand particles. These moments arise in non-spherical particles from the fact that the line of action of the normal contact forces no longer passes through the centre of mass of the particle. Hence rotational moments are generated,[2] introduced here as a couple at particles contacts, which may also oppose particles rotation. In this paper instead, the adopted moment – relative rotation law introduces a couple at contacts, which always opposes particle rotations as proposed by Plassiard *et al.*[3]

To summarise, the mechanical model is composed of a normal, tangential and rolling component. The adopted contact mechanical laws are schematised in Figure 2: a non-linear spring plus a divider act in the normal direction; a non-linear spring plus a slider act in the tangential direction; and a linear rotational spring plus a rotational slider act when one particle rolls over another.

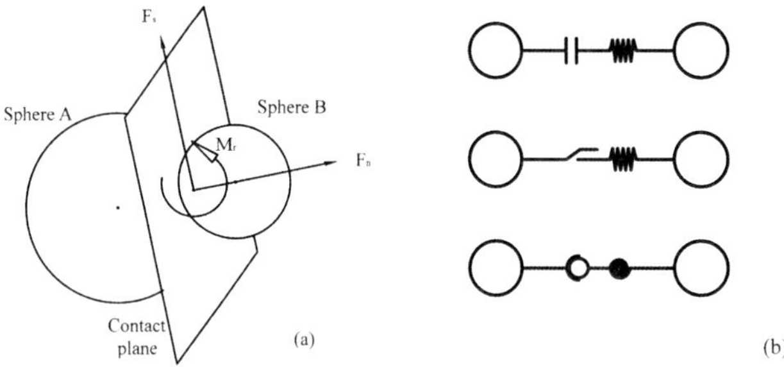

Figure 2 *Contact between two particles: (a) notation and (b) contact model*

The normal force F_n and the incremental tangential component ΔF_s of the contact force are computed from the relative displacements at contacts:

$$F_n = K_n U_n^{3/2} \tag{1a}$$

$$\Delta F_s = -k_s \sqrt{U_n} \Delta U_s \tag{1b}$$

where U_n and ΔU_s are the normal and incremental tangential displacement between the two spheres, K_n is the normal secant stiffness, k_s is the shear incremental stiffness, defined as:

$$K_n = \frac{4}{3} \tilde{E} \sqrt{\tilde{R}} \tag{2a}$$

$$k_s = \frac{2\tilde{G}\sqrt{4\tilde{R}}}{2 - \tilde{v}} \tag{2b}$$

where $\tilde{R}$ is the equivalent radius of the two spheres in contact, $\tilde{v}$ is the Poisson's ratio, $\tilde{E}$ is the Young's modulus and $\tilde{G}$ is the elastic shear modulus. The tangential force is limited by the Mohr-Coulomb condition:

$$|F_s| \leq F_n \tan \phi_\mu \tag{3}$$

The moment – relative rotation contact law was implemented in the code in an incremental fashion as:

$$\Delta M = -k_r \Delta \theta_r \tag{4}$$

where k_r is the rolling stiffness and $\Delta\theta_r$ is the incremental relative rotation between two particles in contact. The rolling stiffness is obtained here assuming that for small displacements $|\Delta F_s| R \cong |\Delta M|$ 4 which leads to:

$$k_s \sqrt{U_n} \Delta U_s R \cong k_r \frac{\Delta U_r}{R} \tag{5a}$$

$$k_r = \beta_r k_s \sqrt{U_n} R^2 \tag{5b}$$

where β_r is the rolling stiffness coefficient. The limit moment is described by the following equation:

$$M_{\lim} \leq \eta R |F_n| \tag{6}$$

where η is a coefficient ruling the amount of M_{lim}.

With such formulation, the model requires only five parameters to be calibrated. These are the intergranular friction angle ϕ_μ, the Young's modulus E, the Poisson's ratio v, the rolling stiffness coefficient β_r and the plastic moment coefficient η.

2.2   Sample Generation

The numerical model was calibrated against experimental tests5 on white Leighton Buzzard Sand (LBS) 14/25, fraction B (Figure 4). This is a very uniform sand ($CU = 1.3$), with a mean grain size $D_{50} = 0.8$ mm, specific gravity of the grains $\rho = 2650$ kg/m³, and minimum and maximum porosities given by $n_{min} = 0.324$ and $n_{max} = 0.436$, respectively.

Triaxial tests were performed on a packing of spherical particles enclosed in a cubic periodic cell. Preliminary tests showed that 5,000 particles are enough to obtain a representative elementary volume. The maximum and minimum porosities of the numerical sample are obtained following:[6] the packing is generated with a very high friction angle ($\phi_{\mu 0} = 80°$) and no contact among particles (Figure 4b). A uniform strain field was applied to all the particles of the periodic cell leading to a gradual increase of the confining pressure up to 1 kPa. In these conditions, the sample is considered to be at its loosest state. Then the friction angle was gradually reduced to 0°, where the numerical sample was considered to be at its densest state. Knowing the maximum and minimum porosities, the sample with a desired numerical relative density can be generated by decreasing the friction angle at constant confining pressure of 1 kPa until the porosity corresponding to the desired relative density is achieved.   Then the samples were isotropically compressed up to 100 kPa, and subsequently sheared (Figure 4c).

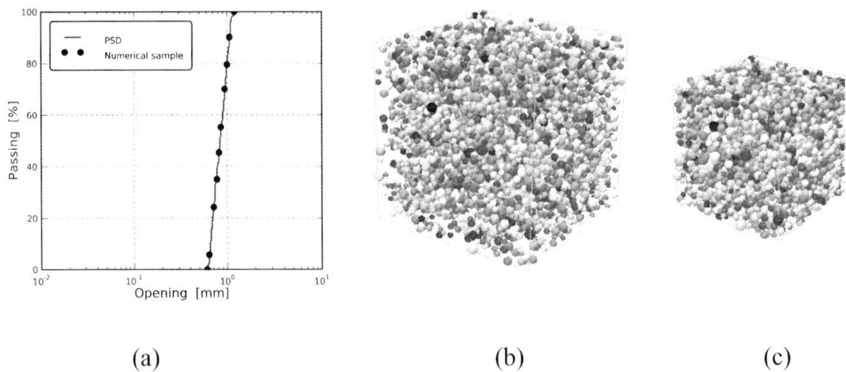

(a)   (b)   (c)

Figure 3 *(a) Particle size distribution of the LBS, (b) sample after generation and (c) sample after compaction.*

3 TRIAXIAL TESTS

3.1   Testing Program

The research program consisted of a parametric analysis of the micro-mechanical parameters, expressed as dimensionless groups, followed by a procedure to calibrate the numerical model against the experimental data. Table 1 lists the material properties used for the parametric analysis, and the value in the numerical model after the calibration. The particles density was scaled up by a factor of 10^9 so that simulations could be completed within a reasonable timescale.[7]

3.2   Sensitivity Analysis

The parameters here analysed are the initial porosity n of the sample, the particle density ρ, the contact law parameters (E, ν, ϕ_μ, β_r and η), and the strain rate $\dot{\varepsilon}$. Triaxial tests were performed first using particles free to roll ($\beta_r = \eta = 0$), secondly including the rolling law at contacts ($\beta_r \neq 0$ and $\eta \neq 0$).

According to dimensional analysis theory, all the material properties can be expressed in terms of dimensionless groups. The use of dimensionless groups reduces the total number of variables necessary to describe the problem. In our case, the response of the granular material can be expressed as:

$$\left(\frac{q}{\sigma_3}, \varepsilon_a\right) = \left(\Pi_I, \phi_\mu, \nu, \Pi_k, \beta_r, \eta, n\right)$$

(7)

where $q = \sigma_1 - \sigma_3$ is the deviatoric stress, and $\varepsilon_a = \varepsilon_1$ is the axial strain (*1* is the vertical direction, *3* is the horizontal one, both are defined positive in compression), Π_I is the inertial number and Π_k the stiffness number, defined as:[8]

$$\Pi_I = R\dot{\varepsilon}\sqrt{\frac{\rho}{\sigma_c}}$$

(8a)

$$\Pi_k = \frac{2R}{U_n} = \left[\frac{E}{3\left(1-v^2\right)\sigma_c}\right]^{2/3}$$

(8b)

Tests conducted in the present study showed that the Poisson's ratio v affects the mechanical response only slightly.[3,8] The same happens to be true for the range of values of intergranular friction angle tested here (from 21 to 36). Hence these two parameters have not been included in the sensitivity analysis presented below.

Table 1 *Micro-mechanical parameters employed in the DEM simulations*

Micromechanical parameter	Sensitivity analysis	Value in the calibrated DEM model
Density ρ (kg/m^3)	2650×10^8, 2650×10^9, 2650×10^{10}	2650×10^9
Strain rate $\dot{\varepsilon}$ (/s)	3.2×10^{-5}, 1.0×10^{-4}, 3.2×10^{-4}	1.0×10^{-4}
Numerical relative density RD (-)	0.25, 0.70, 0.76, 0.86, 0.99	0.70
Porosity n (-)	0.496, 0.429, 0.420, 0.401, 0.396	0.420
Confining pressure σ_c (kPa)	100	100
Young's modulus E (GPa)	0.7, 7, 70	70
Poisson's ratio v (-)	0.25, 0.30, 0.35	0.3
Friction angle ϕ_μ (°)	21, 26, 31, 36	26
Rolling stiffness coeff. β_r (-)	0.5, 1.0, 5.0	1.0
Plastic limit coeff. η (-)	0.5, 1.0, 5.0	1.0

Table 2 *Combinations of particle densities, strain rates and inertial numbers tested*

Test #	Density ρ (kg/m^3)	Strain rate $\dot{\varepsilon}$ (/s)	Inertial number Π_I (-)
TX-01	2650×10^{10}	1.0×10^{-4}	6.5×10^{-4}
TX-02	2650×10^9	3.2×10^{-4}	6.5×10^{-4}
TX-03	2650×10^9	1.0×10^{-4}	2.0×10^{-4}
TX-04	2650×10^8	1.0×10^{-4}	6.5×10^{-5}
TX-06	2650×10^9	3.2×10^{-5}	6.5×10^{-5}

3.2.1 Effect of the inertial number. In order to replicate the loading conditions of experimental triaxial tests, it is necessary to ensure that the numerical samples remain under quasi-static conditions throughout the whole tests. The inertial number Π_I is used here as a measure of the departure from static equilibrium. Triaxial tests were run for different values of Π_I, varying both particle density and strain rate. The input parameters are summarized in Table 2. The results are shown in Figure 4. It turned out that

$\Pi_I =< 2.0 \times 10^{-4}$ (TX-04) warrants rate independent results in terms of critical state and peak friction angles.

3.2.2. *Effect of the stiffness number.* The Young modulus was varied in the range $E = 0.7$, 7 and 70 GPa, corresponding to $\Pi_k = 187$, 870 and 4036, respectively. The results are shown in Figure 5. As expected, the macroscopic Young's modulus increases with the microscopic stiffness (Figure 5a & 5b). However, at larger strains the strength of the system slightly decreases and the dilatancy increases with increasing Young's modulus (Figure 5a & 5c).

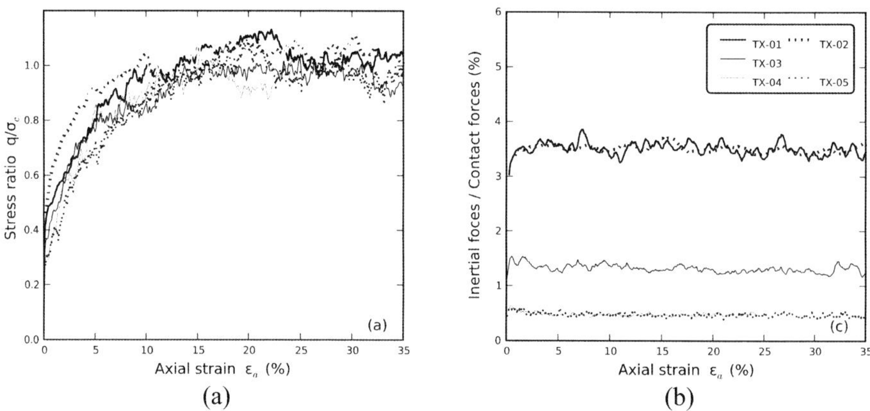

Figure 4 *Evolution of (a) stress ratio, and (b) ratio of inertial over contact forces for samples with various inertial numbers ($\Pi_I = 6.5 \times 10^{-5}$, 2.0×10^{-4}, 6.5×10^{-4}).*

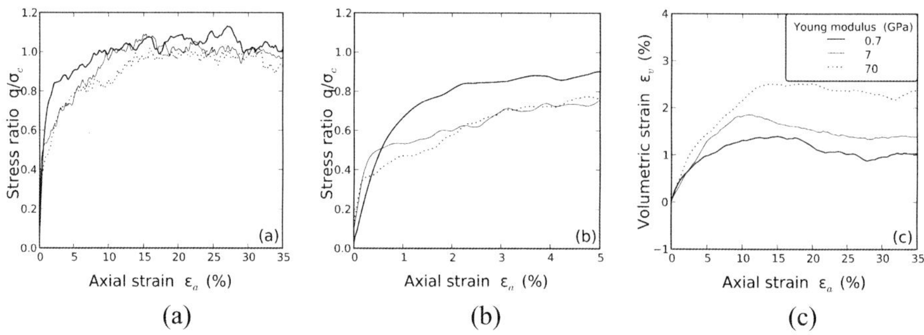

Figure 5 *samples with different values of particle Young's modulus ($\Pi_v = 187$, 870, and 4036): (a) stress ratio vs. axial strain; (b) stress ratio vs. axial strain at small strains; (c) volumetric strain vs. axial strain.*

3.2.3 Effect of the rolling parameters. In Figure 6 the stress-strains curves obtained by the DEM triaxial tests for various values of η and β_r are plotted together with the experimental curves of the Leighton buzzard sand. For $\beta_r = 0$ and $\eta = 0$, particles are free to roll. The curves show that rolling resistance makes the strength of the sample increase considerably. A suitable pair of values for β_r and η was found to make the numerical curve match the experimental one.[9]

It can be noticed that the rolling resistance coefficient β_r has only a slight influence on the initial stiffness, and a strong influence on the critical state friction angle (Figure **6**a). This result is in agreement with those reported in the literature.[3,9] The plastic moment coefficient η, instead, does not exhibit any significant effect on the response.

3.2.4 Effect of the initial porosity. Triaxial tests on samples with different densities were performed. Loose and dense samples are tested, with the porosities investigated varying from 0.396 to 0.496, which correspond to a numerical relative density of 25% and 99% respectively with $\beta_r=\eta=1$. The results are presented in Figure 7, and compared with experimental results for a loose and a dense sample.

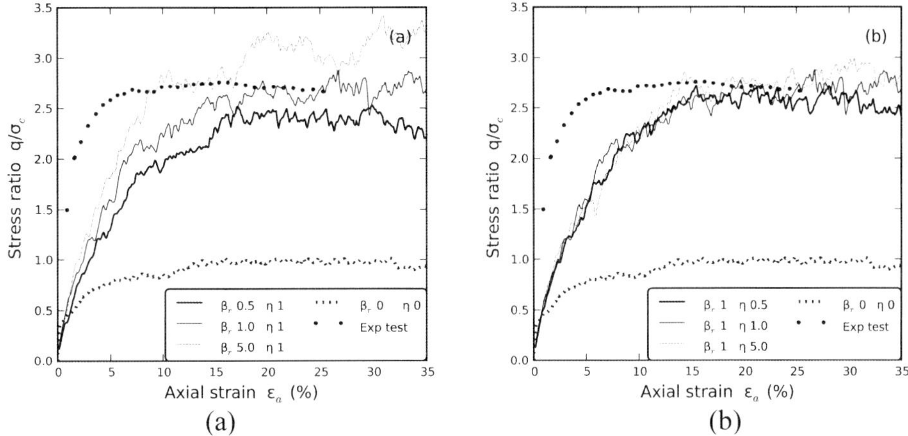

Figure **6** *Evolution of the stress ratios for samples for (a) various rolling stiffness coefficients βr and (b) various plastic moment coefficients η*

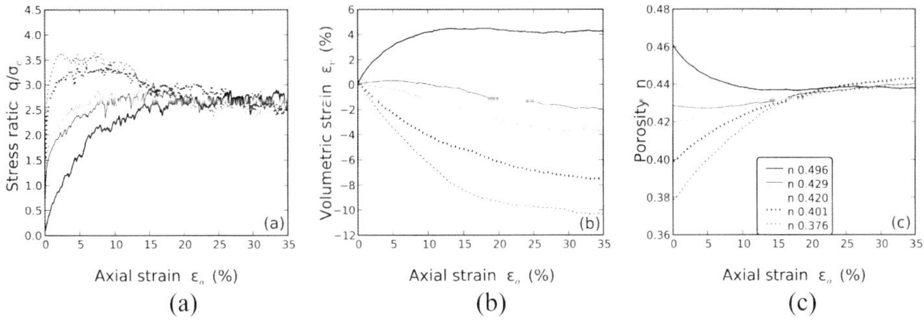

Figure **7** *Evolution of (a) stress ratio, (b) volumetric strain, and (c) porosity for samples with various initial densities.*

The figures show good agreement with the general behaviour of loose and dense sands, and with numerical experiments as well.[6,7] As expected, the initial shear modulus (initial slope of the curve in Figure 7a) is higher for the dense samples, which also exhibit a peak and then a softening behaviour. The volumetric strain shows compression for the loose sample, and expansion for the dense ones. The porosity, different at initial state, approaches to an equal value when the axial strain reaches approximately 20-25%. This means that the critical state is reached.

3.3   The Calibration Methodology

The parameters which need to be calibrated are: the initial porosity of the sample and the coefficients of the rolling law. The calibration procedure undertaken can be summarised in the following steps:

- The Young's modulus E, the Poisson's ratio v, and the friction angle ϕ_μ were taken from the experimental measures on the sand;
- The rolling resistance coefficient β_r was varied in order to match the critical state friction angle of the experimental curve;
- Two experimental stress-strain curves from Schnaid,[5] one for a dense (RD=86%) and one for a loose (*RD*=25%) sample were taken as reference. Numerical samples of different initial porosities were tested to find the values of porosity giving rise to an axial strain - volumetric strain curve matching the experimental curves.
- The η coefficient was changed in the attempt to achieve a peak friction angle ϕ_p near the experimental one.

The values of the micro-mechanical parameters achieved after the calibration exercise are listed in Table 1.

3.3.1 Comparison between the numerical model and the experimental results. The stress – strain curve of the loose numerical sample (see Figure 8) exhibits an excellent agreement with the experimental critical state friction angle and dilatancy angle. The adopted DEM methodology is instead unable to satisfactorily replicate the peak friction angle of the dense sample. A slight discrepancy for the stiffness at small strains for both the loose and the dense sample is observed. Adopting a reduced Young's modulus for the particles would allow matching the experimental stiffness; however, we preferred to keep a realistic value of the Young modulus for the grain stiffness in our simulations.

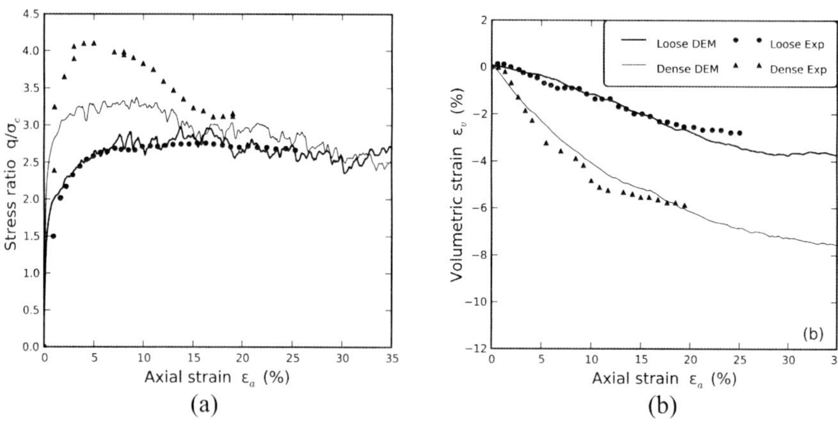

Figure 8 *Evolution of (a) deviatoric stress and (b) volumetric strain for a loose and a dense sample. The dots represent the experimental data.*

4 CONCLUSIONS

DEM numerical simulations of triaxial tests were performed in a 3D periodic cell, employing 5000 spherical particles and adopting a contact law based on the Hertz-Mindlin no-slip solution together with a linear moment – relative rotation law meant to account for the non-sphericity of the real sand grains. The values for the elastic constants and the intergranular friction adopted in the simulations were taken from the elastic and frictional properties of the minerals of the sand grains (mainly quartz) whereas the two parameters of the moment - relative rotation law and the initial porosity of the sample were calibrated.

Following the proposed calibration exercise, the salient features of the stress and strain curves of Leighton Buzzard sand were successfully reproduced for both loose and dense samples with the exception of the peak strength for dense samples. The critical state friction angle and the dilatancy angle (dense samples) of the DEM samples satisfactorily matched the experimental ones.

This calibration exercise was carried out as part of a larger research program aiming to investigate the lateral soil-structure interaction for unburied pipelines lying on sandy seabeds. An extensive program of 3D plain strain tests involving a plane strain section of the pipeline is currently underway making use of the calibrated DEM parameters.

Acknowledgements

The first author acknowledges the Engineering and Physical Sciences Research Council (EPSRC) and BP for the financial support of her PhD.

References

1   P. A. Cundall, in *Micromechanics of granular materials,* Satake & Jenkins (Eds), 1988, 113.

2   C.M. Wensrich and A. Katterfeld, *Powder Technology*, 2012, **217**, 409.

3   J. P. Plassiard, N. Belheine and F. V. Donze, *Granul. Matter,* 2009, **11**, 293.

4   K. Iwashita and M. Oda, *J. Eng. Mech*, **124**, 285.

5   F. Schnaid, *A study of the cone-pressuremeter test in sand.* PhD thesis, University of Oxford, 1990.

6   C. Salot, P. Gotteland and P. Villard, *Granul. Matter*, 2009, **1**, 221.

7   C. Thornton, *Geotechnique*, 2000, **50(1)**, 43.

8   F. Radjaï and F. Dubois, *Discrete-element modeling of granular materials*, ISTE Ltd and John Wiley & Sons Inc, 2011

9   L. Widulinski, J. Tejchman, J. Kozicki and D. Lesniewska, D, *Int. J. of Solids and Structures*, 2011, **48**, 1191.

10 T. G. Sitharam, J. S. Vinod and L. Rothenberg, *Powder and Grains*, 2005, 257.

THE STEADY STATE SOLUTION OF GRANULAR SOLID HYDRODYNAMICS FOR TRIAXIAL COMPRESSIONS

S. Song, Q. Sun and F. Jin

State Key Laboratory of Hydroscience and Engineering, Tsinghua University, China

1 INTRODUCTION

Granular Solid Hydrodynamics (GSH) was proposed for granular materials based on the laws of thermodynamics.[1] GSH was supported and quantified with arguments from physics and soil mechanics, such as the static stress distributions in silos, under point loads, and in sandpiles. When compared with the hypoplastic model, similar results were obtained. In addition, simulations of isochoric deformation cycles using GSH were compared with Wichtmann's experiments.[2] Recently, by using GSH theory, the elastic energy and its relaxation (denoted by the granular temperature) were both calculated and explained.[3]

However, it should be noted that there are up to 12 transport coefficients in GSH; even the granular medium is assumed to be intrinsically isotropic. In reality, due to the complex dissipation mechanisms in granular media, we have to consider a sufficient number of transport coefficients. As a result, it is not a simple matter to ascertain the models because they change with the state. Additional quantifications of GSH are required both from physics and engineering practices. The triaxial test is a conventional measurement in soil mechanics. Many mechanical behaviours, e.g. the stress-strain relationship, are studied on the basis of triaxial tests. The critical state is a very important concept in soil mechanics, and many consequent soil models are developed, such as the cam clay model. For sands, a series of triaxial compression tests were carried out to investigate the critical state of Leighton Buzzard sand.[4] In this study, the GSH was simplified for analysis of the deformational behaviour under the triaxial compression test, and then it was applied to analyze the experimental data obtained by Cheng,[5] and it was found that the steady state of GSH corresponds to the critical state of granular materials.

2 FUNDAMENTALS OF GSH THEORY FOR TRIAXIAL TESTS

GSH is a complete continuum mechanical theory for granular materials. Derivations and explanations can be found in Ref. [1]. GSH consists of (1) five conservation laws for the energy w, mass ρ, and momentum ρv_i; (2) an evolution equation for the elastic strain u_{ij}; and (3) balance equations for two temperatures. Both the true and the granular temperature, T and T_g, respectively, are necessary. In particular, T_g is a new concept introduced in GSH to treat a granular material as a transient elastic object due to particles jiggling and sliding. T_g is similar to the heat-like seismicity.

The elastic strain u_{ij} is defined as the portion of the strain ε_{ij} that deforms the particles and leads to reversible storage of elastic energy. The plastic rest, $\varepsilon_{ij} - u_{ij}$, accounts for rolling and sliding. Because the elastic energy depends on u_{ij} alone, this is a crucial step to retain many useful features of elasticity, particularly in an explicit expression of the stress. Dividing u_{ij} into $\Delta \equiv -u_{ii}$, $u_{ij}^0 = u_{ij} - u_{ii}\delta_{ij}/3$, we define the elastic energy as $E_e = B\sqrt{\Delta}\left(2\Delta^2/5 + u_s^2/\xi\right)$, with $u_s^2 = u_{ij}^0 u_{ij}^0$. The coefficient B accounts for overall rigidity and it increases with the density; ξ relates to the Mohr-Coulomb yield. The elastic energy E_e is convex only for $u_s/\Delta \le \sqrt{2\xi}$ (or equivalently $q/P \le \sqrt{3/\xi}$), coinciding with the Drucker-Prager condition. Then the elastic stress is given as $\pi_{ij} = -\partial E_e/\partial u_{ij}$.

The evolution equation for u_{ij} and T_g and the expression for the Cauchy stress σ_{ij} are briefly introduced here. Considering granular materials as transiently elastic systems, the evolution equation of u_{ij} is $\partial_t u_{ij} = v_{ij} + X_{ij}$, where $v_{ij} \equiv (\nabla_j v_i + \nabla_i v_j)/2$ is the space derivative of velocity v_i and X_{ij} represents a relaxing, dissipative contribution. The Cauchy stress is given as $\sigma_{ij} = \pi_{ij} - \sigma_{ij}^D$, where σ_{ij}^D is the dissipative stress. The evolution equation of T_g is $\rho T_g \partial_t \vartheta = 2\eta_s v_q^2/3 + \eta_v v^2/3 - \gamma T_g^2$. Taking $\eta_s, \eta_v, \gamma \equiv T_g$ for simplifications, we can get the evolution equation of T_g in Equation (7) below.

The energy conservation and entropy production $R = \sigma_{ij}^D v_{ij} + X_{ij}\pi_{ij} + \cdots$, implying that the variables X_{ij} and σ_{ij}^D are written as follows (the viscous terms, large only for high shear rate dense flows, are neglected):

$$X_{ij} = \lambda_{ijkl}\pi_{kl} - \alpha_{ijkl}v_{kl} = \lambda_s \pi_{ij}^0 + \lambda_v \delta_{ij}\pi_{ll}/3 - \alpha_{ijkl}v_{kl}$$
$$\sigma_{ij}^D = \alpha_{ijkl}\pi_{kl}$$

$$(1)$$

The fourth-order transport coefficients in Equation (1) are

$$\alpha_{ijkl} = \alpha_s \delta_{ik}\delta_{jl} + (\alpha_v - \alpha_s)\delta_{ij}\delta_{kl}/3 + \alpha_1(u_{ij}^0\delta_{kl} + u_{kl}^0\delta_{ij})$$
$$\lambda_{ijkl} = \lambda_s \delta_{ik}\delta_{jl} + (\lambda_v - \lambda_s)\delta_{ij}\delta_{kl}/3 + \lambda_1(u_{ij}^0\delta_{kl} + u_{kl}^0\delta_{ij})$$

$$(2)$$

Following the same rules for a triaxial test in which axial symmetry exists (i.e., $\sigma_2 = \sigma_3$), u_{ij}, σ_{ij} and v_{ij} can be rewritten as

$$\Delta = -u_{ii}, u_q = u_{11} - u_{33}$$
$$P = \sigma_{ii}/3, q = \sigma_{11} - \sigma_{33} \tag{3}$$
$$v = -v_{ii}, v = v_{11} - v_{33}$$

The evolution equation for u_{ij} and σ_{ij} takes the following final form:

$$\partial_t \Delta = (1 - \alpha_v)v + 2\alpha_1 u_q v_q - \lambda_0 T_g \Delta$$
$$\partial_t u_q = (1 - \alpha_s)v_q + \alpha_1 u_q v - \lambda T_g u_q \tag{4}$$

$$P = (1 - \alpha_v)B\Delta^{1.5}(1 + u_q^2/3\xi\Delta^2) + 4\alpha_1 B u_q^2 \Delta^{0.5}/3\xi$$
$$q = -2(1 - \alpha_s)B\Delta^{0.5}u_q/\xi - 3\alpha_1 B\Delta^{1.5}(1 + u_q^2/3\xi\Delta^2)u_q \tag{5}$$

The mass conversation equation and entropy increase equations are:

$$\partial_t \rho - \rho v = 0 \tag{6}$$

$$\rho \partial_t \vartheta = v^2/3 + 2v_q^2/3 - T_g^2 \tag{7}$$

$$\vartheta = bT_g \tag{8}$$

where ϑ is the granular entropy per unit mass.

The relaxation of u_{ij} can be described by a simple relaxation time model, where u_{ij} relaxes with relaxation rate $1/\tau$. When T_g is finite and particles vibrate, they briefly lose contact with one another, during which their elastic form will be partially lost. Macroscopically, the relaxation rate increases with T_g and disappears when $T_g=0$. Therefore, for X_{ij} in Equation (1), we assume $\lambda_s \pi_{ij}^0 = -u_{ij}^0/\tau_s = -\lambda T_g u_{ij}^0$ and $\lambda_v \pi_{ll} = \Delta/\tau_v = \lambda_0 T_g \Delta$.

With the conditions of σ_3=const, and v_{11}=const, the stress-strain relationship of triaxial tests could be studied by solving Equations (4) to (8). Then, we evaluated the steady state solution for GSH, by taking $\partial_t \rho \equiv 0, v \equiv 0, \partial_t P \equiv 0, \partial_t q \equiv 0, \partial_t \vartheta \equiv 0$. It was found that $T_g = \sqrt{6}v_q/3$, $u_{qc} = \sqrt{6}(1-\alpha)/2\lambda$ and $\Delta_c/u_{qc} = \sqrt{6}\alpha_1/\lambda_0$. Substituting it into Equation (5), we can get the analytic solutions of the steady state, e.g. the mean stress P and deviatoric stress q as a function of the void ratio e.

3 RESULTS AND DISCUSSIONS

Triaxial compression tests carried out by Cheng were performed on reconstituted samples of Dutch fine sand, which mainly consists of quartz mineral.[5] The classification properties are summarized in Table 1. The dense specimens were prepared in a cylindrical mould by the method of multi-stage vibration, with an average initial void ratio of 0.625. The medium dense specimens were prepared in the same mould with an average initial void ratio 0.725 by the method of air pluviation. Specimens were sheared at confining pressure of 400 kPa.

Table 1 *Properties for Dutch fine sand*

Properties	Value	Properties	Value
Specific gravity, G_s	2.65	Roundness of sand grains	Sub-rounded
Max. void ratio, e_{max}	0.859	Min. void ratio, e_{min}	0.528
Mean grain size, D_{50}	0.156	Uniformity coefficient, D_{60}/D_{10}	1.7

The parameters used in the simulation has been chosen as $B_0 = 7GPa$, $\xi = 5/3$, $\lambda = 80 f_\rho^{0.5}$, $\lambda_0 = \lambda / 3.7$, $\alpha_s = \alpha_v = \alpha = 0.78 f_\rho^{0.15}$ and $\alpha_1 = 25 f_\rho$, where $f_\rho = (1-\rho)/(1-\rho_{lp})$.

3.1 Deformational Behaviour

Two triaxial compression tests with different initial void ratios were simulated, as shown in Figure 1. The dense specimen has a higher peak stress ratio of about 3.0 while the medium dense specimen has a smaller peak stress ratio of about 2.7. The dilatancy of dense specimen is more obvious than the medium one. When sheared, both of them finally reach the same stress ratio and stay the same.

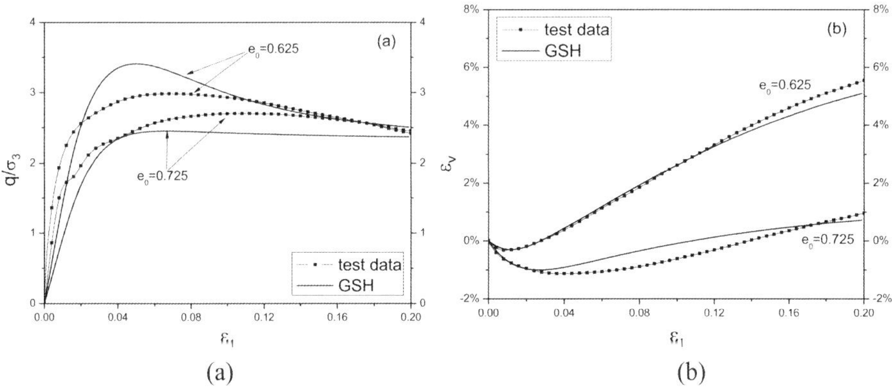

(a) (b)

Figure 1 *Variations in stress ratio q/σ_3 (a) and volumetric strain ε_v (b) with axial strain ε_1 for different initial void ratios*

3.2 The steady state of GSH

When the specimen is continuously sheared with the same shear rate, it finally reaches the critical state, i.e. the state where the material keeps deforming in shear at constant stress and volume.[6] Physically, the steady state solution of GSH has the same meaning as the critical state, and it is necessary to check the properties of the steady state solutions.

As shown in Figure 2, three regions are presented in *lnP-e* phase: (1) the very loose region, the void ratio near the maximum void ratio e_{max}, where the mean stress P is smaller than 100 kPa. The internal structure of sand is unstable. The structural rearrangement among particles will take place and the void ratio will get smaller once the sand is sheared or compressed by a higher stress. (2) The normal dense region (100 kPa<P<5 MPa), containing most of the stress states encountered in engineering and experiments. The usually discussed behaviours of critical state mainly correspond to this region. The relation between void ratio e and the logarithm mean stress lnP is approximately linear. (3) The over-dense region (P>5 MPa), where the void ratio near the minimum e_{min} and the specimen behaves like solids.

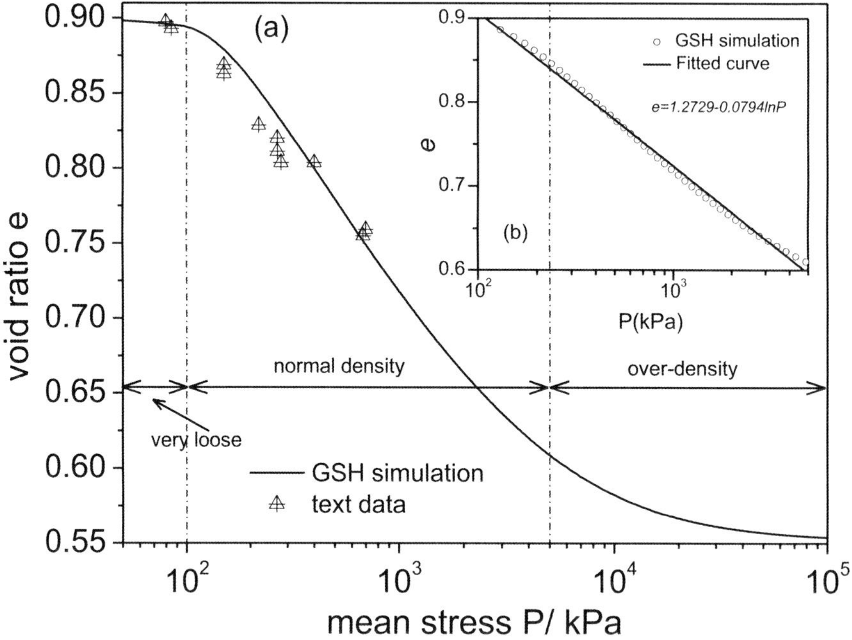

Figure 2 *Void ratio e vs. mean stress P in the steady state of GSH theory*

A series of tests consisting of different confining pressure, void ratio and stress paths were carried out by Cheng.[5] We picked out the void ratio and mean stress in the critical state, shown as triangular symbols in Figure 2a. Most of the tests reached their critical states when the deviatoric strain exceeded 20%. But in some of the tests, the volumetric strain still change after the strain exceeds 20%. However, the stress remained unchanged and the increments of volumetric strain are small and the rates of change are decreasing. So in those cases, the states with a volumetric strain at ε_1=20% are treated as the critical state. As we can see from Figure 2a, most of the test data in the critical state close to the calculated results with the GSH theory. It demonstrates that the critical state in soil mechanics corresponds to the steady state solution of GSH. In the GSH theory, particles with enduring contacts are elastically deformed. During being sheared, they increase the static shear stress. Meanwhile, the deformation is slowly disappeared when grains rattle and jiggle, because they briefly lose contact with one another. Therefore, a constant shear rate not only increases the deformation, as in any elastic medium, but also decreases it due to the jiggle of particles. A steady state would be reached as the two processes are dynamically balanced. Therefore, the elastic deformation and stress remains constant over time, and we can conclude that the critical state may correspond to the steady state solution of GSH theory.

In the normal density region, the mean stress P and deviatoric stress q satisfy the equations $q=MP$ and $e = \Gamma - k \ln P$, which are in fact the definitions of the critical state.[6] In our simulations, we found that M=1.23, Γ=1.27 and k=0.08, as shown in Figures 2b and 3.

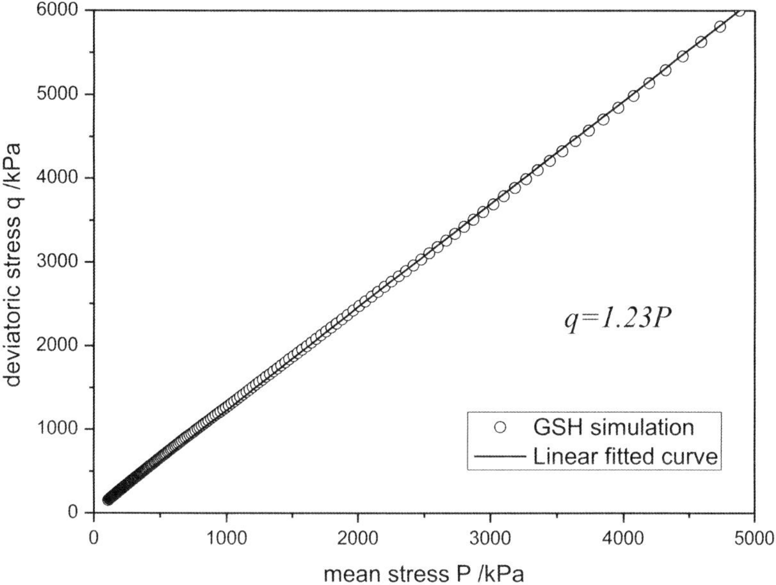

Figure 3 *Deviatoric stress q vs. mean stress P in the steady state of GSH theory*

4 CONCLUSIONS

The GSH equations were simplified for analysimg the deformational behaviour and critical state behaviour under the triaxial tests. It can describe the dilatancy and stress peak of the dense sand and obtain the reasonable results.

The steady state solution of GSH theory is calculated. Three different behaviours in *lnP-e* phase were observed. The very loose region is an unstable configuration; the normal density region is the most important for engineering and experiments, the void ratio and mean stress P has approximately a linear relationship; the over-density region has a solid-like mechanical behaviour. Most of the test data in critical state follow the result of the

GSH simulation. The calculated results demonstrate that the critical state in soil mechanics corresponds to the steady state solution of GSH. It is the result of dynamic balance of the increase of elastic deformation by shear and the loss of elastic deformation by rattling and jiggling.

Since GSH is a complete continuum mechanical theory for granular materials. It provides the theoretical framework to study the macroscopic behaviour of granular matter, so we can use it to study any types of granular materials. Nevertheless, it does not involve the meso-structure of the granular materials; some problems cannot be solved directly. For instance, we cannot directly obtain the function of the free energy and the transport coefficients of the granular materials with GSH. To confirm these, we must compare the GSH with experiments and carry out a large number of calibrations.

References

1 Y. M. Jiang and M. Liu, *Granular Mat.* 2009, **11**, 139.
2 G. Gudehus, Y. M. Jiang and M. Liu, *Granular Mat.* 2011, **13**, 319
3 Q. C. Sun, S. X. Song, F. Jin F and Z. W. Bi, *Granular Mat*, 2011, **13**, 743
4 Z. Y. Cai and X.S. Li, *Chinese J Geotechnical. Engng.* 2004, **26**, 697
5 X. H. Cheng, *Localization in Dutch dune sand and organic clay.* PhD thesis, TU Delft, 2004.
6 A. N. Schofield and C. P. Wroth, *Critical State Soil Mechanics*, McGraw-Hill, 1968.

3D DEM SIMULATIONS OF UNDRAINED TRIAXIAL BEHAVIOUR WITH PRESHEARING HISTORY

G. Gong[1] and A.H.C. Chan[2]

[1] Civil and Environmental Engineering, Shenzhen Graduate School, Harbin Institute of Technology, China. (formerly School of Civil Engineering, The University of Birmingham, B15 2TT).

[2] School of Civil Engineering, The University of Birmingham, B15 2TT

1   INTRODUCTION

Granular material such as sand may experience different stress history before it is used as a foundation and it may have an effect on its behaviour therefore it is instructive to study the effect of such stress history. Finge and co-workers[1] investigated the effects of preloading history on the undrained behaviour of saturated loose Hostun RF sand by performing physical laboratory experiments. They studied the effects induced by isotropic overconsolidation and by drained cycle preshear on the subsequent undrained behaviour of Hostun sand. The analysis of results focused on the evolution of the behaviour according to the deviator stress attained during the drained cycle preshear or as a function of the overconsolidation ratio. An interesting phenomenon as pointed out by Finge and co-workers,[1] is that all the presheared samples have the same initial positive slope in q-p space and the initial stress path for the undrained shear lies on a unique curved line independent of the q_{max} during the preshearing process, which forms a limiting stress boundary in the stress space not crossed by all presheared samples. A similar phenomenon in physical experiments was observed by Gajo and Piffer.[2] With the increase of q_{max} in the drained cycle (the preshear loading process), Finge and co-workers [1] observed that the samples exhibit static liquefaction, temporary liquefaction and complete stability respectively in subsequent undrained compression tests depending on the q_{max} attained during preloading. They attributed the results to the induced anisotropy caused by the preshearing process, which, however, is not in agreement with the findings by Gajo and Piffer.[2]

As far as the authors are aware of, the DEM simulation of undrained behaviour with preshearing history has never been reported before. In the following section, brief details of the simulation are given. The macroscopic and microscopic results and discussions based on DEM simulations are presented in Sections 3 and 4, respectively. In Section 5, the main conclusions are summarised.

2   SIMULATION DETAILS

Three-dimensional DEM simulations have been performed on polydisperse systems of 3600 elastic spheres using the same TRUBAL code as used by Thornton.[3] Contact interactions are calculated using algorithms based on the theories of Hertz for normal interaction and Mindlin and Deresiewicz[4] for tangential interaction. The undrained condition is modelled in 'dry' constant volume tests without considering the interstitial fluid. Further details can be found in Refs [5] and [6]. The Young's modulus and Poisson's ratio for each particle were specified as $E = 70$ GPa and $v = 0.3$, respectively. During the shear stage, the interparticle friction coefficient was set to $\mu = 0.5$. Nine different sizes of spheres were used (the actual number of particles is given in brackets): 0.25 mm (2), 0.26 mm (20), 0.27 mm (220), 0.28 mm (870), 0.29 mm (1376), 0.30 mm (870), 0.31 mm (220), 0.32 mm (20), 0.33 mm (2), with an average particle diameter of 0.29 mm. The notional density of each particle is 2650 kg/m^3, which is scaled up by a factor of 5×10^{12} in order to perform quasi-static simulations within a reasonable timescale. The time step used in the simulations is based on the minimum particle size and the Rayleigh wave speed.[7]

A loose sample with a porosity of 0.419 was obtained by adjusting the inter-particle friction values during isotropic compression stages and the inter-particle friction was changed back to 0.5 when the mean stress (p) is nearly 100kPa. Detailed preparation procedures can be found in Ref. [8]. From an initial isotropic stress state ($p = 100$kPa) the sample was presheared by applying conventional drained compression at constant "radial" stress $\sigma_2 = \sigma_3 = 100$ kPa until the deviator stress reached the desired value. The sample was then unloaded back to the initial isotropic state of stress. In this manner the loose sample was presheared to $q = 20$ kPa, 40kPa, and 60kPa to provide three presheared samples for subsequent simulations of constant volume (undrained) tests, and the samples are called preshear20, preshear40 and preshear60, respectively. Table 1 shows the deviator fabric ($\phi_1 - \phi_3$), indicating that the systems are essentially isotropic, at the end of preshearing process,. Therefore, the preshearing process will not induce initial anisotropy, which is in agreement with Gajo and Piffer.[2]

All the simulations start from an initial stress state that is almost isotropic with all the normal (principal) stresses approximately equal to 100 kPa. The strain-rates in the three principal strain directions were set to be $1.0 \times 10^{-4} s^{-1}$ (compression), $-5.0 \times 10^{-5} s^{-1}$ and $-5.0 \times 10^{-5} s^{-1}$ (extension), respectively.

The fabric tensor is defined by equation (1) below. The deviator fabric ($\phi_1 - \phi_3$) is used to describe induced anisotropy for axisymmetric fabric conditions, where ϕ_1, ϕ_3 are the major and minor principal values of ϕ_{ij}.[9]

$$\phi_{ij} = \frac{1}{C} \sum_1^C n_i n_j \tag{1}$$

where n_i is the unit vector normal and C is number of contacts.

Table 1 *Sample data at the end of preshearing*

Preshear q_{max} (kPa)	Porosity	Z_m	$\phi_1 - \phi_3$	Z_m at the end of preshear loading
0	0.4190	5.03	0.004	5.03
20	0.4185	4.95	0.004	4.92
40	0.4177	4.91	0.002	4.84
60	0.4143	4.83	0.006	4.79

Note: the definition of Z_m is given by equation (2). The definitions of ϕ_1 and ϕ_3 are from equation (1)

The mechanical coordination number Z_m is calculated using Equation (2) below:[3]

$$Z_m = \frac{2C - N_1}{N_p - N_1 - N_0}, \quad Z_m \geq 2 \tag{2}$$

where N_p is number of particles, N_1 and N_0 are the number of particles with one and no contacts respectively.

3 MACROSCOPIC BEHAVIOUR

In this section, macroscopic behaviour based on DEM simulations is presented in terms of undrained stress path (deviator stress against mean stress) and evolution of deviator stress.

Figure 1 shows the stress paths (deviator stress $q = (\sigma_1 - \sigma_3)$, mean stress $p = (\sigma_1 + \sigma_2 + \sigma_3)/3$) for the undrained behaviour of the loose sample with and without preshearing history, where in the figure porosity0.419 refers to the original sample without preshearing history. It can be seen that the undrained behaviour with preshearing history is different from that without preshearing history. It is interesting to note that the samples follow the same initial stress path prior to reaching the maximum deviator stress. The same phenomenon was reported in Refs [1] and [2].

Figure 2 shows the deviator stress plotted against deviator strain for the presheared and unpresheared samples. It can be seen that the initial slope is identical for all the samples. The maximum deviator stress increases with the increase of q_{max} obtained during preshearing process. The two samples presheared to deviator stresses of 20 kPa and 40 kPa exhibit steady state[10] at large strains but both samples exhibit a higher deviator stress at steady state compared with that of the unpresheared sample. Such steady state behaviour is also termed "liquefaction" in soil mechanics literature. The sample presheared to a deviator stress of 60 kPa exhibits quasi-steady state[11] since it can be seen that the deviator stress is increasing gradually at larger strains. Such quasi-steady state behaviour is also termed "temporary liquefaction" in soil mechanics literature.

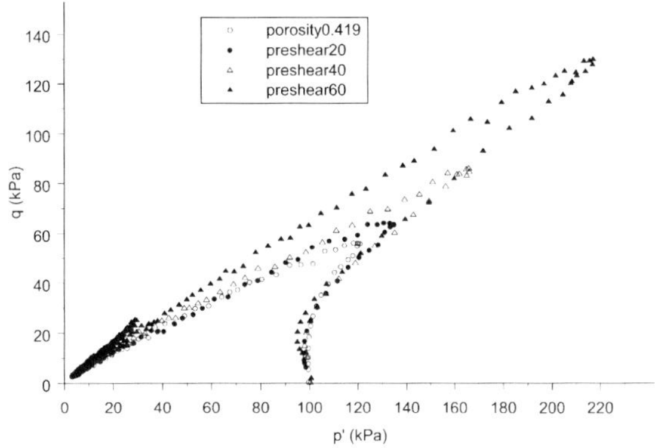

Figure 1 *Undrained stress path*

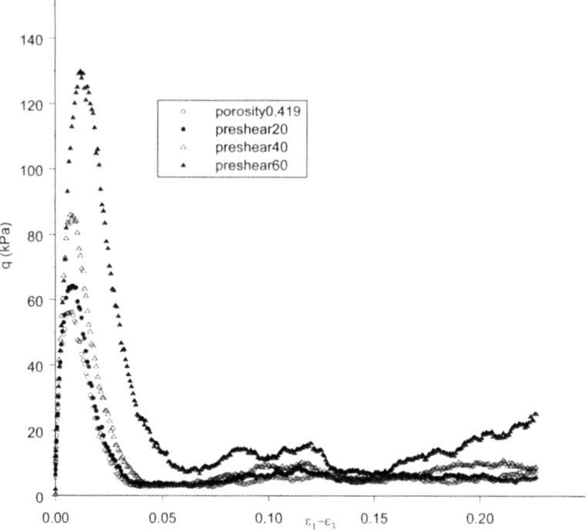

Figure 2 *Evolution of deviator stress*

4   MICROSCOPIC BEHAVIOUR

In this section, microscopic behaviour based on DEM simulations is presented in terms of mechanical coordination number and redundancy index.

Gong and co-workers [12] proposed a redundancy index, which is defined as the ratio of the number of constraints to the number of degrees of freedom in the system and can be expressed by

$$I_R = Z_m \left(\frac{3-2f}{12} \right) \tag{3}$$

where f is the fraction of sliding contacts.[8]

If $I_R > 1$, the system is a redundant one and in this case the macroscopic behaviour can be said to be 'solid-like'. If $I_R < 1$, the system becomes a mechanism and is unstable, which can lead to further loss of contacts with the consequence that the macroscopic behaviour can be said to be 'liquid-like'. The condition with $I_R = 1$ defines the limiting condition between the 'solid-like' and 'liquid-like' and can be used to define the onset of liquefaction (temporary liquefaction).

Figure 3 shows the evolution of the mechanical coordination number Z_m against deviator strain for the presheared and unpresheared samples. It is observed that with the increase of the amount of the preshear during the preshearing process, higher mechanical coordination number at a given deviator strain is observed in subsequent undrained shearing. Lade[13] found that the locus of the stress state associated with initial peak values of q on the different effective stress paths fall onto a unique line, which can be termed the Lade instability line. It is of interest to observe that for all the presheared and unpresheared systems, the initial peak states of q (Lade instability) occurs corresponding to a unique value of Z_m (=4.5) irrespective of the preshearing history as shown in Figure 4, and it has also been shown in Ref. [12] that the unique value of Z_m (=4.5) corresponding to Lade instability is independent of the samples' initial porosities.

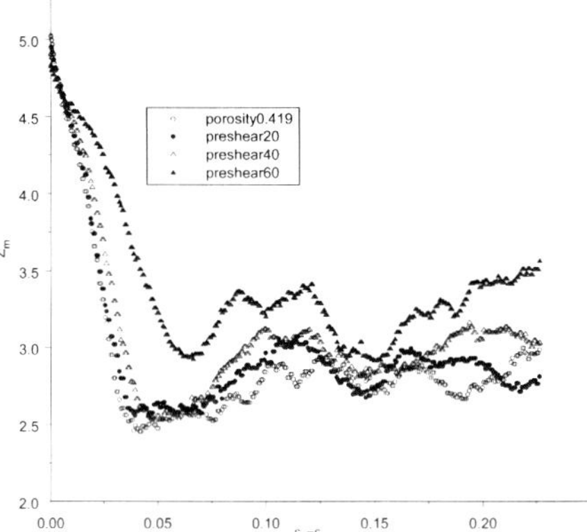

Figure 3 *Evolution of the mechanical coordination number*

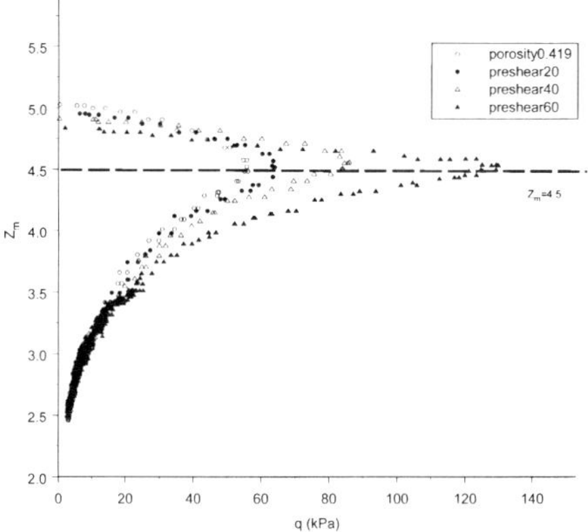

Figure 4 *Mechanical coordination number against deviator stress*

Figure 5 shows the evolution of redundancy index for the presheared and unpresheared samples under undrained axisymmetric compression as well as for the 'drained' sample. The 'drained' in Figure 5 means constant mean stress axisymmetric compression for the sample with a porosity of 0.419, used for reference only, and it can be seen that the 'drained' sample exhibit complete stability ($I_R>1$ always) microscopically, which is consistent with the known common fact in soil mechanics that a loose sample does not exhibit liquefaction (temporary liquefaction) under drained condition. In Figure 6, the total region is divided into solid-like region ($I_R>1$) and liquid-like region ($I_R<1$) with the $I_R=1$ as the limiting phase-transition line. It can be seen that the presheared and unpresheared samples all drop to the liquid-like region, i.e. exhibit liquefaction (temporary liquefaction), at small deviator strains. With increase of the amount of maximum deviator stress obtained during the preshearing process the corresponding sample exhibits higher redundancy index for a given deviator strain, which explains why the presheared samples exhibit higher resistance to liquefaction (temporary liquefaction) than the unpresheared and why the presheared sample with larger preshearing deviator stress exhibits higher resistance from a microscopic point of view.

Figure 6 shows redundancy index against mechanical coordination number for the presheared and unpresheared samples under undrained axisymmetric compression. In the

Figure, the condition Z_m=4.5 identifies the Lade instability, i.e. initial peak states of deviator stress under undrained condition. The condition I_R=1 identifies the onset of liquefaction (temporary liquefaction). It can be seen from Figure 6 that the Lade instability (Z_m =4.5) corresponds to I_R>1, i.e. in solid-like region, and from Figure 6 it can be easily deduced that the Lade instability occurs at a smaller deviator strain than the deviator strain corresponding to I_R=1 for a given sample, indicating that the onset of liquefaction (temporary liquefaction) lags behind the Lade instability. The relationship between I_R and Z_m is described by Equation (3), which is dependent of the fraction of sliding contacts (f). For reference, the evolution of f is shown in Fig. 7, where it can be seen that the value of f is generally smaller than 0.08 during all the evolution for all the samples.

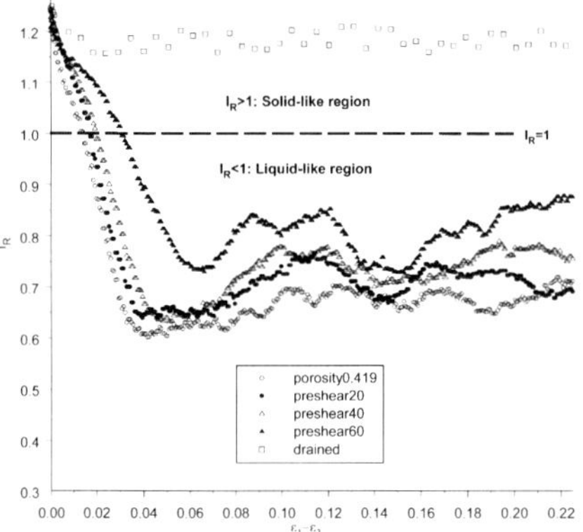

Figure 5 *Evolution of redundancy index*

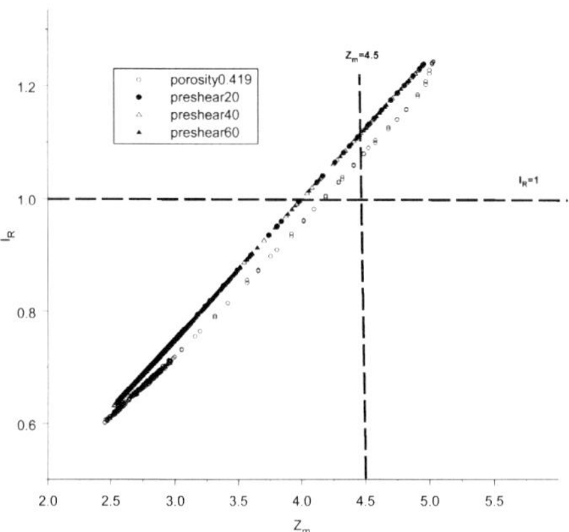

Figure 6 *Redundancy index against mechanical coordination number*

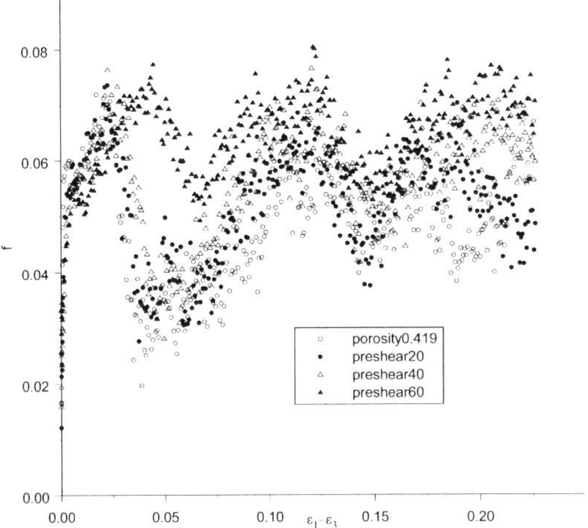

Figure 7 *Evolution of fraction of sliding contacts*

5   CONCLUSIONS

This paper reports the results of undrained axisymmetric/triaxial simulations with and without preshearing history. It is demonstrated that DEM can mimic preshearing history effect on undrained behaviour and is an ideal tool for simulating and examining preshearing history effect at macroscopic and microscopic levels.

The presheared simulation results show that the presheared samples are denser compared with the original unpresheared sample, and exhibit higher resistance to liquefaction and from a microscopic point of view such higher resistance can be attributed to the evolution of redundancy index. The DEM simulation also captures the limiting boundary in q-p space in published experimental observations, which cannot be crossed by any of the presheared samples.

The DEM simulations indicate the preshearing process will not cause induced anisotropy, which is in agreement with Gajo and Piffer.[2]

It is found that the Lade instability is found to correspond to a unique mechanical coordination number (=4.5), independent of preshearing history. It also found that the onset of liquefaction (temporary liquefaction) in terms of redundancy index lags behind the onset of the macroscopic strain softening in terms of the Lade instability for the presheared and unpresheared samples under undrained conditions.

Acknowledgements

The first author would like to thank the Engineering and Physical Sciences Research Council, UK (Grant No. GR/R91588) for providing the funding for his PhD study, as well as Dr. Colin Thornton's guidance on this study that formed part of his PhD research. Both authors would like to wish Colin all the best wishes as he enters his eighth decade of life and hope that he continues to contribute significantly to the particulate research community.

References

1   Z. Finge, S. Boucq and T. Doanh. In *Proc. 3ʳᵈ international symposium on deformation characteristics of geomaterials*, Swets & Zeitlinger, Lisse, Lyon, France, 2003, 729.

2   A. Gajo and L. Piffer. *Soils and Foundations*, 1999, **39**, 43.

3   C. Thornton. *Geotechnique*, 2000, **50**, 43.

4   R.D. Mindlin and H. Deresiewicz. *Trans. ASME, J. Appl. Mech.*, 1953, **20**, 327.

5   C. Thornton and K.K. Yin. *Powder Technology*, 1991, **65**, 153.

6   C. Thornton, C. In *Mechanics of Granular Materials – an introduction* (eds. M. Oda and K. Iwashita), Balkema, Rotterdam, 1999, 207.

7   C. Thornton and C.W. Randall. *Micromechanics of granular materials*, Satake and Jenkins (eds), Elsevier, Amsterdam, 1988, 133.

8   G. Gong. *DEM simulations of drained and undrained behaviour.* PhD thesis, University of Birmingham, UK, 2008.

9   C. Thornton and G. Sun. *Powers and Grains 93*, Thornton (ed.), 1993, 129.

10   S.J. Poulos. *Journal of Geotechnical Engineering, ASCE*, 1981, **17**, 553.

11   A. Alarcon-Guzman, G.A. Leonards and J.L. Chameau. *Journal of Geotechnical Engineering, ASCE*, 1988, **114**, 1089.

12   G. Gong, C. Thornton and A. Chan. *Journal of Engineering Mechanics, ASCE*, 2012. (doi:10.1061/(ASCE)EM.1943-7889.0000366)

13   P.V. Lade. *Journal of Geotechnical Engineering, ASCE*, 1992, **118**, 51.

STRONG FORCE NETWORK OF GRANULAR MIXTURES UNDER ONE-DIMENSIONAL COMPRESSION

N.H. Minh and Y.P. Cheng

Department of Civil, Environmental and Geomatic Engineering, University College London, Gower Street, WC1E 6BT, London, UK

1 INTRODUCTION

The behaviour of granular systems is highly dependent on the size distribution of the constituent particles. Force transmission through the assemblies of discrete particles of different sizes can be characterized by a high degree of force inhomogeneity. This happens when a small number of contacts in the system attract forces of a significantly higher magnitude than the other contacts. The force transmission can be represented by two distinct classes - the strong force and the weak force contact networks.[1,2] The strong forces were defined as contact forces contributing primarily to the macroscopic deviatoric stress, and hence they determine the shear strength of the sample. These strong contacts are constituted to a shear-strength related load-bearing mechanism. Column-like structures[3] carrying strong forces were observed inside granular systems, and the buckling of these structures (i.e. the strong force chains) marks the onset of the failure condition.[4] The weak forces, on the other hand, tend to be oriented in perpendicular to the loading direction, and support the stability of the strong force chains. The weak forces only contribute to the mean stress but have a negligible contribution to the deviatoric stress. The weak network behaves like an interstitial liquid, whereas the load-bearing strong network behaves like a solid.[1] For slightly polydisperse systems,[1,2] the characteristic force that divides the strong forces and the weak forces is approximately the mean contact force of the whole system. The relative proportion of the strong contacts in a three dimensional system[2] is about 30% of the total number of contacts, as compared to the value of 70% for the weak contacts. This bimodal character of the force network is an important feature of granular media; it reflects the high level of redundancies in the system, when only some of the contacts are required to resist the overall applied shear forces.

Unlike granular systems in the powder industry where the properties of the particles are usually consistent, soils are natural materials whose properties vary greatly by location and geological history. Soils are normally mixtures of particles of different sizes and their particle size distribution has a significant impact on the mechanical behaviour of soils. In this study, we investigate the effect of particle size distribution on the compression behaviour of imperfect binary mixtures of a sand-like material and a finer uniform silt-size material; the maximum size ratio in the assembly is about 10. The force transmissions in the binary mixtures were studied in detail. We found that the characteristic force, above which defines the strong forces, is not necessarily the mean force. This however depends on the particle size distribution and the stress level. The existence of different contact-types plays a major role in governing the contact force distributions and the load-bearing mechanism for these mixture systems.

2 SIMULATION PROCEDURE OF ONE-DIMENSIONAL COMPRESSION

The Itasca discrete element method (DEM) package,[5] PFC3D, was used to simulate the one-dimensional compression behaviour of granular mixtures. Numerical samples were created by mixing two component materials, one has the particle size distribution (PSD) of a real sand material and the other (hereinafter referred to as silt) is a material purely composed of particles 0.1 mm in diameter. The PSD(s) of these two materials are shown in the inset of Figure 1. Mixtures were created with different silt contents (by mass) in the range of 0%, 10%, 20% and up to 100%. To prepare the numerical samples, spherical particles of different sizes were randomly generated inside a cubical space bounded by rigid frictionless walls (wall stiffness is 100 times of the average particle stiffness) to form gas-like assemblies at an initial solid fraction of 0.5. The input parameters are listed in Table 1. The linear elastic contact model[5] with Coulomb friction was used for the inter-particle interaction; the contact particle stiffness was calculated based on a scaling relationship with the constant particle elastic modulus and the particle diameter: $k_n = k_s = 4E_cR$ (where $E_c = 1$ GPa is the elastic modulus; R is the particle radius; k_n and k_s are the normal stiffness and the shear stiffness of the particle, respectively). The normal contact stiffness (k^n) and the shear contact stiffness (k^s) are then defined based on the stiffness of the two contacting entities (e.g., A and B) as:

$$k^n = \frac{k_n^{[A]}k_n^{[B]}}{k_n^{[A]} + k_n^{[B]}}\,; k^s = \frac{k_s^{[A]}k_s^{[B]}}{k_s^{[A]} + k_s^{[B]}} \tag{1}$$

Table 1 *Input parameters for 3D DEM simulation*

Input Parameters	Value
Particle density	2650 kg/m^3
Particle friction, μ	0.5
Contact elastic modulus, E_c	1×10^9 N/m^2
Particle stiffness ratio, k_s/k_n	1.0
Wall friction, μ_{wall}	0.0
Wall stiffness, k_{wall}	100 * average particle stiffness

The initial particulate systems were brought to a mean pressure of approximately 100 kPa, with no gravity, by applying a location-dependent velocity to each particle and to each boundary wall, in order to move them towards the centroid of the sample at a constant strain rate. During the process, the systems were allowed to relax periodically by zeroing the assigned velocity, and letting particles rearrange until the unbalanced forces became negligible. A temporary inter-particle friction of 0.0 was used in this stage to create packing of the possible densest state. Once the system reached the specified mean stress, the preparation stage was finished and the simulation configuration was changed into one-dimensional compression. Lateral walls were now fixed, the inter-particle friction was set equal to 0.5 and the samples were compressed vertically in the same constant-strain-rate manner, such that the particles and the horizontal walls were moved toward the horizontal mid-plane with periodic relaxation. The stress and strain were calculated from the force and the displacement obtained at the boundary walls. The sample size was chosen so that a consistent behaviour could be achieved and the simulation could be run within a reasonable time scale. Summary of the sample size and the number of particles are given in Table 2 for the different cases of silt content; note that d ≥ 0.4mm are particles of the (large) sand material, whereas d = 0.1mm are particles of the (small) silt material in the mixture. Due to the very large numbers of particles in these samples, the differential density scaling[5] was used to reduce the computational time.

Table 2 *Sample size and the number of particles for DEM samples*

Silt content	Initial sample size	d ≥ 0.4 (mm)	d = 0.1 (mm)	Total particles
0%	15^3 mm^3	10680	0	10680
10%	7^3 mm^3	973	32754	33727
20%	7^3 mm^3	857	66294	67151
30%	7^3 mm^3	751	98950	99701
40%	7^3 mm^3	643	131605	132248
50%	7^3 mm^3	536	164261	164797
100%	3^3 mm^3	0	25783	25783

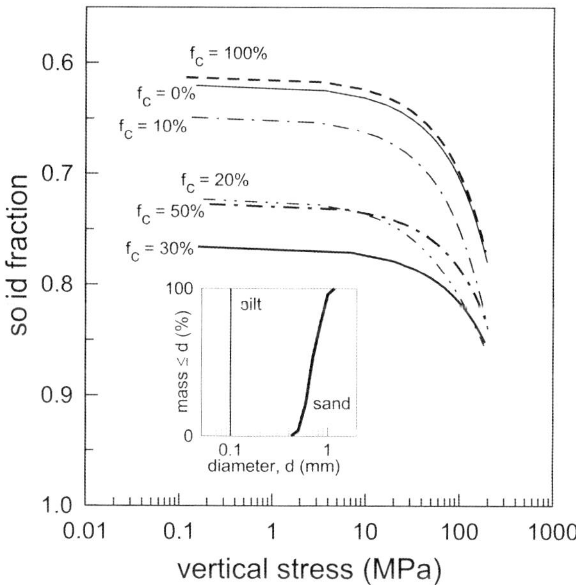

Figure 1 *Compression curves and PSD of two component materials*

3 SIMULATION RESULTS OF ONE-DIMENSIONAL COMPRESSION

Figure 1 shows the compression curves of the solid fraction plotted against the vertical stress for different silt content (f_c). As the samples were compressed and the vertical stress increased, the void space inside the granular systems reduced its volume, leading to a higher solid fraction. In general, samples of the mixture materials have higher initial solid fraction than those of the uniform gradations (i.e. the 0% and the 100% cases). For the mixture materials, the packing efficiency was improved when the small particles filled the void space between the larger grains. Figure 2 shows an approximately linear relationship between the horizontal stress and the vertical stress during one-dimensional compression. A linear fitting line was fitted to each data set, the slope of which yields the value of the at-rest lateral earth pressure coefficient K_0 as described in the field of soil mechanics. Note that samples of the uniform gradations (f_c = 0% and f_c = 100%) behaved similarly in both Figures 1 and 2, which was due to the scaling relationship between the particle stiffness and the particle size used in this study. Any difference in the behaviour of the mixture materials is purely the result of the interactions of different particle sizes.

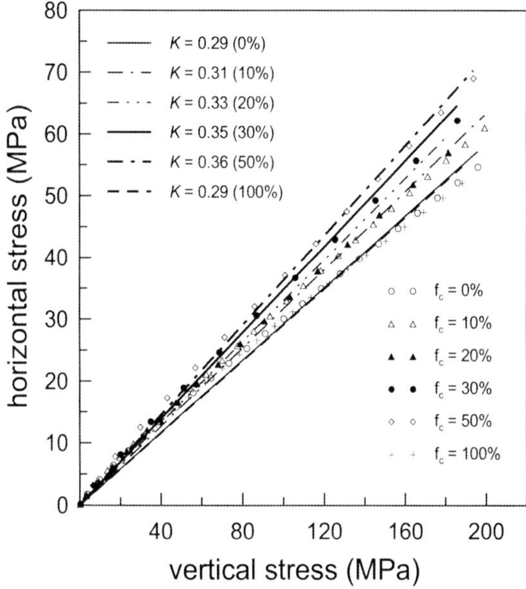

Figure 2 *K_0 - behaviour during one dimensional compression*

To study the contribution of the individual contact forces to the overall stress tensor of the whole system, we applied the formulation of the ensemble average stress tensor[6] σ_{ij} as:

$$\sigma_{ij} = \frac{1}{V}\sum_1^C (R_1 + R_2)Nn_i n_j + \frac{1}{V}\sum_1^C (R_1 + R_2)Tn_i t_j \tag{2}$$

where C is the total number of ball-ball contacts, V is the volume of the sample, R_1 and R_2 are the radii of two spheres in contact, N and T are the magnitudes of the normal and the tangential contact forces, n_i and t_i are the normal unit vector and the tangential vector to the contact plane respectively. In quasi-static simulations, calculating the stress tensor from the internal contact forces and from the boundary forces would yield similar results.[7] For the mixture materials, the total number of contact C in Equation (2) can be presented as the summation of the numbers of three different contact types defined as C_{b_b},

C_{b_s}, C_{s_s} where the subscripts b(large) and s(small) specify the size of two contacting entities (e.g., C_{b_s} is the number of sand-silt contacts in the system). Consequently, the stress tensor σ_{ij} in Equation (2) can also be presented as the summation of three contact-type-related stress tensors as:

$$\sigma_{ij} = \sigma_{ij}^{b\_b} + \sigma_{ij}^{b\_s} + \sigma_{ij}^{s\_s} \tag{3}$$

Calculating the overall stress tensor (σ_{ij}) and the three component stress tensors ($\sigma_{ij}^{b\_b}, \sigma_{ij}^{b\_s}, \sigma_{ij}^{s\_s}$), we can determine their corresponding deviatoric stress values $\sigma_d = \sigma_1 - \sigma_3$ where the principal stress components (σ_1 and σ_3) are approximately the vertical and the horizontal stresses. Furthermore, to consider the cumulative contribution of contact forces of different magnitudes to the deviatoric stress, we considered contacts with normal force smaller than a cut-off value. The deviatoric stress calculated using only these contact forces were plotted on the vertical axis against the horizontal axis of the normalized normal force by the mean normal force of the whole system, $<f_n>$. One of such plots for the case of pure sand ($f_c = 0\%$) is shown in Figure 3 for the vertical stress level of 5 MPa; the range of the abscissa shows that the maximum normal force in the system is more than 6 times higher than the average contact force. The ordinate at any point on the curve gives the value of the deviatoric stress calculated using only contacts with normal force less than the abscissa times the average force.

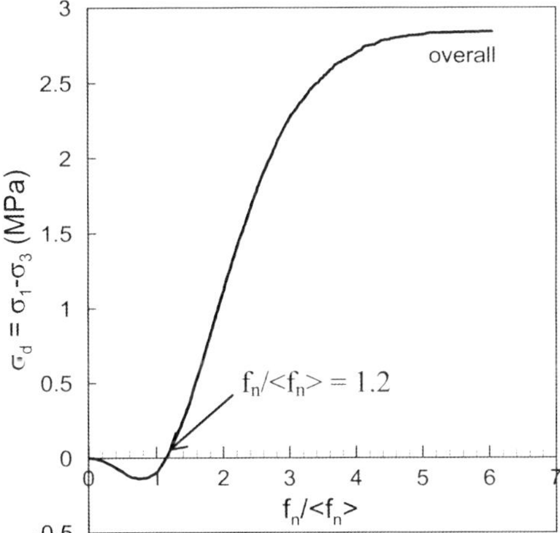

Figure 3 *Cumulative contribution of the contact network to the overall deviatoric stress ($f_c = 0\%$)*

Figure 3 shows that the contacts with normal force less than $1.2*<f_n>$ contribute negatively to the deviatoric stress; these are weak forces[1,2] acting in the horizontal direction, whereas those with a magnitude satisfying $f_n \geq 1.2*<f_n>$ contribute primarily to the positive deviatoric stress, and are hence the load-bearing strong forces described in the literature.[1,2] Figure 4 shows particles transmitting at least one strong force that are located within a vertical 1.2 mm thin slice at the centre of the sample; the picture gives an approximate 2D image of the strong force chains of spherical particles percolating from the top to the bottom walls, as similar to the simulation results based on two dimensional disks by Radjai *et al.*[1] The line connecting particles in Figure 4 describes the transmitted normal forces, with its thickness indicating their relative magnitudes. The force value of $1.2*<f_n>$ can be considered as the characteristic force that differentiates the strong from the weak forces. The number of strong forces was found to be approximately 30% of the total number of contacts in the system, which agrees with the results by others.[2]

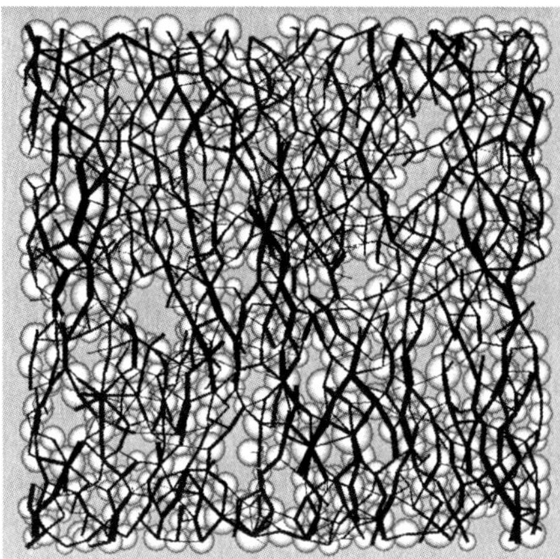

Figure 4 *Particles transmitting normal forces $\geq 1.2*<f_n>$ within the thin central vertical slice ($f_c = 0\%$)*

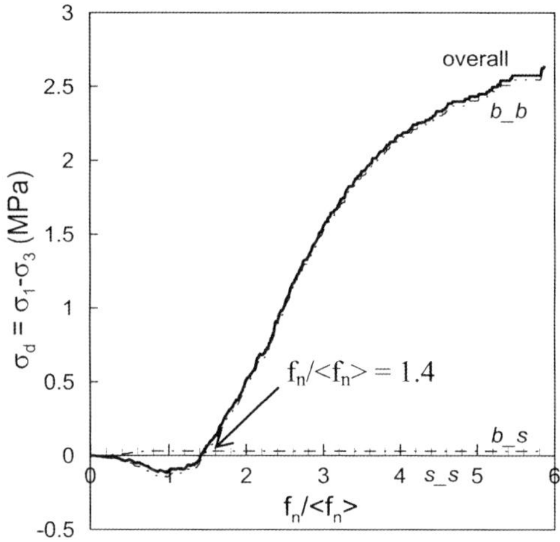

Figure 5 *Cumulative contribution of the contact networks to the overall deviatoric stress ($f_c = 10\%$)*

Figure 5 and Figure 6 show the results of the deviatoric stress contribution and the particles transmitting strong force for the case of 10% silt content when $\sigma_v = 5$ MPa. The overall deviatoric stress is mostly due to the contribution of the *b_b* contact network as compared to the negligible contributions of the other two *b_s* and *s_s* networks. This result is consistent with Figure 6 where only sand particles are shown to transmit strong force. Although silt particles exist in the system, they are lost in the void space and are not involved in the load-bearing mechanism. The characteristic force in this case is equal to $1.4*<f_n>$. For the case of $f_c = 20\%$ (not shown), the characteristic force is equal to $0.6*<f_n>$; in this case, the b_s network has a slightly higher contribution to the overall deviatoric stress, but it is still insignificant as compared to the contribution of the b_b network. The behaviour of the mixture is hence still dominated by the larger sand particles. This agrees with the experimental study of sand-silt mixture[8] where it was found that for the silt content less than a transitional value, the mixture behaviour is dominated by the sand particles.[9] The transitional silt content was obtained when the mixture had the maximum packing efficiency in terms of a maximum solid fraction; the silt particles were densely packed within the void space of the larger sand grains. Note that in Figure 1, the highest initial solid fraction was obtained when $f_c = 30\%$, which can be considered as the transitional silt content for the DEM samples in this study.

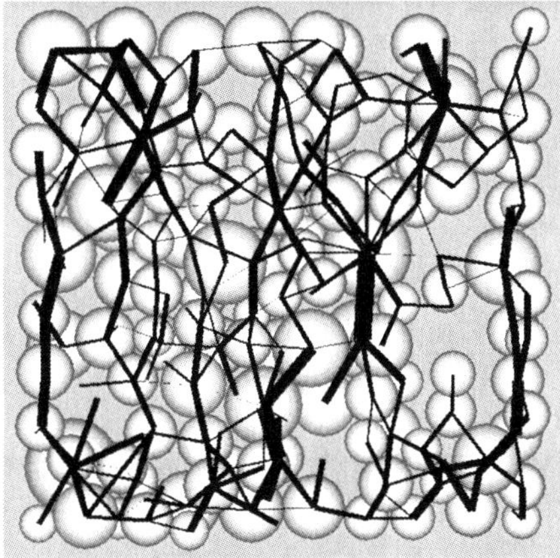

Figure 6 *Particles transmitting normal forces $\geq 1.4*<f_n>$ within the thin central vertical slice ($f_c = 10\%$)*

For the case of 30% silt content under the same vertical stress level, Figure 7 shows a very different picture about the contribution of different contact networks to the overall deviatoric stress. With the optimal packing efficiency, both the b_b and the b_s networks have a comparable contribution, and for the first time, the s_s network also has a visible contribution to the deviatoric stress. As different contact types were involved in the force transmission and the contact stiffness was scaled with the size of the contacting entities (Equation (1)), the range of the magnitude of the contact forces varies greatly, and the maximum force in the system can be as large as over 150 times that of the average contact force. A blown-up plot (not shown) of the curves in Figure 7 showed however, that the characteristic force - where the overall curve crosses the horizontal axis - is only equal to $1.2*<f_n>$. This is because at $f_n/<f_n> = 1.2$, a positive contribution of the s_s network is balanced by a negative contribution of the b_s network which results in the neutral contribution. The contribution of the b_b network in this range is negligible, the crossing point of this curve is delayed until $f_n/<f_n>$ is equal to about 20, as shown in Figure 7. Figure 8 shows that the force transmission through this system goes through a combined network of both large and small particles. Some percolated large-large force chains still exist in the system, but many others are diffused into the smaller contact forces belonging to the b_s and the s_s networks.

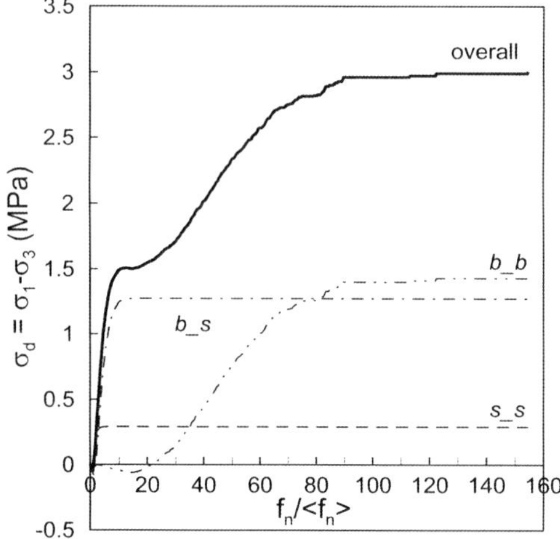

Figure 7 *Cumulative contribution of the contact networks to the overall deviatoric stress ($f_c = 30\%$)*

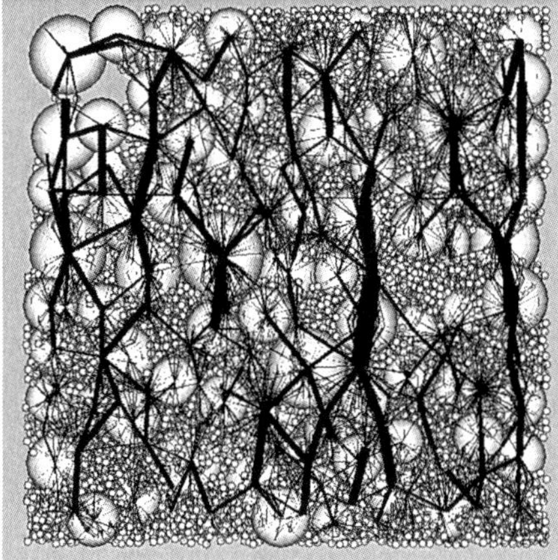

Figure 8 *Particles transmitting normal forces $\geq 1.2*<f_n>$ within the thin central vertical slice ($f_c = 30\%$)*

We found that the observed behaviour in the case of $f_c = 30\%$ can be linked to the connectedness of the different contact networks.[9] As the contacts of all three contact types are abundant in this system, their networks are well connected and all can transmit forces. In reality, the shape of the contribution curve of each contact network is similar, the respective crossing point from negative to positive contribution however is different for different networks, and this value can be considered as the individual characteristic force related to each contact type. Furthermore, the percentage of the strong contacts in each contact network is about the same 30% of the total number of contacts of that network, which is similar to the case of uniform assemblies with only one contact type. Simulation results at a higher stress level show that the contact-type-related characteristic force may depend on stress level, especially for the sand-dominated systems ($f_c < 30\%$) where more small particles can be recruited into the force transmission at a higher stress level, and hence their corresponding b_s and s_s networks can develop a greater relative contribution to the overall deviatoric stress.

4 CONCLUSIONS

We studied the force transmission of imperfect binary mixtures under one-dimensional compression. Three types of particle-particle contacts were categorized depending on the size of the two spheres in contact. The contribution of each contact type network to the deviatoric stress was calculated, and the simulation results showed that these contributions were dependent on the particle size distribution of the mixture (or the silt content), and the stress level. Silt particles were not involved in the strong force transmission for the sample of 10% silt content, and the overall deviatoric stress was primarily due to the contact network between sand-sand particles. For the sample of the maximum packing efficiency when $f_c = 30\%$, all three contact networks transmitted forces and were together involved in the shear-strength related load-bearing mechanism. The internal view of the sample showed that the extremely high magnitude of contact forces between sand particles were diffused into the many smaller magnitudes of b_s and s_s contact forces. This means that the force percolation from the top to the bottom wall goes through a combined network of both large and small particles. The characteristic force, which differentiates strong forces from weak forces, depends on the contact-type, PSD and the stress level. The value of the overall characteristic force however is quite close to the mean force of the system, although the range of variation in the force magnitude of the mixture systems can be much greater than that in the uniform to slightly polydisperse systems.

Acknowledgements

The authors gratefully acknowledge the financial support from EPSRC (Grant No. EP/F036973/1) and thank Dr. Colin Thornton for his valuable suggestions for this work.

References

1 F. Radjai, D. Wolf, M. Jean and J.J. Moreau, *Physical Review Letters*, 1998, **80**, 61.
2 C. Thornton and S.J. Antony, *Philosophical Transactions of the Royal Society A: Mathematical, Physical and Engineering Sciences*, 1998, **356**, 2763.
3 A. Drescher and G. de Josselin de Jong, *J. Mech. Phys. Solids*, 1972, **20**, 337.
4 P.A. Cundall and O.D.L. Strack, in *Mechanics of Granular Materials: New Models and Constitutive Relations*, edited by J.T. Jenkins and M. Satake, 1983.
5 Itasca, *Particle flow code in 3 dimensions (PFC3D) version 4*, Minnesota, USA, 2008.
6 C. Thornton and L. Zhang, *Géotechnique*, 2010, **60**, 333.
7 X. Tu and J.E. Andrade, *International Journal for Numerical Methods in Engineering*, 2008, **75**, 1581.
8 P.V. Lade and J.A. Yamamuro, *Canadian Geotechnical Journal*, 1997, **34**, 918.
9 N.H. Minh and Y.P. Cheng, *Géotechnique*, 2012, in press.

VERIFICATION OF THE DOUBLE SLIP AND ROTATION RATE MODEL FOR
ELLIPTICAL GRANULAR FLOW USING THE DISTINCT ELEMENT METHOD

L.Q. Li,[1,2] M.J. Jiang[1,2] and Z.F. Shen[1,2]

1. Key Laboratory of Geotechnical and Underground Engineering of Ministry of
Education, Tongji University, Shanghai 200092,China;
2. Department of Geotechnical Engineering, Tongji University, Shanghai 200092,China

1 INTRODUCTION

Non-coincidence of the principal stress tensor and the principal plastic deformation rate
tensor is called non-coaxiality in geomechanics. From the theoretical and practical points
of view, non-coaxial models for granular materials are particularly important since
granular materials show evident non-coaxial behavior in a variety of conditions.[1-2] The
non-coaxiality is an important feature of numerous plasticity models describing plastic
flow by means of kinematic theories.[3-10] The averaged micro-pure rotation rate (APR) have
been deduced based on a micro-analysis of the kinematics of disc particles in contact, and
the double slip and rotation rate model (DSR^2 model) have been proposed by introducing
the variable APR into the unified double-sliding plasticity model.[8-10] from the observation
and analysis in the undrained simple shear tests using the discrete element method (DEM)
code NS2D, it has been concluded that the DSR^2 model could be used to depict the
non-coaxiality of circular particle assemblages.

However, considering that DEM simulations of circular particle assemblies lead to
typically low internal friction angles compared with real sands, and the particle shape of
sands is mainly non-circular, there is a need to further study whether the DSR^2 model is
applicable to non-circular particle assemblages. For this purpose, first, the averaged
micro-pure rotation rate (APR) based on the kinematics of elliptical particles in contact
have been deduced, then a developed DEM code NS2D was used to generate assemblies
that were composed of elliptical particles with aspect ratios A_m= 1.4 and 1.7, respectively,
and then the assemblies were subjected to undrained simple shear tests to verify the DSR^2
model. The work in this paper is carried out under 2-D planar strain condition.

2 APR OF ELLIPTICAL PARTICLES

We shall consider a 2-D assembly of rigid ellipses incrementally deforming during a time
interval from t to $t+dt$. In the time interval, two elliptical particles 1 and 2 are assumed to
remain in contact as shown in Figure 1. At time t, let o_1 and o_2 denote the ellipse centers,
R_1^t and R_2^t denote the contact radius. $\dot{u}_{1i}^t$, $\ddot{u}_{1i}^t$, $\dot{u}_{2i}^t$ and $\ddot{u}_{2i}^t$ denote the translational
velocity and acceleration at the centre of elliptical particles 1 and 2, respectively. $\dot{\theta}_1^t$, $\ddot{\theta}_1^t$,

$\dot{\theta}_2^t$ and $\ddot{\theta}_2^t$ denote the rotational velocity and acceleration of elliptical particles 1 and 2, respectively. Take counterclockwise rotation as positive. At time t, let $\mathbf{n}_1^t$ and $\mathbf{n}_2^t$ denote two unit vectors, which are normal and tangential to the common plane of the contact c that could be obtained using the Common Normal Method.[11] Let α^t denote the angle between the common plane and the x-axis, $\dot{\alpha}^t$ denote its corresponding rotation rate. r_1^t and r_2^t are the corresponding rotation radius of ellipse 1 and 2 at the contact c, which can be derived by

$$r_1^t = R_1^t \cos \beta_1^t \tag{1a}$$
$$r_2^t = R_2^t \cos \beta_2^t \tag{1b}$$

where β_1^t and β_2^t are the included angle as shown in Figure 1a.

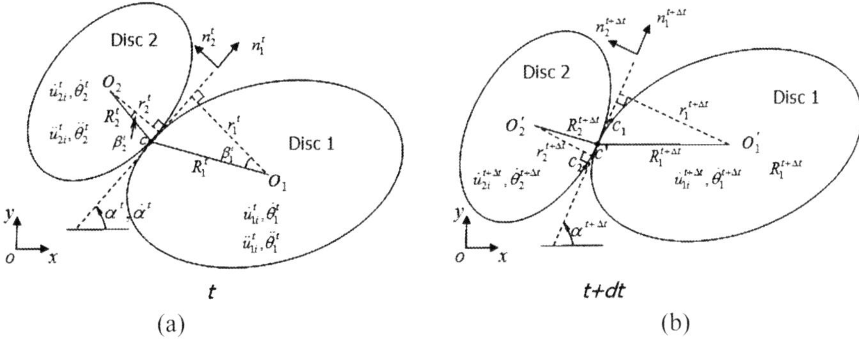

Figure 1 *Kinematics of two ellipses in contact at times t to t+dt*

At time $t+dt$, let o_1' and o_2' denote the current ellipse centers, c' denote the current contact. The arcs c_1c' and c_2c' represent the displacement of the contact c on ellipse 1 and 2 during dt, respectively. Summation of the two arcs yields the relative sliding distance of the two ellipses and can be obtained by

$$\Delta u_s = \int_t^{t+\Delta t} [(\dot{u}_{1i} - \dot{u}_{2i})n_{2i} - (\dot{\theta}_1 r_1 + \dot{\theta}_2 r_2)]dt \tag{2}$$

where $i = 1, 2$ represents x and y direction, respectively, and the tangential vector $n_2 = (\cos\alpha, \sin\alpha)$. The Einstein's summation convention is adopted here.

Let $\dot{u}_s = (\dot{u}_{1i} - \dot{u}_{2i})n_{2i} - (\dot{\theta}_1 r_1 + \dot{\theta}_2 r_2)$, according to the mean value theorem for integrals, Equation (2) can be rewritten as

$$\Delta u_s = \dot{u}_s^{t+k\Delta t} \Delta t \tag{3}$$

where $0 < k < 1$.

According to Taylor series expansion, $\dot{u}_s$ at $t+dt$ can be expressed as

$$\ddot{u}_s^{t+k\Delta t} = \dot{u}_s^t + \ddot{u}_s^t k\Delta t \tag{4}$$

where the second and higher orders are neglected. In Equation (4), $\ddot{u}_s^t$ can be obtained by

$$\ddot{u}_s^t = (\ddot{u}_{1i}^t - \ddot{u}_{2i}^t)n_{2i}^t + (\dot{u}_{1i}^t - \dot{u}_{2i}^t)\dot{n}_{2i}^t - (\ddot{\theta}_1^t r_1^t + \ddot{\theta}_2^t r_2^t + \dot{\theta}_1^t \dot{r}_1^t + \dot{\theta}_2^t \dot{r}_2^t) \tag{5}$$

$$\dot{u}_{1i}^{t+\Delta t} = \dot{u}_{1i}^t + \ddot{u}_{1i}^t \Delta t \tag{6a}$$
$$\dot{u}_{2i}^{t+\Delta t} = \dot{u}_{2i}^t + \ddot{u}_{2i}^t \Delta t \tag{6b}$$

$$\dot{\theta}_1^{t+\Delta t} = \dot{\theta}_1^t + \ddot{\theta}_1^t \Delta t \tag{7a}$$
$$\dot{\theta}_2^{t+\Delta t} = \dot{\theta}_2^t + \ddot{\theta}_2^t \Delta t \tag{7b}$$

$$r_1^{t+\Delta t} = r_1^t + \dot{r}_1^t \Delta t \tag{8a}$$
$$r_2^{t+\Delta t} = r_2^t + \dot{r}_2^t \Delta t \tag{8b}$$

where $\dot{r}_1^t$ and $\dot{r}_2^t$ are the change rate of rotation radius of ellipses 1 and 2 at t, respectively. $\dot{\mathbf{n}}_2^t$ is the change rate vector of $\mathbf{n}_2^t$,

$$\dot{\mathbf{n}}_2^t = \dot{\alpha}^t(-\sin\alpha^t, \cos\alpha^t) = \frac{\alpha^{t+\Delta t} - \alpha^t}{\Delta t}(-\sin\alpha^t, \cos\alpha^t) \tag{9}$$

Based on Equations (4)-(9), Equation (3) can be rewritten as

$$\begin{aligned}
\Delta u_s &= \dot{u}_s^t \Delta t + \ddot{u}_s^t k\Delta t^2 \\
&= [(\dot{u}_{1i}^t - \dot{u}_{2i}^t)n_{2i}^t - (\dot{\theta}_1^t r_1^t + \dot{\theta}_2^t r_2^t)]\Delta t + [(\dot{u}_{1i}^{t+\Delta t} - \dot{u}_{1i}^t) - (\dot{u}_{2i}^{t+\Delta t} - \dot{u}_{2i}^t)]n_{2i}^t k\Delta t + \\
&\quad [(\dot{\theta}_1^{t+\Delta t} - \dot{\theta}_1^t)r_1^t + (\dot{\theta}_2^{t+\Delta t} - \dot{\theta}_2^t)r_2^t]k\Delta t + [(\dot{u}_{1i}^t - \dot{u}_{2i}^t)\dot{n}_{2i}^t + \frac{r_1^{t+\Delta t} - r_1^t}{\Delta t}\dot{\theta}_1^t + \frac{r_2^{t+\Delta t} - r_2^t}{\Delta t}\dot{\theta}_2^t]k\Delta t^2
\end{aligned} \tag{10}$$

It is known that the friction at inter-particle contacts is a source of elastic/dissipated energy, which is associated physically with Δu_s, for granular materials.[10] Here, we define a variable, the sliding rotation $\Delta\theta^s$, as follows:

$$\Delta\theta^s = \frac{\Delta u_s}{r} \tag{11}$$

where $r = 2r_1 r_2 / (r_1 + r_2)$.

Dividing Equation (11) with dt and taking the limit as $dt \to 0$, we can obtain the corresponding sliding rotation rate $\dot{\theta}^s$

$$\dot{\theta}^s = \lim_{\Delta t \to 0} \frac{\Delta\theta_s}{\Delta t} = \frac{1}{r}[(\dot{u}_{1i} - \dot{u}_{2i})n_{2i} - (\dot{\theta}_1 r_1 + \dot{\theta}_2 r_2)] \tag{12}$$

We also define the second kinematic variable, the pure rotation rate $\dot{\theta}^p$, as follows:

$$\dot{\theta}^p = \frac{1}{r}(r_1\dot{\theta}_1 + r_2\dot{\theta}_2) \tag{13}$$

Equation (12) could hence be expressed as

$$\dot{\theta}^s = -\dot{\theta}^p + \frac{1}{r}(\dot{u}_{1i} - \dot{u}_{2i})n_{2i} = -\dot{\theta}^p + \dot{\alpha} \tag{14}$$

Equation (14) indicates that the sliding rotation rate $\dot{\theta}^s$ at a contact is composed of two terms. The first one is associated to particle rotations and particle sizes and the second one is associated to the translational motion of particle center. Conventionally, the motion and deformation of the granular material, which is regarded as continuum, can be determined solely by the velocity and/or displacement, and is hence represented by $\dot{\alpha}$.[12] The effect of particle slip and rotation on energy dissipation is thus neglected in continuum mechanics. It is believed that the resistance at a contact includes rolling resistance and sliding resistance, and only sliding resistance was focused by considering that the rolling resistance at contacts in the assembly is very small.[10] Therefore, the effect of sliding resistance $\dot{\theta}^p$ on energy dissipation is taken into account in this study.

If the variable indicating the elastic/dissipated energy for the ellipse contacts in the assembly is defined as the averaged pure-rotation rate (APR), denoted by ω_3^c, it can be expressed as

$$\omega_3^c = \frac{1}{N}\sum_{k=1}^{N}\dot{\theta}^p = \frac{1}{N}\sum_{k=1}^{N}\left[\frac{1}{r^k}\left(\dot{\theta}_1^k r_1^k + \dot{\theta}_2^k r_2^k\right)\right] \tag{15}$$

where the summations are over the N contacts in the assemblies.

3   THE DOUBLE SLIP AND ROTATION RATE MODEL

For completeness, some formulae proposed by Harris[6] are reproduced here. The unified kinematic equations governing the velocity field for the incompressible materials are

$$(d_{11} + d_{22}) = 0 \tag{16a}$$

$$2(\vartheta + w_{12})\sin\varphi = (d_{11} - d_{22})\sin 2\alpha_\sigma - 2d_{12}\cos 2\alpha_\sigma \tag{16b}$$

where ϑ is an angular velocity which may be given different physical interpretations in different kinematic models, φ is the angle of internal friction, and α_σ is defined as the angle of major principal stress axis. d_{ij} and w_{ij} are the deformation rate tensor and the spin tensor as shown in Equation (17).

$$d_{ij} = \frac{1}{2}\left(\frac{\partial v_i}{\partial x_j} + \frac{\partial v_j}{\partial x_i}\right), \quad w_{ij} = \frac{1}{2}\left(\frac{\partial v_i}{\partial x_j} - \frac{\partial v_j}{\partial x_i}\right) \qquad (17)$$

where v_i denotes the velocity components, with i or j=1, 2 in the 2-D case.

In the DSR^2 model, $\vartheta = \omega_3^c$, i.e. the APR, which can be obtained from Equation (15) by numerical experiments. In the next section, we shall verify the DSR^2 model using DEM simple shear tests on elliptical particle assemblies with aspect ratio A_m=1.4 and 1.7, respectively.

4   DEM VERIFICATION OF THE DSR^2 MODEL

4.1 DEM Numerical Experiments

The developed DEM code NS2D[13], which is in essence similar to that proposed by Cundall and Strack,[14] was used to carry out numerical experiments to verify the DSR^2 model. The two kinds of elliptical particle assemblies with A_m=1.4 and 1.7 respectively, used in the experiments, have the same distribution of particle size, and their grading curve is expressed as the geometric mean of the long and short axes of elliptical particles as shown in Figure 2a. Undrained simple shear tests were conducted for the verifications, in which the DEM samples are regarded to be 'incompressible', i.e. the volume change is zero. In this paper, the incompressibility is targeted by two requirements: the 'material requirement' by selecting the samples that can satisfy the condition of perfect plasticity with the contractancy/dilatancy close to zero, and the 'test requirement' in which the volumetric change is zero. These targets can be achieved by using several DEM samples of different compactness, such as loose, medium-dense and dense samples, which provide the upper and lower bounds for incompressible materials. And these samples were prepared by the Under-Compaction technique (UCM) proposed by Jiang *et al.*[15]

Two stages in the simple shear tests are carried out on the DEM samples after they are generated: consolidation and simple shear stage. At the consolidation stage, the samples, which are allowed to deform vertically, were vertically loaded and kept stable under σ_v =200 kPa/400 kPa. At the shear stage shown in Figure 2b, the top and bottom walls were vertically fixed but moved horizontally by following the left and right walls rotating with the rotation rate $\dot{\theta}$. Hence, the volume of these DEM samples keeps constant during the shear stage. Monotonic and cyclic simple shear tests have been conducted with $\dot{\theta}$ controlled respectively by

$$\dot{\theta} = \dot{\theta}_0 \qquad (18a)$$

$$\dot{\theta} = \dot{\theta}_0 \cos\left(\frac{2\pi t}{T_0}\right) \qquad (18b)$$

where $\dot{\theta}_0$ is the initial rotation rate, t is the rotation time and T_0 is the period.

Table 1 provides all the test and material parameters used in the DEM experiments. 8 groups of undrained simple shear tests were carried out on one kind of elliptical particle assemblies with A_m=1.4 and 1.7 respectively: two monotonic shear tests on the dense samples and the loose samples respectively, one monotonic shear test on the medium-dense sample; one cyclic shear test on the dense, medium-dense and loose sample, respectively.

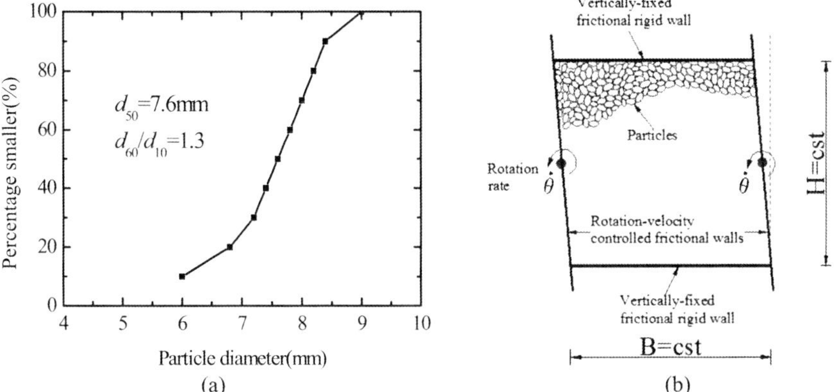

Figure 2 *Distribution of grain size used in DEM verifications (a) and boundary conditions in the simple shear stage (b)*

Table 1 *Test and material parameters used in the DEM verifications*

Specimen size (mm×mm)	*ca.* 325×340
Initial rotation rate $\dot{\theta}_0$ (rad/min) (monotonic, cyclic shear tests)	0.5, 0.15
Period in cyclic simple shear test T(min)	2
Density of particles (kg/m^3)	2, 600
Total number of particles	2032
Incremental time for DEM computation Δt (s)	1.0×10^{-4}
Normal, tangential spring stiffness (N/m)	1.5×10^{9}, 1.0×10^{9}
Desired planar void ratio (dense, medium-dense, loose samples with A_m=1.4)	0.18, 0.20, 0.23
Desired planar void ratio (dense, medium-dense, loose samples with A_m=1.7)	0.16, 0.19, 0.22

4.2. Verification 1

The first verification is that the motion and deformation of bodies in the continuum mechanics is defined only according to the velocity and/or displacement and can be represented by $\dot{\alpha}$ in Equation (14). In the simple shear tests, the velocity field of the DEM samples can be expressed as

$$v_x = -\dot{\theta}\cdot y \tag{19a}$$
$$v_y = 0 \tag{19b}$$

where y is the vertical distance between one point and the rotation center of the side wall, and $\dot{\theta}$ is the rotation rate of the corresponding wall in Equation (18).

In the DEM tests, the average velocity can be obtained based on the band method, which has been described in the literature,[10] and then compared to Equation (19). Figure 3 shows the distributions of the average velocity and the corresponding theoretical values during the monotonic simple shear tests on the medium-dense specimen (A_m=1.4, σ_v=400 kPa). The number of bands in Figure 3 is 5, 7, 9, 11, 13, 15, 17, 19 and 21, respectively. Figure 3a shows that although the number of bands is different, the horizontal average velocity is almost linearly distributed, and is in good agreement with the theoretical values calculated using Equation (19a). Figure 3b demonstrates that the vertical average velocity is basically zero, which is in agreement with Equation (19b). The results in Figure 3 show that Equation (19) can be used to describe the velocity fields defined at the particle centers. This confirms the statement in Section 2 that the motion and deformation of bodies in the continuum mechanics can be described by the motion of grain centers in the discrete granular mechanics.

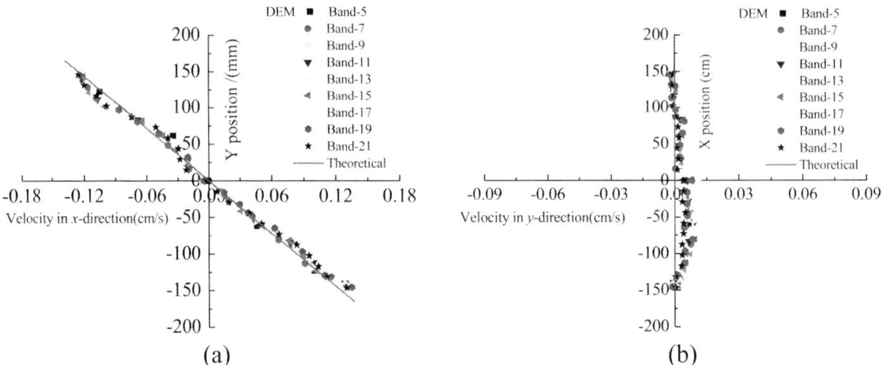

Figure 3 *Distribution of the average velocity in the DEM monotonic shear test on the medium-dense specimen (A_m=1.4) with different numbers of bands: (a) Horizontal velocity; (b) Vertical velocity*

4.3. Verification 2

Using Equations (16b), (17) and (19), the theoretical rotation rates measured by the DSR[2] model in the simple shear tests may be yielded as

$$\vartheta = -\frac{d_{12}\cos 2\psi_\sigma}{\sin\varphi} - w_{12} = \left(\frac{\cos 2\psi_\sigma}{2\sin\varphi} + \frac{1}{2}\right)\dot{\theta} \tag{20}$$

where the internal friction angle φ=30°, the principal stress direction ψ_σ can be measured in the DEM tests, and $\dot{\theta}$ is the rotation rate of the rigid walls defined in Equation (18). Note that the APR can be measured in the DEM tests through Equation (15).

Figure 4 shows the theoretical rotation rates measured, and the APRs predicted by the DSR[2] model for the medium-dense specimen (200 kPa) with A_m=1.4 and 1.7, respectively, and their respective average values for all specimens deduced from all the monotonic DEM tests. Figure 4 indicates that although there is a little difference between the theoretical rotation rates measured and the APRs predicted, there appears to be good agreement between these quantities for all the specimens with aspect ratios A_m=1.4 and 1.7. It should also be mentioned that good agreement is observed between the theoretical values measured and the APR predicted in the DSR^2 model in the tests on all the other specimens of different relative densities. Therefore, it can be concluded that in the monotonic simple shear tests, the variable APR in the DSR^2 model can be used for the prediction of the angular velocity in the kinematic model.

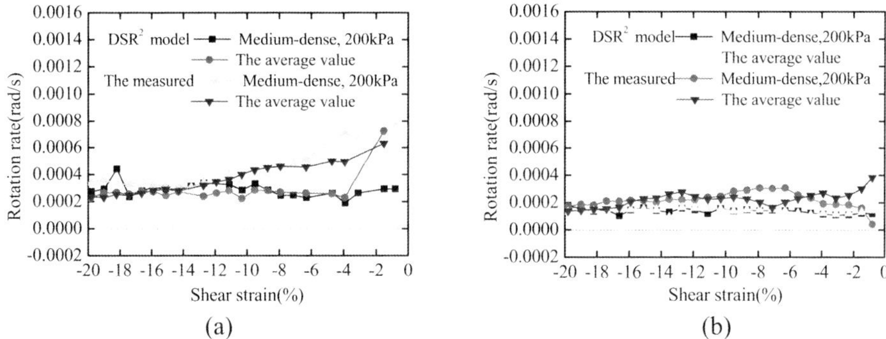

(a)   (b)

Figure 4 *The APRs measured and predicted by the DSR^2 model for the medium-dense specimen (200 kPa), and their respective average values for all specimens in the all monotonic DEM simple shear tests: (a) A_m=1.4; (b) A_m=1.7*

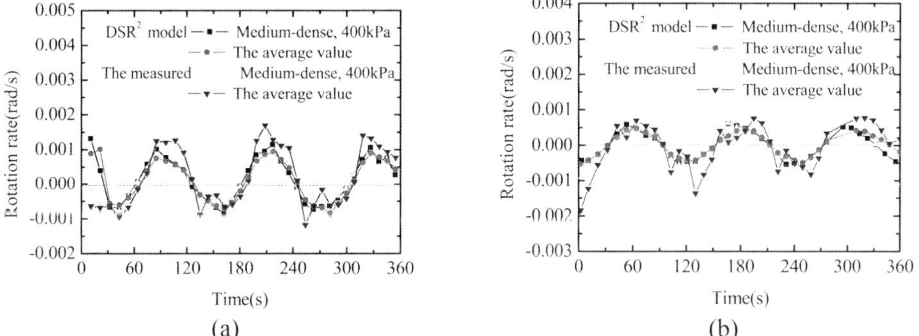

(a)   (b)

Figure 5 *The APRS measured and predicted by the DSR^2 model for the medium-dense specimen (200 kPa), and their respective average values for all specimens in the all cyclic DEM simple shear tests: (a) A_m=1.4 and (b) A_m=1.7*

Figure 5 shows the predicted theoretical rotation rates, and the APRs predicted by the DSR^2 model during the cyclic simple shear tests on the medium-dense specimen (400 kPa) with A_m=1.4 and 1.7, respectively, and their respective average values for all specimens deduced from all the cyclic DEM tests. It can be seen that there is also good agreement between these predicted and measured quantities in the DEM tests. They are both periodic, varying between -0.0015 rad/s and 0.0015 rad/s with the same period during cyclic shear tests on the medium-dense specimen and the average values for all the specimens with A_m=1.4, and varying between -0.0012 rad/s and 0.0012 rad/s for the specimens with

A_m=1.7. Hence, considering the numerical results in the literature,[10] and those presented in Figures 4 and 5, it is clear that the angular velocity in kinematic models could be well predicted with the variable APR in the DSR^2 model in the DEM simple shear test on circular and elliptical particles assemblies.

5. CONCLUSIONS

This paper presents an evaluation of the DSR^2 model using the distinct element method. Strain rate-controlled simple shear tests were carried out on the elliptical particles assemblage with A_m=1.4 and 1.7, respectively, using the developed code NS2D. It can be concluded from the observation and analysis in the monotonic and cyclic DEM experiments that the DSR^2 model can give a good prediction of the angular velocity in the kinematic model, and the variable APR in the DSR^2 model has been deduced based on the relative motion of contact of elliptical particles, and is a rational and important parameter in the kinematic model.

References

1   K. Ishihara, I. Towhata, *Soils Found.*, 1983, **23**, 11.
2   M. Gutierrez, K. Ishihara and I. Towhata, *Soils Found.*, 1991, **31**, 121.
3   Mandel. J. Sur les lignes de glissement et le, *Comptes Rendus del' Academie des Sciences*, 1947, **225**, 1272.
4   De Josselin de Jong G., *Geotechnique*, 1971, **21**, 155.
5   A. J. M. Spencer, *J. Mech Phy Solids*, 1964, **12**, 337.
6   D. Harris, *Proceedings of the Royal Society A*, 1995, **450**, 37.
7   S. Nemat-Nasser, *J. Mech Phy Solids*, 2000, **48**, 1541.
8   M.J. Jiang, D. Harris and H.S. Yu, *Int. J. Numer. Anal. Geomech*, 2005, **29**, 643.
9   M.J. Jiang, D. Harris and H.S. Yu, *Int. J. Numer. Anal. Geomech*, 2005, **29**, 663.
10 M.J. Jiang, D. Harris and H.S. Yu, *Mech Res Commun*, 2006, **33**, 651.
11 X. Lin and T.T. Ng, *Int. J. Numer. Anal. Geomech*, 1995, **19**, 653.
12 S. Khan and S.J. Huang, *Contimuum Theory of Plasticity*, John Wiley & Sons, New York, 1995.
13 L.Q. Li, M.J. Jiang and X.F. Wu, *Rock and Soil Mechanics*, 2011, **32**, 713. (in Chinese)
14 P.A. Cundall and O.D.L. Strack, *Geotechnique*, 1979, **29**, 47.
15 M.J. Jiang, J.M. Konrad and S. Leroueil, *Comput Geotech*, 2003, **30**, 579.

MICROMECHANICS OF SEISMIC WAVE PROPAGATION IN GRANULAR MATERIALS

J. O'Donovan[1], C. O'Sullivan[1] and G. Marketos[1]

[1]Department of Civil & Environmental Engineering, Imperial College, London, SW7 2AZ

1 INTRODUCTION

Understanding seismic wave propagation is important in geotechnical engineering, for example the shear wave velocity can be used to determine the soil stiffness. Typically the continuum theory of body wave propagation is used to model this wave propagation even though granular materials are inherently discrete. There are two types of seismic body waves, compressional and shear waves. The speed of the compressional wave is a function of the constrained modulus, M, of the material. The shear wave speed, Vs, is a function of the shear modulus, G, of the material. Both M and G are functions of the Young's modulus, E, and the Poisson's ratio, v.   Seismic waves are used to infer soil parameters both in the field and laboratory. By propagating the waves in different directions and with different polarisations the degree of anisotropy of soil stiffness can be measured. Bender elements are used to generate seismic waves through laboratory samples, such as a triaxial cell.[1] Prior research has shown how bender elements can be used to ascertain the complete cross-anisotropic stiffness matrix for a pluviated sand sample.[2] The wave that is recorded at the receiver bender element is very different to the wave transmitted complicating clear identification of the arrival of the shear and compressional waves. This is due to the geometric spreading of the wave from the transmitter, reflections off the boundaries and dispersion of the wave due to energy loss during propagation. This leads to uncertainty regarding the V_s values obtained from bender element tests and to variation of results for identical samples depending on the technique used to estimate the travel time. As a consequence of this uncertainty, bender element tests are often coupled with small strain stress probes to guide the interpretation of values from received signal traces. As wave propagation in granular material is composed of a series of individual particle-particle interactions a discrete element method (DEM) simulation is ideal to isolate the particle scale fluctuations in stress and velocity which arise due to the propagation of the seismic wave. The commercial DEM program used in this research is Particle Flow Code 3D (PFC^{3D}) Version 4.0 available from Itasca Consulting Engineering.[3]

2 DISCRETE ELEMENT METHOD SIMULATION

The granular sample considered here is a dense, regularly packed sample with monosized particles. This simple simulation was chosen to act as a bridge between previous continuum studies and this novel discrete study. The simplicity also leads to easier interpretation of the results and a framework can be built up which will be used to guide the examination of more complex, randomly packed samples which are currently being analysed.

The particle size is 1.10 mm and the packing is face-centred cubic, where the majority of the particles contact 12 other particles with the exception of the outer particles. The average coordination number of the sample is therefore 11.54. The sample size is 13.20 mm in both the x and y directions and 12.44 mm in the z direction and is composed of 2,673 particles. The complete sample properties are listed in Table 1 and include the inputs used in the Hertz-Mindlin contact model considered here.

Initially, in what can be considered to be the sample preparation step, the particles were specified to have no friction. The sample boundaries were rigid walls and these were used with a servo-control algorithm to isotropically compress the sample to 100 kPa. A value for friction of 0.35 was added to the simulation at this point. At this point the walls were also removed and a stress controlled boundary was applied.[4] Using this numerical membrane, an area was assigned to each of the outer particles in the sample, which was then multiplied by the desired stress, 100 kPa, and the resulting force was applied to the centroid of the particle considered. The direction of the force was orthogonal to the principal axis and towards the centre of the sample. The sample was allowed to settle as the change in boundary conditions resulted in a brief, small amplitude dynamic oscillation in the stresses in the sample. The resulting sample was a regularly packed sample isotropically compressed to 100 kPa and contained by flexible boundaries. This is analogous to the flexible boundary present in the true triaxial cubical cell apparatus (CCA).[2]

Table 1 *The sample properties used for a representative DEM simulation*

Particle Size	1.10 [mm]
Particle Density	2.57×10^{-3} [g/mm^3]
Interparticle Friction	0.35 [-]
Contact Model	Hertz-Mindlin
Particle Shear Modulus [G]	28.69×10^9 [Pa]
Particle Poisson's Ratio [v]	0.22 [-]
Viscous Damping at Contacts	0.10 [-] (reducing to 0.01 [-] for BE test)
No. of Particles	2673 [-]

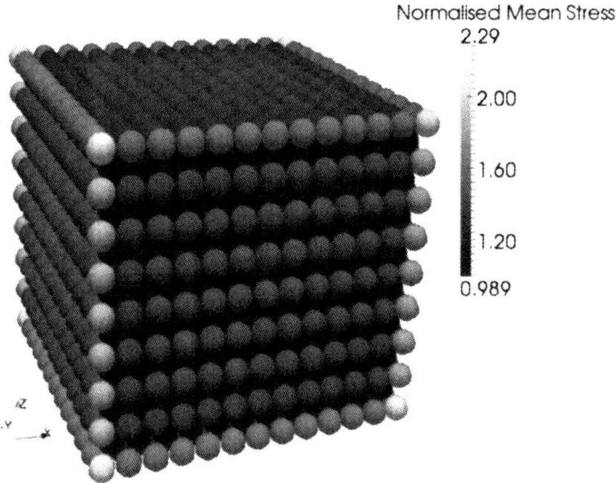

Figure 1 *Illustration of DEM test sample composed of monosized particles in a face-centred cubic packing*

To create a seismic wave in a soil sample a disturbance must be created. In a laboratory test this is done using a piezoceramic bender element. In the simulation presented here a disturbance was created by translating an individual particle. This particle was chosen to be mid-way along the z-axis, mid-way along the y-axis and near zero on the x-axis as shown on Figure 2. A particle inside the boundary particles was chosen as a transmitter as the bender elements would be inserted into the sample a given penetration length. This approach has been used several times before in DEM analysis of bender elements.[5,6,7] A shear wave that propagated in the x-direction and oscillated in the y-direction was simulated. The shear wave was chosen as this is repeatedly reported in the literature as a difficult wave to determine the travel time. The frequency of the bender element (100 kHz) was chosen to be quite high to create a seismic wave of low wavelength. This keeps the value of R_d high (R_d is the number of full shear waves occurring in the sample before arrival). It has been shown that a high value of R_d leads to better quality received signal and avoids near field effects.[8,9] The amplitude of the sine wave was also specified to be very low and was 0.000125 mm. Different amplitudes were simulated and the largest amplitude that did not produce a change in coordination number or significant particle slippage was chosen. Although there was some slippage around the transmitter particle during transmission all contacts were regained and there were no permanent plastic deformations. Figure 2 also illustrates the position of the receiver particle. The displacement of the receiver particle was tracked at a high sampling rate in order to obtain a clear and highly sampled received signal.

3 WAVE PROPAGATION

The propagation of the disturbance was tracked through the simulation using particle scale metrics uniquely available in DEM. As in the previous two-dimensional analysis[5] the particle velocities and representative shear stresses were considered. The representative shear stresses were taken to be as in Equation 1.

$$s^p = \frac{\overline{\sigma}^p_{ij}}{2} = \frac{\overline{\sigma}^p_{xy} + \overline{\sigma}^p_{yx}}{2} \tag{1}$$

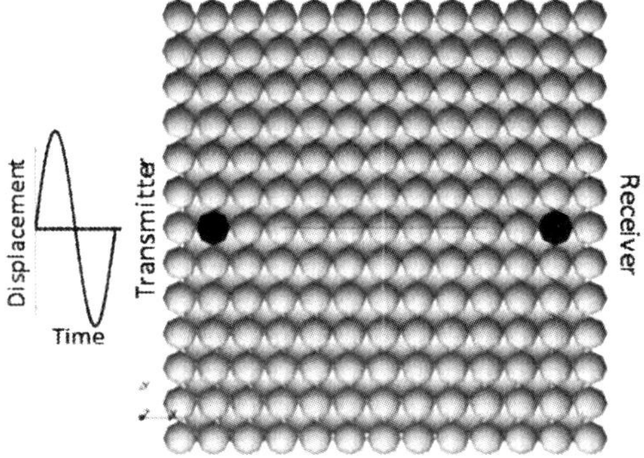

Figure 2 *Position of transmitter and receiver particles for bender element test in DEM simulation*

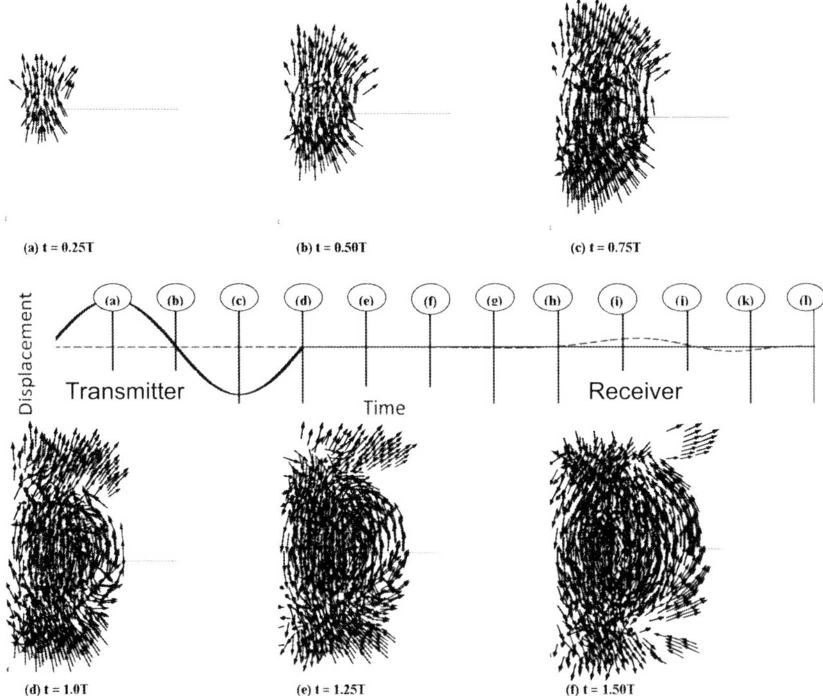

Figure 3 *Particle velocity vectors for time points (a) to (f) (8.33 and 50 µs). Velocities with a magnitude less than 1 mm/s are discarded*

Figure 3 illustrates the individual velocities of each particle in the system. The velocity vectors are represented as arrows which are located at the particle centroid and only appear once their magnitude is greater than 1.0 mm/s. This threshold removes smaller velocities which would cloud the system response to the initial transmitter particle motion. Figures 3a, 3b and 3c show that the movement of the transmitter particle causes particles to move with the transmitter particle on all sides of the particle. A three-dimensional response to a one-dimensional motion was observed when examined in an elevation view. Figures 3d and 3e illustrate the formation of the central shear wave lobe, as predicted previously.[10] The shear wave is observed as the propagation of the wave is perpendicular to the particle motion indicated by the direction of the velocity arrows. From Figure 3d to 3e the shear wave is seen to grow in size as a central lobe in the system. Again it is observed as a lobe that expands in all directions. Figure 3f shows the formation of the compressional wave at the sides of the sample. The compressional wave is indicated by the particles which have velocity arrows that are parallel to the direction of the wave propagation, i.e. parallel to the x-axis. The central shear wave lobe is also seen to be propagating towards the receiver disk. By the time that Figure 3f is captured the transmitter particle has stopped moving, however, there is still motion around the transmitter particle due to the elasticity of the system.

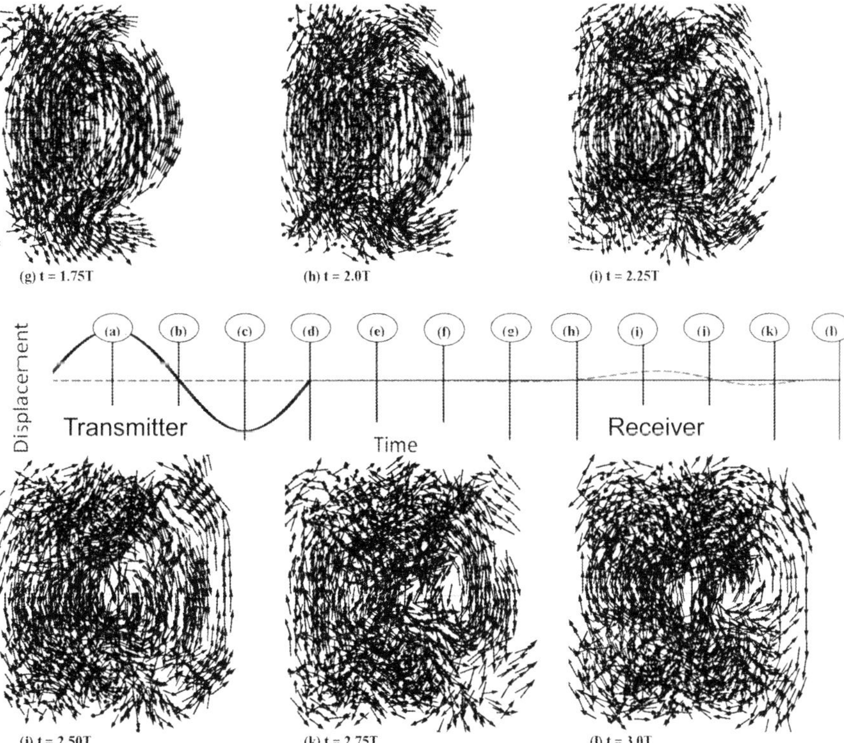

Figure 4 *Particle velocity vectors for time points (g) to (l) (58.33 and 100 µs). Velocities with a magnitude less than 1 mm/s are discarded*

Figures 4g and 4h show that the shear wave is composed of components with alternating oscillation direction. The leading shear wave is seen to move close to the receiver position. The P-wave at the edge of the sample is also observed to be composed of waves of alternating oscillation direction and the direction of particle motion is seen to change over time and space. Figure 4i shows the arrival of the shear wave at the receiver position. On the received signal trace the time at which snapshot (i) occurs there is movement of the receiver particle in the y-direction. Figure 4j illustrates the second shear wave arriving at the receiver particle and the change in direction of the particle motion is seen on the received signal trace. Figures 4k and 4l show the complicated movement following the leading shear and compressional waves arriving at the receiver particle. The complex motion of the particles in the system is clearly observed in these snapshots. Figures 3 and 4 illustrate a particle scale response similar to the two-dimensional system.[5] The vortex behaviour of particle velocities is evident in both two and three-dimensional cases. This vorticity in particle velocity arrows has been previously documented in wave propagation.[11]

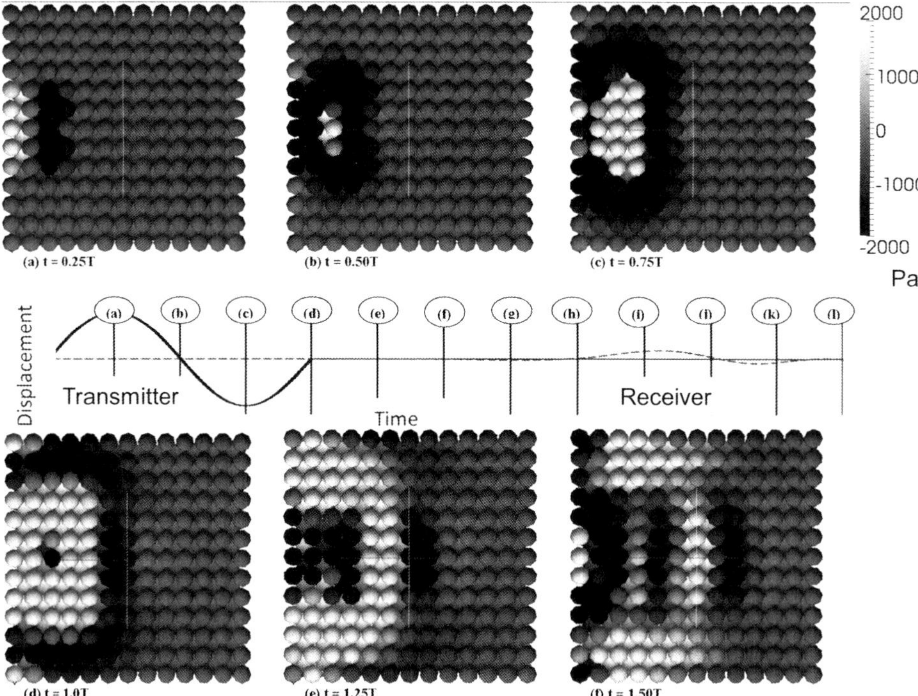

Figure 5 *Particles coloured according to representative particle shear stress for time points (a) to (f)*

Figure 5 illustrates relative representative particle shear stress for the simulation and these plots are prepared in a similar way to a previous 2D investigation.[5] Figure 5a show the shear stresses as the transmitter particle moves in one direction. Interestingly particles directly in front of and behind the transmitting particle experience no change in shear stress due to the transmitter particle motion at this stage. Figures 5b and 5c show the changes in relative representative shear stress due to the reversal of the transmitter particle motion. This is a similar behaviour to the normalised representative particle mean stress where alterations in response are seen due to alterations in the direction of the motion of the transmitter particle. Figures 5d and 5e show a final change in shear stress around the transmitter particle due to its final change in direction before coming to a stop. Figure 5f shows the relative representative particle shear stress propagating from the transmitter particle in close correlation with the previous plots showing particle velocity vectors. The shear wave can be observed to propagate as a change in relative particle shear stress. It is observed to propagate as a central lobe through the sample.

The alternating shear waves are seen as alternating patterns in shear stress shown in Figures 6g, 6h and 6i. This agrees with the alternating direction of the particle velocities seen in Figure 4 for the same snapshots. The shear stress wave is seen to arrive at the receiver disk at Figure 6i which is again in agreement with Figure 4i. There also appears to be little or no change in representative particle shear stress at the edge of the sample indicating that there is no shear wave propagating at the edge. Figures 6j, 6k and 6l show the complex behaviour of the particles after the initial shear wave has passed through the system. This, again, correlates well with the snapshots shown in Figure 4j, 4k and 4l where the system response is seen to grow steadily more complicated.

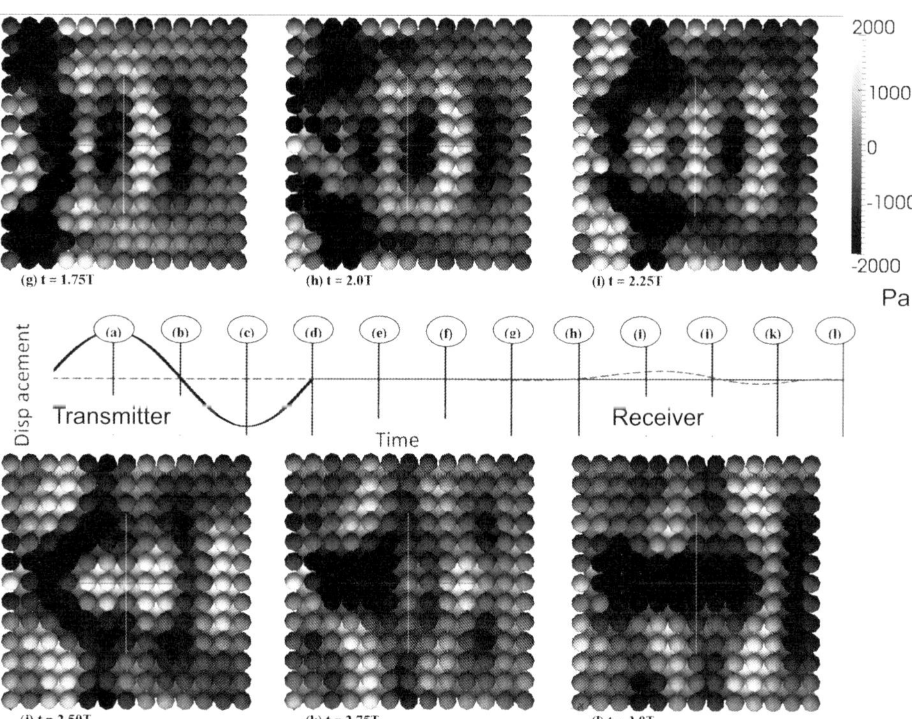

Figure 6 *Particles coloured according to representative particle shear stress for time points (g) to (l)*

4 RECEIVED SIGNAL ANALYSIS & G vs. σ_0

The received signal is plotted with the transmitted signal in Figure 7. In this figure displacements in the x- and y-directions are shown for the received signal while only the input y-direction displacement is shown for the transmitted signal as it is not displaced in any other direction and after input the displacement is set to zero. The amplitudes of the received signal are plotted ten times larger than the transmitted signal to make them observable on the same figure. The amplitude of the signal was reduced as the wave spreads geometrically. The vertical line plotted in Figure 7 is the estimated arrival time. This was calculated following the equations presented in the literature[12] and is a function of the particle material properties and sample confining pressure. The equation presented in the paper calculates a value for sample shear modulus, G_{pack}, based on the type of lattice packing, particle Poisson's ratio, particle shear modulus and confining stress. This value of G_{pack} can be substituted into Equation 2 as G_{max}:

$$G_{max} = \rho V_S^2 \tag{2}$$

to calculate a shear wave speed, V_S, using the sample density, ρ. The shear wave speed can be used with the shear wave travel distance to calculate a shear wave travel time the vertical line is plotted at this travel time. The shear wave travel distance is taken to be the straight line distance between the centroids of the transmitter particle and receiver particle as shown in Figure 2. The arrival time predicted by the analytical equations[12] agrees very well with the point of first inflection on the received signal in the y-direction. This is often taken to be the arrival time in experimental bender element tests.

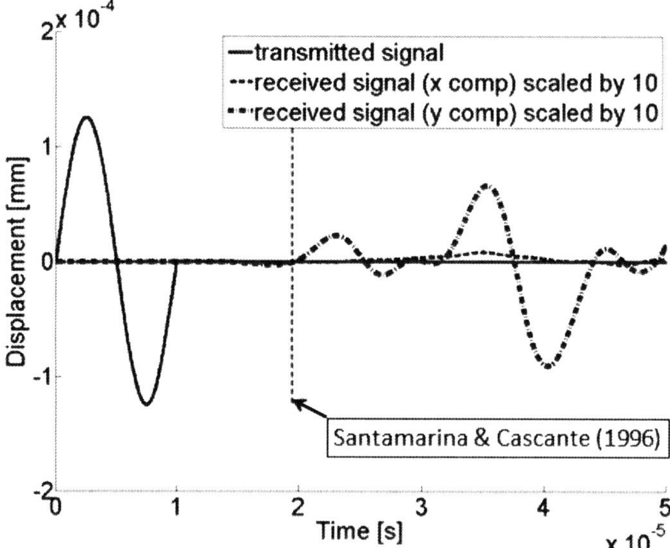

Figure 7 *Transmitted signal, received signal and analytical solution plotted on a graph of displacement vs. time*

A parametric study was carried out using results from numerical bender element tests on samples with different confining pressures. It can be analytically predicted[12] that the sample shear modulus should be proportional to the confining pressure to the power of 1/3. Figure 8 shows different G values for different confining pressures using different travel time determination techniques. While there is some scatter among the results from different travel time determination techniques a relationship between G and confining pressure that is close to $G \propto \sigma_0^{0.3}$ is observed. The different confining pressures were reached during the servo-controlled wall compression stage. Once the specified confining pressure had been achieved the walls were removed and a flexible boundary with the specified confining pressure was applied. This finding is important as it shows that DEM can successfully simulate the predicted response of a lattice packed granular medium to seismic wave propagation. This response is different to that observed in experimental tests and this will be explored in more detail in further work.

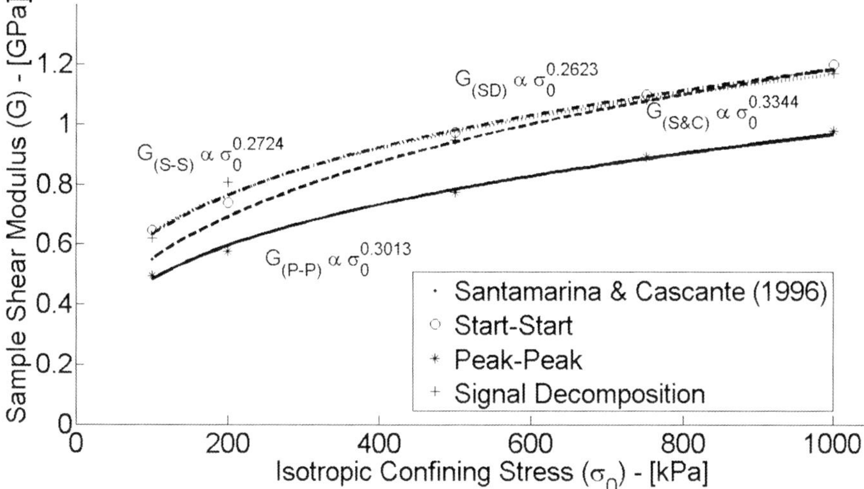

Figure 8 *Sample shear modulus, G, vs. isotropic confining stress, σ_0, for different values of σ_0 and different travel time determination techniques*

5 CONCLUSIONS

DEM is shown here to be capable of capturing wave propagation in granular material by providing easy access to important micromechanical parameters. Two ways of studying the particle scale responses are presented here, namely individual particle velocity vectors and representative particle shear stresses. The shear wave produced is seen to propagate as a spherical lobe through the centre of the sample and compressional waves are seen to travel along the edges of the sample but are of much smaller amplitudes. Individual particle velocity vectors and representative particle shear stress were used to improve our understanding of the mechanism of wave propagation through a highly idealised granular medium. The arrival time of the shear wave produced in the DEM wave propagation simulation agreed well with the arrival time predicted by an analytical equation in the literature. It showed that the first movement of the DEM received signal corresponded to the arrival of the shear wave. The increase in amplitude after the initial oscillation was due to reflections off the boundaries arriving at the receiver particle. Finally, a plot of G vs. σ_0

showed that a DEM simulation incorporating a Hertz-Mindlin contact model produced a relationship of G α $\sigma_0^{0.3}$ which is not far from the predicted G α $\sigma_0^{1/3}$ relationship. This was true of most of the travel time determination techniques used, however, the position of the trendline varied between different techniques. Work is ongoing to ascertain how to reconcile these differences in results for different travel time determination techniques. It is hoped that by analyzing the micromechanical data produced in a DEM simulation travel time determination in experimental bender element tests can be improved.

Acknowledgments

The authors would like to acknowledge funding provided by EPSRC grants: EP/G064180 and EP/G064954, and our research partners at the University of Bristol: Dr. E. Ibraim, Dr. M. Lings & Mr. S. Hamlin.

References

1. D. J. Shirley and L. D. Hampton, *Journal of the Acoustical Society of America*, 1978, **63**, 2.
2. T. Sadek, *The multiaxial behaviour and elastic stiffness of Hostun sand.* PhD thesis, University of Bristol, 2006.
3. Itasca, PFC^{3D} *Version 4.0 User Manual*, Itasca Consulting Group, Minneapolis, 2007.
4. G. Cheung and C. O'Sullivan, *Particuology,* **6**, 2008.
5. J. O'Donovan, C. O'Sullivan, and G. Marketos, *Granular Matter,* [in press], 2012.
6. K. Carter, *Simulation of Bender Element Tests in the Cubical Cell Apparatus Using the Distinct Element Method*, MSc Dissertation, Imperial College London, 2010.
7. S. Clement, *Simulating Bender Element Tests using the Distinct Element Method*, MSc Dissertation, Imperial College London, 2006.
8. G. Alvarado, *Influence of Late Cementation on the Behaviour of Reservoir Sands*, PhD Thesis, Imperial College London, 2007.
9. S. Hardy, *The implementation and application of dynamic finite element analysis to geotechnical problems*, PhD Thesis, Imperial College London, 2003.
10. J. –S. Lee and J. C. Santamarina, *Journal of Geotechnical and Geoenvironmental Engineering*, 2005,**131**, 9.
11. L. Li and R. M. Holt, *Oil & Gas Science and Technology*, 2002, **57**, 5.
12. J. C. Santamarina, G. Cascante, *Canadian Geotechnical Journal*, 1996, **33**, 5.

MICROMECHANICAL STUDY ON SHEAR WAVE VELOCITY OF GRANULAR MATERIALS USING DISCRETE ELEMENT METHODS

X. Xu[1,2], D. Ling[1], Y. P. Cheng[2] and Y. Chen[1]

[1]Key Laboratory of Soft Soils and Geoenvironmental Engineering of MOE, Zhejiang University, China
[2]Department of Civil, Environmental and Geomatic Engineering, University College London, UK

1 INTRODUCTION

Shear wave velocity is a key parameter to evaluate the engineering properties of granular materials, like sand. It is generally obtained using field seismic testing techniques in situ, and piezoelectric transducers in various laboratory tests.[1] These extensive studies have provided us a promising understanding on the mechanical properties of granular materials in macro-scale.

With the recognition that granular materials are discrete in nature, rather than continuous, micromechanical study on shear wave velocity using Discrete Element Methods (DEM) has become increasingly important to understand their response in micro-scale,[2,3] as well as to bridge the relationships between micro- and macro-scale mechanisms. And this is just unfolded.[4~6] Two different methods for the measurement of shear wave velocity based on both shear vibration and torsional vibration are presented in this paper.

2 DISCRETE ELEMENT MODEL

2.1 Assembly Generation

A three-dimensional assembly consisting of 21,495 sphere particles with diameters ranging from 0.15 to 0.20 mm is generated in a cylindrical region by radius expansion (Figure 1a). Each particle is prescribed properties including a radius, density, coefficient of contact friction *etc*. The Hertz-Mindlin contact model is employed because it is suitable for simulating the pressure-dependent behaviour at small-strain. The parameters adopted in this study are listed in Table 1.

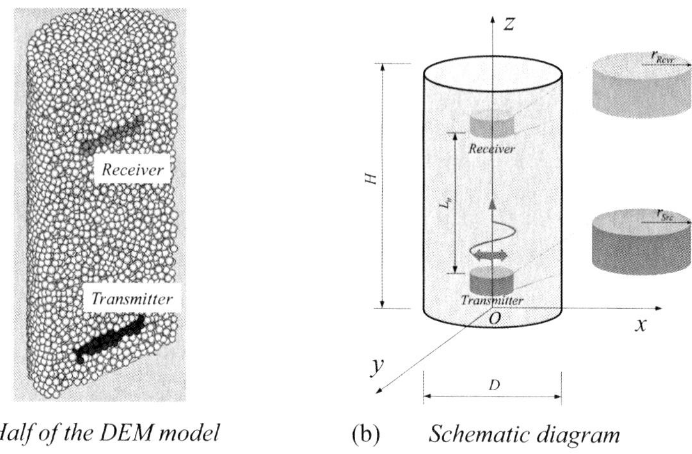

(a) *Half of the DEM model*   (b)   *Schematic diagram*

Figure 1 *The model for shear wave propagation*

Table 1 *Input parameters used in the DEM model*

Parameters	Value	Parameters	Value
Model diameter D (mm)	4	Particle diameter d (mm)	0.15~0.20
Model height H (mm)	8	Particle shear modulus G_g (GPa)	1.0
Confining pressure σ_c (kPa)	100	Particle Poisson's ratio v_g	0.20
Particle density ρ_g (kg/m^3)	2630	Particle friction coefficient μ_g	0.50
Particle number N	21,495		

2.2 Test Procedures

Four basic steps for shear wave propagation modelling are described as follows: (1) Initial consolidation. The DEM specimen is isotropically consolidated at a confining pressure of 100 kPa in this study; (2) Noise elimination. To ensure that particles inside the sample come close to equilibrium, a number of more cycles are usually needed after isotropically consolidation; (3) Transmitter and receiver installation. In the model, a group of particles in a certain region (e.g., round-block shaped) were selected as a transmitter and another one as a receiver (Figure 1). (4) Transmitting and monitoring. The excitation is then created by applying a velocity pulse to the transmitter in a certain direction. The averaged velocity of the receiver in the same direction is then monitored in order to detect the arrival of the disturbance.

3 MEASUREMENT OF SHEAR WAVE VELOCITY BASED ON SHEAR VIBRATION

Enlightened by the basic principles of bender element, measurement of shear wave velocity on the basis of shear vibration is firstly given in this section. A single sine velocity pulse is applied to the transmitter in x direction. After the excitation being created, the averaged velocity of the receiver in this direction is then monitored (Figure 1b).

3.1 Shear Wave Propagation

The propagation of shear wave in granular materials is plotted in Figure 2. Each diagram is a longitudinal section drawing of the velocity field at a certain time after the excitation being created, with the positions of both transmitter and receiver. The velocity field only appears within the transmitter at the very beginning (Figure 2a). The vibration of the particles inside the transmitter brings movements to the particles surrounded, and the velocity field gradually move towards the receiver. The maximum value of velocity reaches the receiver at 22.5 microsecond (Figure 2d).

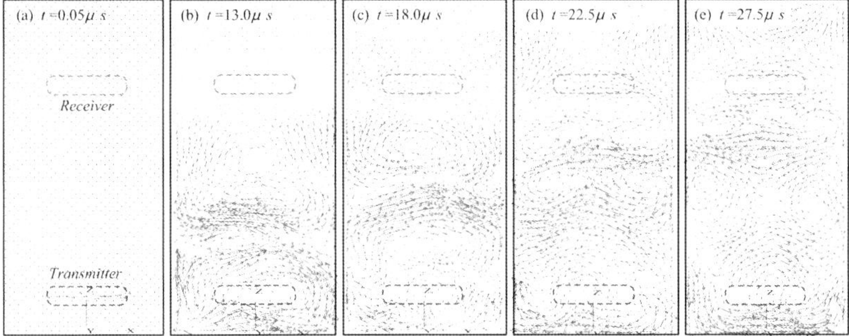

Figure 2 *Shear wave propagation*

3.2 Determination of Travel Time

Cross-correlation analysis is adopted in this study due to its superiority of both determining the travel time and identifying similarities between two signals. An example is given in Figure 3. The excitation wave and received wave are plotted in the upper plot, and the calculated cross-correlation coefficient is displayed in the lower plot. The travel time t is determined by taking the time of point D corresponding to the maximum of cross-correlation coefficient, as shown in Figure 3. The shear wave velocity is calculated using the wave travel time and the distance of the travel path, in exactly the same way as in laboratory tests.

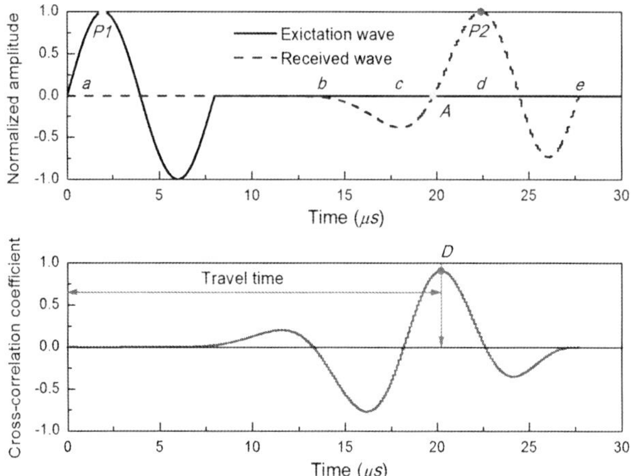

Figure 3 *Determination of travel time via cross-correlation analysis*

3.3 Parametric Analysis

Systematic parametric analysis on shear wave modelling is carefully examined by conducting a number of series of DEM simulations, including the influence of excitation frequency, excitation amplitude, travel distance, size of the transmitter and receiver, damp etc. Taking the size of transmitter and receiver for an example, it is indicated that half the sample diameter for both the transmitter and receiver will achieve the best results. The selection of excitation frequency also need to satisfy the condition of $L_{tt}/\lambda > 2$ and $\lambda/d_{50} > 10$, where L_{tt} and λ are travel distance and wave length, respectively.

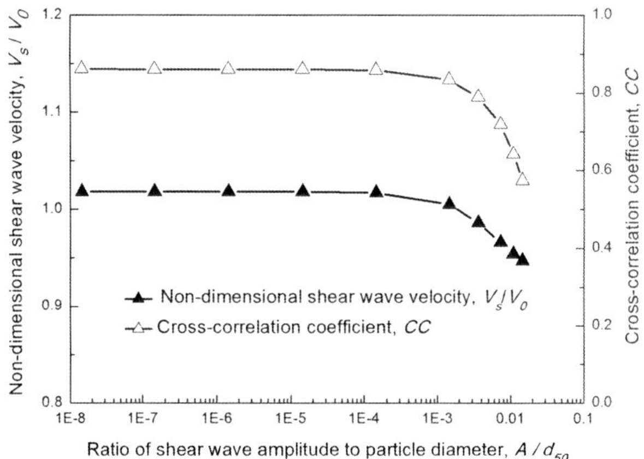

Figure 4 *Influence of excitation amplitude on shear wave velocity*

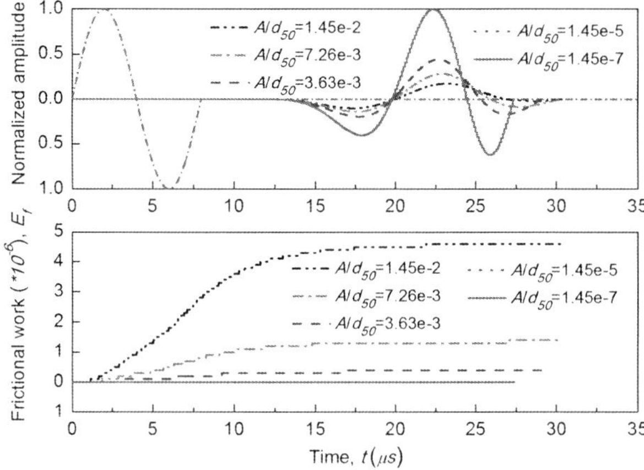

Figure 5 *Influence of excitation amplitude on received wave and frictional work*

Herein we just present the influence of excitation amplitude as shown in Figure 4, which is a key parameter for shear wave modelling. The reference value V_0 on the left vertical axis is acquired by calculating the small-strain shear-modulus first as the shear stress divided by the engineering shear strain. As observed from Figure 4, the non-dimensional shear wave velocity V_s/V_0 gradually converged to a stable value of approximate 1.0 with the diminution of the ratio of excitation amplitude to particle diameter A/d_{50}, and there was a very similar trend for the corresponding cross-correlation coefficient. It can be concluded from Figure 4 that a value of 10^{-4} or less for A/d_{50} is sufficient to obtain a stable and correct shear wave velocity.

The shape of the received wave can also be taken as a significant indicator for identifying the proper choice of excitation amplitude, as shown in the upper plot of Figure 5. With the increment of A/d_{50}, the shape of the received wave gradually distorted comparing with that of excitation wave, by reducing the amplitude and increasing frequency. It indicated that the applied excitation amplitude also ensured a stable and reasonable shape of received wave.

The distortion of the received wave can be explained by the generation of frictional work, i.e. sliding in particle contacts, as emphasized in the lower plot of Figure 5. The large value of A/d_{50} corresponds to the high frictional work, and it disappears when A/d_{50} is low enough. The above discussion provides clear evidence to support the shear wave modelling in this study.

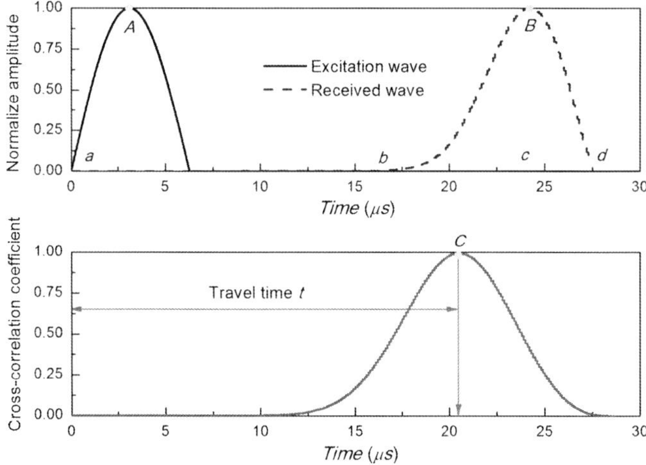

Figure 6 *Determination of travel time for torsional wave*

4 MEASUREMENT OF SHEAR WAVE VELOCITY BASED ON TORSIONAL VIBRATION

Another method for measuring the shear wave velocity based on torsional vibration is briefly discussed in this section, also named as torsional wave for short. It is an unpolarized transverse wave, with the advantages of anti-interference, large radiation and stable waveform, comparing to the shear wave mentioned in Section 3.

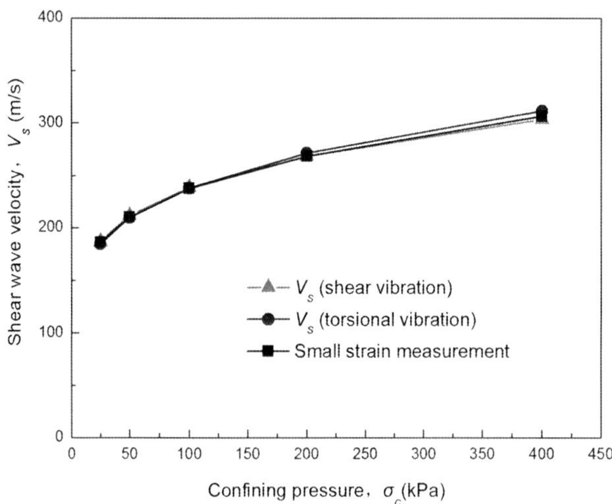

Figure 7 *Comparison of shear wave velocity obtained using different methods*

A semi-cycle of angular velocity pulse is applied to the transmitter, and the corresponding disturbance of the receiver is monitored afterwards. Figure 6 presents the determination of travel time for torsional wave through cross-correlation analysis, very similar to the discussion on Figure 3. It is indicated that the torsional wave has a perfect performance, with the cross-correlation coefficient almost equalling to 1.0. Moreover, it is also found that the torsional wave velocity shows good agreement with the result from shear vibration, as shown in Figure 7.

5 VERIFICATION

Numerical simulations on both regular and random packing assemblies are then conducted to verify the reliability of two measuring methods on shear wave velocity of granular materials. Figure 8 presents the comparison among the results from analytical solution, shear wave modelling, and homogenization techniques[7] with various hypotheses for simple cubic packing assembly. It is shown that different methods give similar results, and the relative errors are within 5%.

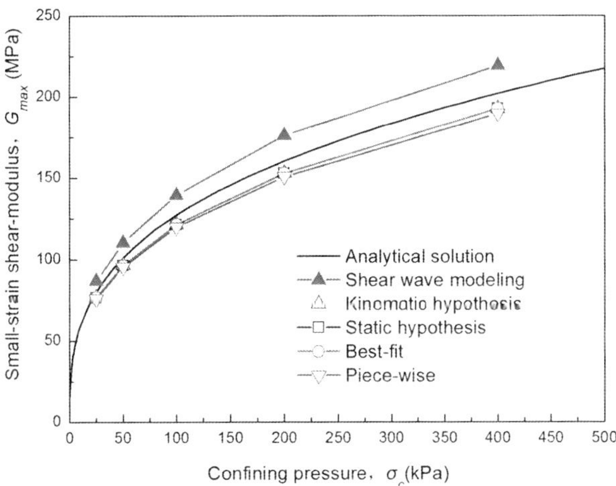

Figure 8 *Verification of shear wave modelling for simple cubic packing assembly*

For random packing assembly, verification of shear wave modelling is then given in Figure 9. It shows that the results from shear wave modelling almost lie in the middle, between the results of infinitely rough and perfectly smooth in Effective Medium Theory (EMT). The reason for the difference is that EMT assumes particle evenly stressed and it is always not the case in granular materials. The value of strong force can be as much as five times that of mean force. Meanwhile, it is found that homogenization techniques overestimate the real stiffness of granular materials.

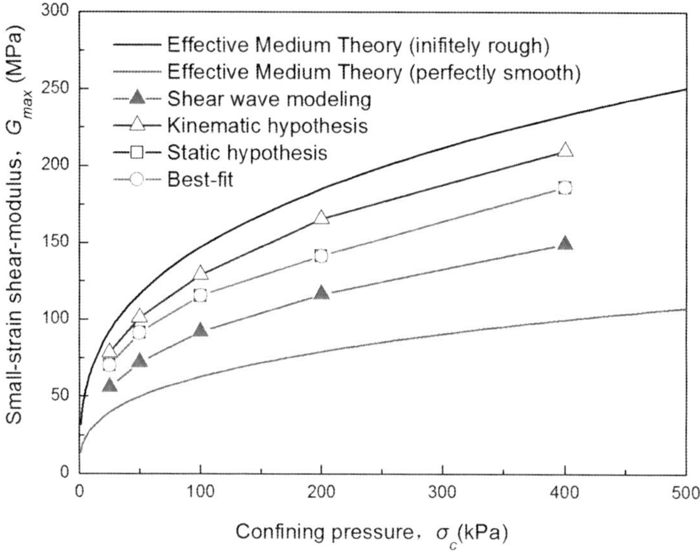

Figure 9 *Verification of shear wave modelling for random packing assembly*

6 CONCLUSIONS

The principal conclusions of this study are as follows:

(1) Shear wave modeling in DEM is implemented by applying a velocity pulse to the transmitter in a certain direction and monitoring the corresponding average velocity of the receiver.

(2) Reasonable parameters are the key to the successful shear wave modeling in DEM, including $L_{tt}/\lambda > 2$, $\lambda/d_{50} > 10$, and no frictional work generation for both shear and torsional vibrations.

(3) The results of homogenization techniques overestimate the shear wave velocity of granular materials in random packing.

Acknowledgements

Much of the work described in this paper was supported by the National Basic Research Program of China (973 Project) (Grant No. 2007CB714203), the National Natural Science Foundation of China (Grant Nos. 50778163 and 50708095), and ERASMUS MUNDUS EXTERNAL COOPERATION WINDOW (EM ECW) scholarship. These financial supports are gratefully acknowledged. The authors thank Prof. Kenichi Soga at University of Cambridge, Prof. M A Curt Koenders at Surrey University, and Prof. J Carlos

Santamarina at Georgia Institute of Technology, for their valuable exchanges of ideas and help with this study.

References

1 C.R.I. Clayton, *Géotechnique*, 2011, **61**, 5.
2 C. Thornton, *Géotechnique*, 2000, **50**, 43.
3 K. Soga and C. O'Sullivan, *Soils and Foundations*, 2010, **50**, 861.
4 L. Li and R.M. Holt, *Oil & Gas Science and Technology - Rev.* IFP, 2002, **57**, 525.
5 E. Somfai, J.N. Roux, J.H. Snoeijer, M. van Hecke and W. van Saarloos, *Physical Review E*, 2005, **72**, 021301.
6 X. Xu, D. Ling, B. Huang and Y. Chen, *Chinese Journal of Geotechnical Engineering*, 2011, **33**, 1462 (in Chinese).
7 C.L. Liao, T.P. Chang, D.H. Young and C.S. Chang, *International Journal of Solids and Structures*, 1997, **34**, 4087.

MECHANICAL BEHAVIOUR OF METHANE HYDRATE SOIL SEDIMENTS USING DISCRETE ELEMENT METHOD: PORE-FILLING HYDRATE DISTRIBUTION

Y. Yu[1], Y. P. Cheng[1] and K. Soga[2]

[1]Department of Civil, Environmental and Geomatic Engineering, University College London, London, WC1E 6BT, United Kingdom
[2]Geotechnical and Environmental Research Group, Department of Engineering, University of Cambridge, Trumpington Street, Cambridge, CB2 1PZ, United Kingdom

1 INTRODUCTION

Methane gas is the predominant element composing natural gas, which is assumed to be a relatively clean-burning energy resource with immense worldwide deposits. A large amount of methane is trapped within a crystal structure of water, forming a methane hydrate that can only exist under the conditions of low temperature and high pressure. Normally, methane hydrate develops in the pores of soil sediments under deep seabeds or permafrost regions.[1] This highly compacted methane hydrate soil sediment has attracted interest as a potential energy resource by extracting methane gas from hydrate-bearing sediment dissociation, but it also has an impact on climate change and geotechnical issues, such as seafloor instability and near well conditions.[2]

In the pores of methane hydrate soil sediments there are three typical hydrate formation patterns that describe the microscopic hydrate distribution: pore-filling, load-bearing and cementation. In this paper, the research is focused on pore-filling hydrate distribution in the sediment pores (Figure 1).[1,3,5]

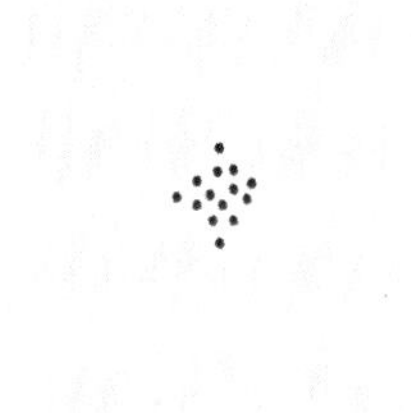

Figure 1 *Pore-filling: hydrates (black) nucleate on sediment grain (grey) boundaries and grow freely into pore spaces without bridging two or more particles together.*

Triaxial shear tests are usually employed to conduct research on mechanical behaviours of hydrate-bearing soil in laboratory studies. It was found that the strength and volume change behaviour of the sediments are affected by hydrate formation and distribution pattern, as well as hydrate saturation.[2] However, it is difficult for laboratory studies to control the formation, distribution and saturation of the hydrate. Hence, in this paper, the numerical simulations with Discrete Element Methods (DEM) were used to provide insights into the mechanical response of hydrate-bearing soil sediments with pore-filling hydrate distribution. The discrete element code PFC3D 4.0 was used.[4] It was observed that the results of the DEM simulations had similar trends to the laboratory results reported by Soga et al.[5]

2   DEM SAMPLE PREPARATION OF PORE-FILLING HYDRATE-BEARING SOIL SEDIMENTS

Two sample preparation methods were used and compared by Brugada et al.:[1] "Soil + Hydrate" case and "Soil → Hydrate" case. "Soil + Hydrate" is usually used in laboratory studies by mixing soil particles with synthetic hydrates. Hydrate particles were randomly placed into a sample together with the soil particles at low confinement, and then the sample was isotropically consolidated to a desired effective stress. However, in the "Soil → Hydrate" case, the soil sample was first isotropically consolidated to the desired effective stress. Following this, hydrate particles were put into the void space of the sample to reach different hydrate saturations. Through the comparisons of the triaxial simulation results, the mechanical behaviours were independent of the two different preparation methods.[1] Hence, in this research, the "Soil → Hydrate" sample preparation method was used throughout the simulations, which were similar to the natural hydrate formation in sediments under deep seabeds and permafrost regions.

The parameters of DEM sample preparation are summarized in Table 1. The particle sizes of the soils followed the Gaussian distribution. But there is a lack of data in the literature and the experimental results about the elastic modulus (E_c) of hydrates developed inside soil pores.[1] In this research, simulations were performed at varying hydrate elastic moduli (E_c) from 0.286 MPa (very soft value) to 286 MPa (the same value as that of soil particles). So the hydrate-soil elastic modulus ratios ($E_{c\text{-hyd}}/E_{c\text{-soil}}$) were varied between 0.001 and 1. For these DEM simulations, the normal contact stiffness k_n is proportional to the particle elastic modulus (E_c) and the particle diameter (D), and $k_n = 2DE_c$. That is, when $E_{c\text{-hyd}}/E_{c\text{-soil}}$ of 0.1 is used, the hydrate-soil contact stiffness ratio $k_{n\text{-hyd}}/k_{n\text{-soil}}(D_{50})$ becomes 0.023.

Table 1 *Parameters in the DEM simulations*

Property	Soil	Methane hydrate
Particle size D (mm)	0.1-0.25 *Gaussian distribution*	0.04
Density (kg/m³)	2600	900
Elastic Modulus E_c (MPa)	286	0.286 – 286
Normal contact stiffness k_n (N/m)	$2DE_c$	$2DE_c$
Shear contact stiffness k_s (N/m)	$0.7k_n$	$0.7k_n$
Inter-particle friction μ	0.75	0.75

The samples were cylinders of 1.75 mm (diameter) × 3.5 mm (height). The spherical soil particles were first generated at half the value of the target sizes, and they were then enlarged to the final sizes. The total number of soil particles was 1956. Gravity was not applied to the particles. The soil sample (with initial porosity n=0.43) was isotropically consolidated to isotropic effective stresses of 1 MPa, 2 MPa or 3 MPa. After that, the half sized hydrate particles were randomly generated inside the pore space of the cylindrical soil sample, and then they were expanded to the target diameter of 0.04 mm, in order to reach the desired hydrate saturation (Figure 2). The radii expansion process was used to avoid overlapping the particles and to shorten the time it takes to reach equilibrium. Note that the simulated hydrate saturation (S_h) should correspond to a higher hydrate saturation of the natural hydrate-bearing soil sediments for several reasons, including the assumed hydrate particle size and the existence of some other materials in the pores of natural sediment. The total number of hydrate particles varied from 10,114 and 40,459 at hydrate saturation of 10% and 40%, respectively.

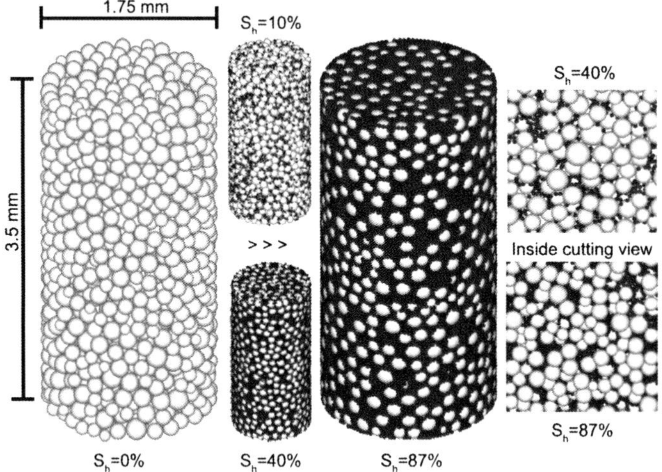

Figure 2   *DEM hydrate-bearing soil sediments cylindrical samples*

After the sample preparation, a series of drained triaxial shear tests were systematically performed at different conditions of hydrate saturations, confining pressures and hydrate-soil elastic modulus ratios. During the sample preparations and triaxial tests, the top, bottom and lateral walls were frictionless. The radial pressures of 1 MPa, 2 MPa or 3 MPa were kept constant to the cylindrical wall throughout the triaxial tests while the axial deviatoric loading was applied by displacing the top boundary at a constant speed.

3   RESULTS AND ANALYSES OF MECHANICAL BEHAVIOUR IN THE SIMULATIONS

3.1 Effect of Hydrate Saturation on Shear Strength

Under the confining pressure of 1 MPa and a hydrate-soil contact stiffness ratio $k_{n\text{-hyd}}/k_{n\text{-soil}}(D_{50})$ of 0.023, the stress-strain relationships of hydrate-bearing samples are shown in Figure 3 at different hydrate saturations (S_h= 0%, 10%, 20%, 30%, 40%). The initial stiffness increased as there was higher hydrate saturation. When the hydrate saturation was 10%, the maximum deviator stress was similar to that of the hydrate-free sample. This means that hydrates with pore-filling distribution do not contribute to the strength of the sediments when hydrate saturation is low. Hydrates were formed inside the pore space rather than at the soil particles' contacts.[1] When the hydrate saturation was above 20%, the peak shear strength obviously increased from 1.4 MPa (S_h=0%) to 2.75 MPa (S_h=40%). At large axial strain levels, the deviator stresses began to reduce to some constant values. There was a reduction in the critical state shear strength (softening behaviour) when the hydrate saturation was higher than 20%. The hardening effect in the small strain and the softening behaviour in the large strain of the simulated results were similar to the experimental data obtained by Masui *et al.*[2] However, this same pattern of behaviour was observed in the DEM simulations when the samples were confined by the uniform cylindrical wall boundaries.

Figure 4 shows the contact force chains on the middle cutting plane of the hydrate-bearing soil sample (S_h = 40%) throughout the triaxial test. The thickness of the black lines in Figure 4 represents the value of the contact force. It is indicated that the strong forces (thick black lines) were mainly transmitted through the soil skeleton at the initial state. When the sample reached the peak strength, more hydrate particles contributed to the transmission of the contact force, which was depicted by thin black lines. Accordingly, the number of the strong forces between soil particles (thick black lines) decreased obviously. The distribution of the contact forces tended to be uniform at the critical state.

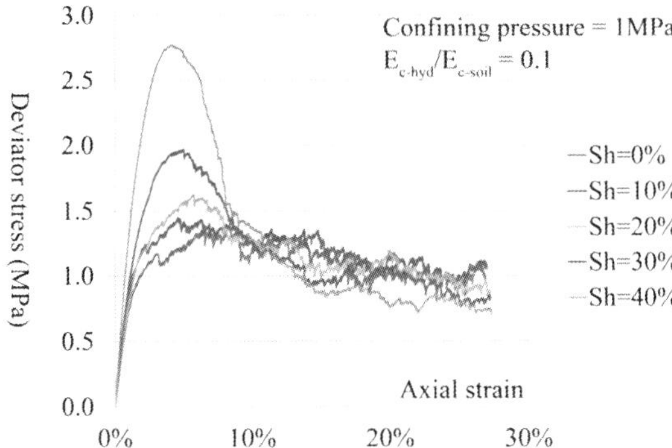

Figure 3   *Effect of hydrate saturation on shear strength of hydrate-bearing samples*

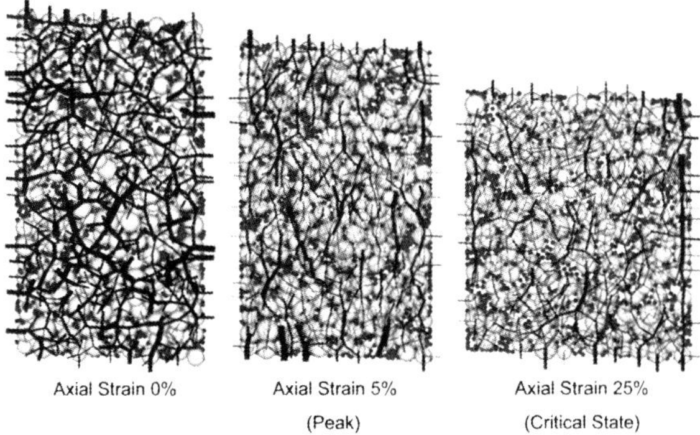

Figure 4   *Contact force chains on the middle cutting plane of the hydrate-bearing sample (S_h = 40%) throughout the triaxial test (f_{max} = 0.1147 N)*

The peak and critical state failure envelopes were plotted on the mean effective stress versus deviator stress (p'-q) space in Figure 5 and Figure 6, respectively. The shear strength increased with the mean effective stress. At S_h =30%, the increase in the peak friction angle (ϕ_p) became evident, demonstrating the hardening effect of the hydrate saturation in peak strength discussed above (Figure 5). While the critical state lines plotted on the p'-q space (Figure 6) revealed the softening behaviour with an increase in hydrate saturation. At S_h =20%, the critical state friction angles (ϕ_c) started to decrease with an increase in hydrate saturation. This may be due to inclusion of fine hydrate particles in soil matrix.

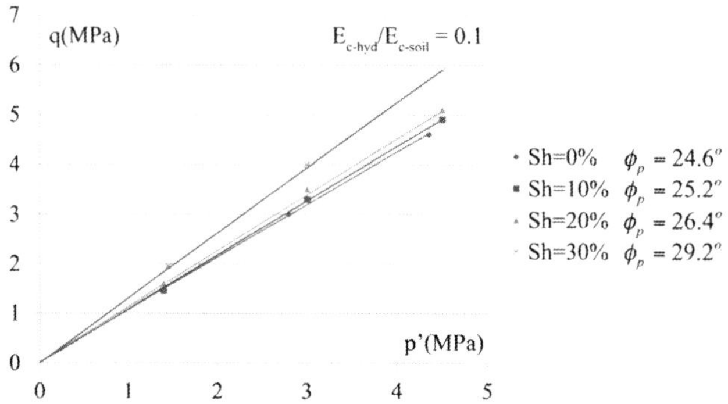

Figure 5   *Peak failure envelope projected on p'-q plane*

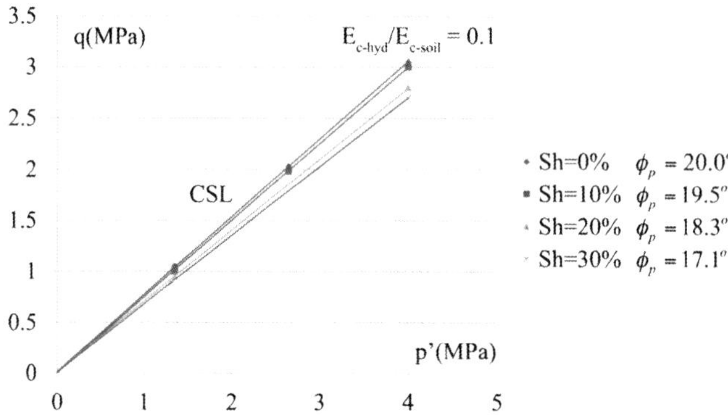

Figure 6   *Critical state lines project on p'-q plane*

3.2 Effect of Hydrate Saturation on Volumetric Response

At the same conditions mentioned in Session 3.1, the volumetric response at different hydrate saturations (Figure 7) showed that contractive behaviour was followed by the dilative tendency, which was enhanced when hydrate saturation increased. In Figure 7, the volumetric strain (ε_v) and axial strain (ε_a) relationship was plotted at different hydrate saturations. It is clear that the dilatancy rate ($d\varepsilon_v / d\varepsilon_a$) increased when the hydrate saturation increased.

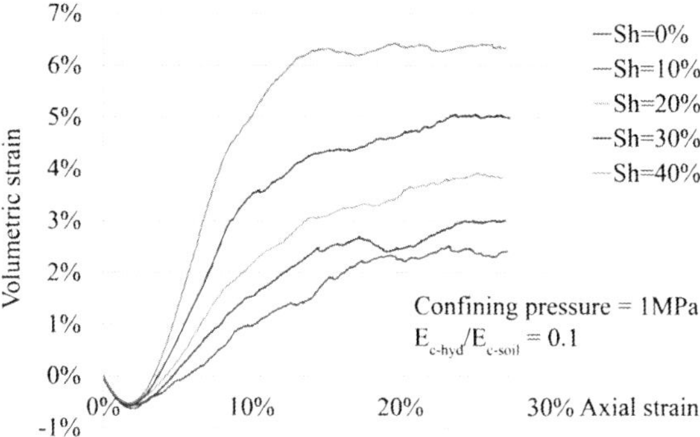

Figure 7 *Effect of hydrate saturation on volumetric response*

In the critical state, the dilation reduced to zero and the samples maintained constant volumes. Furthermore, the critical state granular void ratio e – *ln*p' projections were plotted in Figure 8. The simulations showed that the resulting critical state granular void ratios may not be unique indicating possible effect of fine hydrate particles on the critical state behaviour. In addition, the hydrate induced dilatancy was less obvious at high pressure. Note that granular void ratio *e* was obtained by the following equation:

$$e = \frac{V_v + V_h}{V_s}$$

where V_v is the void volume, V_h is the hydrate volume and V_s is the soil volume.

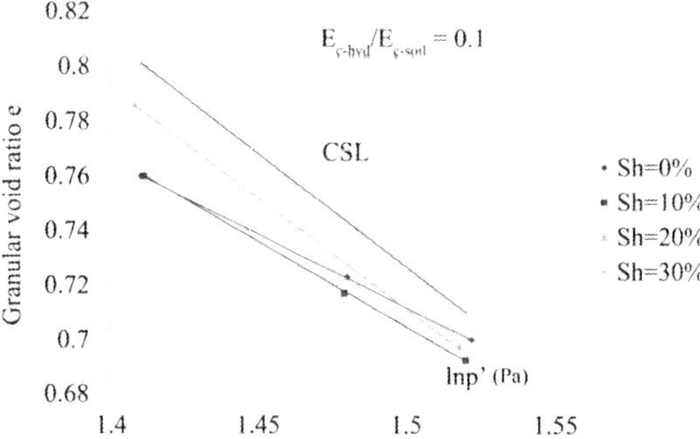

Figure 8 *Critical state lines projected on granualar void ratio e –lnp' plane*

4   Conclusions

For pore-filling hydrate distribution, the hydrate contributions to the strength of the sediments were of a frictional nature. When the hydrate saturation was above 20%, the hydrate particles' contribution to the enhanced shear strength became increasingly obvious.

The hydrate particles, however, contributed to reducing the critical state angle of friction when saturation was above 20%. The simulations also showed that the resulting critical state granular void ratios may not be unique. These results indicate that the presence of fine hydrate particles in soil matrix may influence the critical state behaviour, which requires further study

Further investigations are being conducted on the hydrate-soil contact stiffness ratio and micro-mechanism of hydrate-bearing samples throughout the triaxial shear tests. Simulations are being performed on the methane hydrate soil sediments with cementation hydrate distribution, in order to explore the mechanical behaviour and the associated micro-mechanisms.

Acknowledgments

The author would like to thank the financial support through the UK-China Scholarships for Excellence from the government of China, the government of UK and University College London.

References

1   J. Brugada, Y. P. Cheng, K. Soga and J. C. Santamarina, *Granular Matter*, 2010, **12**, 517.
2   A. Masui, H. Haneda, Y. Ogata and K. Aoki, in *Proceedings of the 5th Int. Conf. on Gas Hydrates*, Trondheim, Norway, 2005, 657, Paper No. 2037.
3   W. F. Waite, J. C. Santamarina, D. D. Cortes, B. Dugan, D.N. Espinoza, J. Germaine, J. Jang, J. W. Jung, T. Kneafsey, H. S. Shin, K. Soga, W. Winter and T. S. Yun, *Rev. Geophys.*, 2009, **47**, 38.
4   Itasca, PFC3D: *Particle flow code. User's guide version 4.0*, Minneapolis, USA, 2008,.
5   K. Soga, S. L. Lee, M. Y. A. Ng and A. Klar, *Characterisation and Engineering Properties of Natural Soils,* Taylor and Francis,, London, 2006, **4**, 2591.

ON THE EFFECT OF SOIL MODIFICATION WITH LIME USING GRADING ENTROPY

E. Imre[1,2], J. Szendefy[2], J. Lőrincz [1], P.Q. Trang[2] and Vijay P. Singh[3]

[1]SZIE, Budapest, Hungary
[2]BME, Budapest, Hungary
[3]Texas A and M University, USA

1 INTRODUCTION

The entropy path of lime modification and particle breakage was investigated and compared in terms of the entropy coordinates using some measured data. According to the results, the lime modified soil has an overall state improvement in terms of non-normalized entropy coordinates, the change is in the opposite direction as in the case of breakage (or degradation). The path in terms of normalized entropy coordinates is basically the same if the fraction number is unchanged, however, the discontinuity of the entropy path due to the fraction number changes may convert it as un-determinate.

2 GRADING ENTROPY[1]

2.1 Statistical Cells (Fractions and Elementary Cells)

The sieve aperture diameters have a scale factor of two while uniform cell system is needed for the statistical entropy theory of a discrete distribution. That is why a double statistical cell system is used: a uniform cell system in which the width, d_0, is arbitrary, and a non-uniform sized cell system ('fraction system') which is defined on the pattern of the classical sieve aperture diameters. The diameter range for fraction j (j =1, 2...see Table 1) is as follows:

$$2^j d_0 \geq d > 2^{j-1} d_0 \tag{1}$$

where d_0 is the elementary cell width.

Table 1 *Definitions of fractions and elementary cells*

j	1	...	23	24
Limits	d_0 to $2 d_0$	...	$2^{22} d_0$ to $2^{23} d_0$	$2^{23} d_0$ to $2^{24} d_0$
S_{0j} [-]	0	...	22	23

2.2 Space of the Possible Grading Curves

The probabilities are computed from the fraction information exclusively, assuming that the distribution of the grading curves is uniform within a fraction. As a result, the distribution of the double cell system is known and can be computed if the relative frequencies of the fractions x_i (i = 1, 2, 3...N) are specified. The relative frequencies of the fractions x_i (i = 1, 2, 3...N) of each grading curve satisfy the following equation:

$$\sum_{i=1}^{N} x_i = 1, \quad x_i \geq 0, \ N \geq 1 \tag{2}$$

Equation (2) defines an N-1 dimensional, closed simplex (which is the generalised N-1 dimensional analogy of the 2 dimensional triangle or, 3 dimensional tetrahedron) if the relative frequencies x_i are identified with the barycentre coordinates of the simplex points. The vertices can be numbered 'continuously' from 1 to N, and, therefore, the simplex is referred to as continuous. The sub-simplexes of a simplex are partly continuous, partly gap-graded.

2.3 The Grading Entropy and the Entropy Coordinates

The grading entropy S can be given as the sum of two parts:[1]

$$S = S_0 + \Delta S \tag{3}$$

$$S_0 = \sum_{x_i \neq 0} x_i S_{0i}, \tag{4}$$

$$\Delta S = -\frac{1}{\ln 2} \sum_{x_i \neq 0} x_i \ln x_i. \tag{5}$$

which are referred to as base entropy S_0 and entropy increment ΔS. S_{0i} is the grading entropy of the i-th fraction (see Table 2) and is given as follows:

$$S_{0i} = i - 1 \tag{6}$$

Table 2 *The fractal dimension D for a fixed A (= 2/3 in the function of N)*

N [-]	2	3	4	5	6	7
D [-]	2	2.25	2.40	2.49	2.56	2.62

The normalized base entropy, i.e., the so-called relative base entropy A, is defined as:

$$A = \frac{S_o - S_{o\min}}{S_{o\max} - S_{o\min}}, \tag{7}$$

where S_{0max} and S_{0min} are the base entropies of the largest and the smallest fractions in the mixture, respectively. The normalized entropy increment B is defined as follows:

$$B = \frac{\Delta S}{\ln N} \tag{8}$$

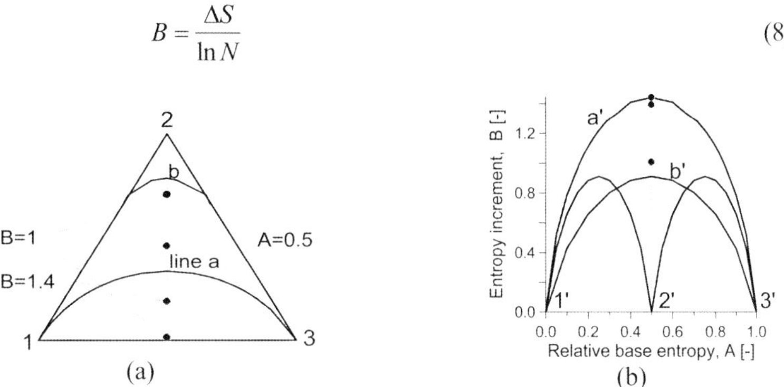

Figure 1 *The normalized entropy map for a simplex with N =3. (a) The 2D simplex and its hyper-plane section with A=0.5. (b) The normalized entropy diagram. Note that the inverse image of points A = 0.5, B = 1 and B = 1.4 are shown in the hyper-plane section with A=0.5 in the 2D simplex (a) with the inverse of the maximum B line that is the optimal line.*

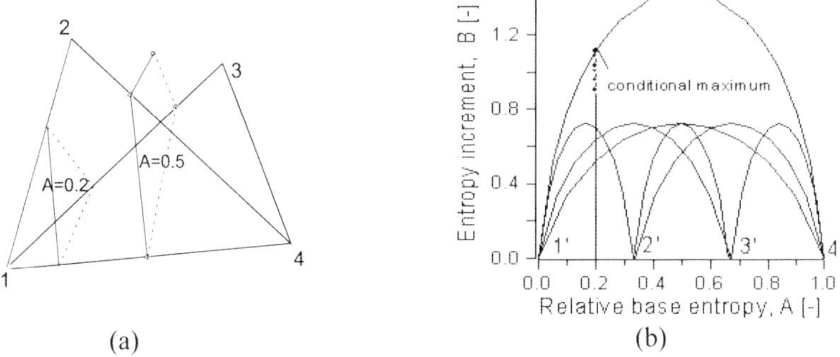

Figure 2 *N=4, the image of the 3D simplex in the entropy space. (a) The simplex and its two parallel hyper-plane sections with A=0.2 and 0.5. (b) The entropy diagram with the indication of the image of the hyper-plane section with A=0.2.*

2.4 Entropy Diagram and Inverse Image

Two maps can be defined for a specified simplex Δ: the non-normalized entropy map $\Delta \rightarrow [S_0, \Delta S]$ and the normalized entropy map f_n: $\Delta \rightarrow [A, B]$.[2] These maps between the N-1 dimensional simplex and the two dimensional space of the entropy coordinates are continuous on the open simplex and can continuously be extended to the closed simplex. The image of the maps - the normalized and non-normalised entropy diagrams, respectively – are compact, like the simplex (Figs 1 to 3). It can be proved that the inverse image of the regular values of the map is similar to the parts of an N-3 dimensional circle.[2]

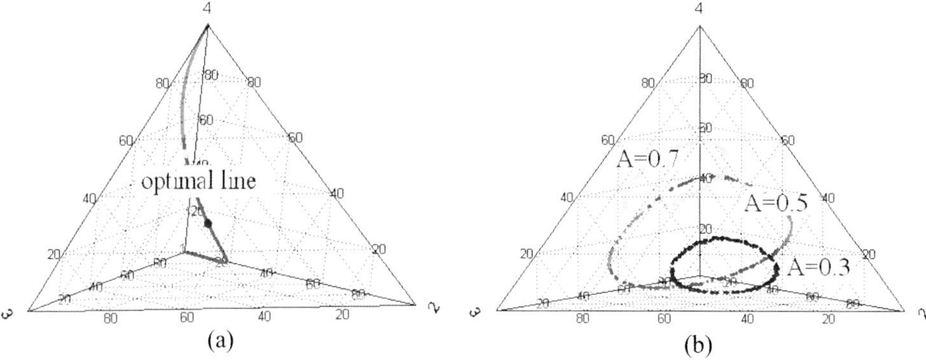

Figure 3 *The 3-simplex of the grading curves for N=4, indicating the inverse of (a) the maximum B line (optimal line), (b) three inner entropy diagram points (A=0.7 B=1.2; A=0.5 B=1.2, A=0.3 B=1.2) which are essentially three circles.*

2.4.1 Maximum and Minimum Entropy Increment Lines. For a given number of fractions N and a given relative base entropy A, there is a maximum entropy increment. This can be best computed with conditional optimization.[1] According to the solution, for a fixed N and A, the following grading curve or point of the simplex maps into this point:

$$x_1 = \frac{1}{\displaystyle\sum_{j=1}^{N} a^{j-1}} = \frac{1-a}{1-a^N}, \tag{9}$$

$$x_j = x_1 \, a^{j-1} \tag{10}$$

where parameter *a* is the root of the following equation :

$$y = \sum_{j=1}^{N} a^{j-1}[j-1-A(N-1)] = 0. \tag{9}$$

Following from the Descartes rule of signs, polynomial *y* has one and only one positive root for *a*.[2] The corresponding points of the simplex constitute a continuous line called optimal line that can be seen in Figures 2(a) and 4(a). The corresponding grading curves are with finite fractal distribution, while the fractal dimension is:

$$D = 3 - \frac{\log a}{\log 2} \tag{10}$$

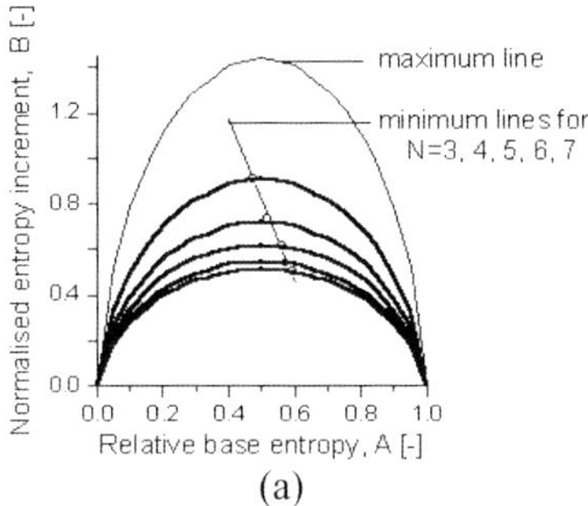
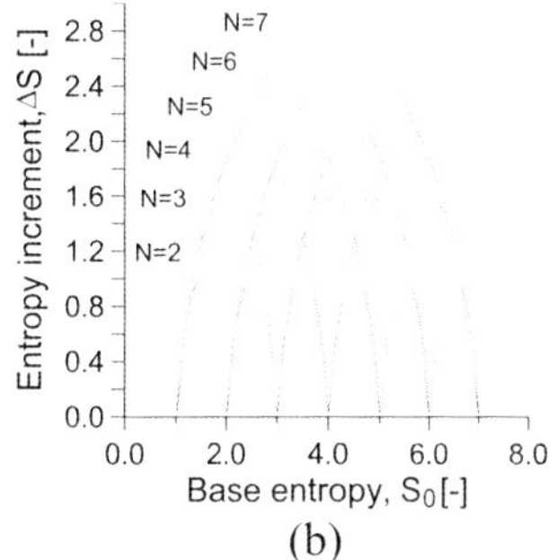

Figure 4 *The multiple diagram. (a) The normalized entropy map for simplexes with N varying from 2 to 7. (b) The non-normalized entropy diagram (the maximum ΔS lines) for the continuous sub-simplexes of a simplex with N = 7.*

As A varies between 0 and 1 then positive root a varies between 0 and ∞, extreme values represent the extreme fractions 1 and N.[2] The minimum bounding line has a smaller significance than the maximum bounding line, and the image of edge $1 - N$ is used as an approximate minimum B (or ΔS) bounding line.

2.4.2 *Multiple Diagram.* The grading curve changes during lime modification or breakage. It is not enough to consider the continuity of the entropy map with respect to x_i (i = 1, 2, 3...N) but it is necessary to consider it with respect to N that is an integer variable changing discontinuously. Since the formulae of the normalized entropy coordinates contain the integer N, these coordinates will not be continuous with respect to the change in N.

In a multiple diagram, a finite number of simplexes (with various N, d_{min} values) are represented through the entropy map. At the confluence of the individual simplexes, N changes due to the emergence or disappearance of zero or non-zero fractions. In this case the non- normalized entropy map is continuous and, therefore, the single and the multiple diagrams coincide. The global maximum value of the maximum ΔS lines is ln$N$/ln2 and the global maximum of the approximate minimum ΔS lines is 1.

In the case of the multiple normalized entropy diagram the range of the normalised coordinates and the location of the global maximum point of B are independent of N (B = 1/ln2). At the symmetry point of the diagram (A=0.5), B=1/ln2 and the fractal dimension D =3, independently of the value of N, whereas in any other point both B and D depend on N (Table 2). For small N values, the maximum B lines nearly coincide with, and can be approximated to, the simple specific entropy formula valid for N =2.

Table 3 *Illustration for the discontinuity of the normalised entropy path by adding zero fractions: various entropy coordinates for the 'same' actual collection of particles.*

N [-]	2	3	3	4
	(50% of grains in each fraction)	(a smaller fraction added with zero frequency)	(a larger fraction added with zero frequency)	(larger and smaller zero fractions added)
S_{0min}[-]	2	1	2	1
S_{0max}[-]	3	3	4	4
A [-]	0.5	0.75	0.25	0.5
B [-]	1/ln2	1/ ln3	1/ ln3	1/ln(4)

The normalized entropy map is not continuous with respect to N. The discontinuity is illustrated in a simple example. Let us consider a two-fraction soil, with relative frequencies $x_1 = x_2 = 0.5$. We can embed it into a two dimensional simplex (i.e. triangle) in such a way that a larger or a smaller zero fraction is added, as illustrated in the data in Table 3.

The discontinuity can be qualified using the entropy coordinate A as a stability measure. The structure of the large grains is unstable and erosion may occur if $A<2/3$ (in Zone I) and is stable otherwise.[1]

3 LIME MODIFICATION

The effect of lime stabilization on grading has been investigated since 2004 at the Budapest University of Technology and Economics.[3] The soils were prepared with various lime and water contents. The grading curve was determined at each stage of lime content increment. The results are presented here using three samples: two clays (Bánokszentgyörgy and Szarvas) and a loess (Adony) soil, as shown in Figure 5 and Table 4.

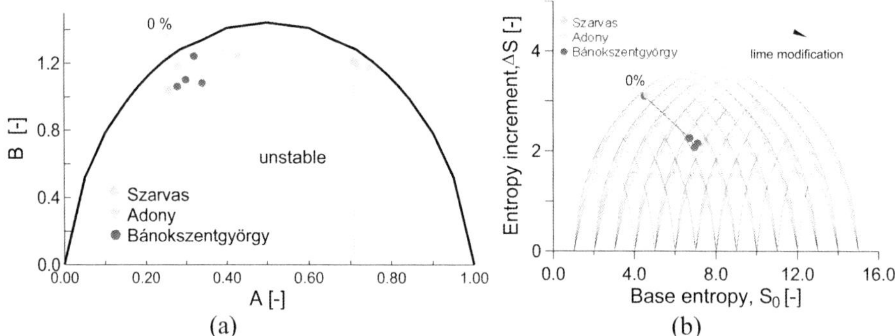

Figure 5 *Entropy path of lime treatment (a) in the normalized diagram (b) in the non-normalized diagram.*

Table 4 *Lime modified soils – classification test results*

	Adony	Szarvas	Bánokszentgyörgy
Clay [%]	7,0	34,5	25,2
Silt [%]	55,6	52,7	62,6
Sand [%]	29,5	3,9	12,2
Gravel [%]	7,9	8,9	0,0
I_P [%]	6,4	24,8	29,1

With the addition of a few percent of lime the grading curve did not change significantly. The non-normalised entropy path is clear: the base entropy increases, the entropy increment decreases, hence the soil has a more uniform grading. the larger the base entropy increase is ,the more plastic the clay is.

The S_0, the weighted mean of the fraction entropies, is monotonically and uniquely related to the mean grain size and it follows that an increase in the mean grain diameter should cause an increase in the base entropy S_0. The entropy increment ΔS is a measure of the disorder of the grain system, which originates from the mixing of the fractions. It can be related to the entropy principle.

The normalized entropy path has the same behaviour provided that the number of fractions remains constant. However, it has a discontinuity if the number of fractions increases due to the appearance of smaller fractions. In this situation, the discontinuity is with 'opposite sign' (A decreases, B decreases) if a new smaller fraction appears. Figure 5 shows that the normalised entropy path that is un-determinate in terms of A can either be stabilizing or not (possibly due to the decrease in the fraction number N due to the addition of lime).

4 DISCUSSION

4.1 Entropy Principle and Breakage

The directional properties of the variation of the grading curve due to particle breakage (and degradation) is connected to the grading entropy through the entropy principle as follows.[4] The non-normalized entropy path during breakage is characterized by a monotonic increase in the entropy increment ΔS (entropy principle) and decrease in the base entropy S_o (due to the decrease in the mean grain diameter). The results are shown in Figure 6 where the results of test RS3 reported by Coop et al.[5] is also presented (they conducted a series of ring shear tests to investigate the development of particle breakage with shear strain for a carbonate sand).

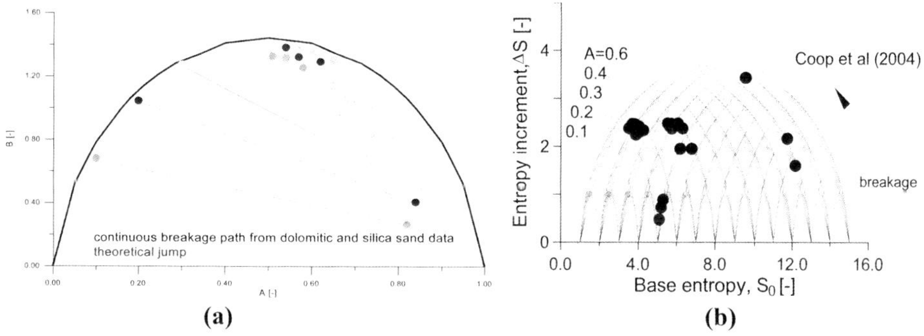

Figure 6 *The entropy path during breakage.[7] (a) Normalized entropy path, five samples. Open circles - silica sand, full circles - carbonate sand, the solid lines - the computed discontinuitiesby adding some zero fractions. (b) Non-normalized entropy paths for six samples.*

However, it is a discontinuity in the entropy path if the number of fractions increases due to the appearance of smaller fractions. In this situation, the discontinuity is with 'opposite sign' (the normalised base entropy A increases, the normalised entropy increment B decreases). Since the base entropy A is a continuous stability measure, it follows that during breakage the appearance of a new fraction increases the entropy stability measure.

4.2 Entropy Principle and Lime Modification

The entropy path of lime modification was investigated in terms of the entropy coordinates using the entropy principle.[6] The non-normalized entropy path is characterized by a monotonic decrease in the entropy increment ΔS (entropy principle) and an increase in the base entropy S_o (due to the increase in the mean grain diameter). The lime modified soil has an overall state improvement in terms of non-normalized entropy coordinate ΔS, the change is in the opposite direction as in the case of breakage (or degradation).

The path in terms of normalized entropy coordinates is basically the same except that it depends also on the fraction number changes. However, there is a discontinuity in the entropy path if the number of fractions decreases due to the addition of lime. In this situation, the discontinuity is with 'opposite sign' and stability decrease in the normalised base entropy A for the less plastic soils, only the plastic clay sample show a stability increase.

The structure of the large grains is unstable and erosion may occur if $A<2/3$ (in Zone I, Figure 7) and is stable otherwise. One could expect that lime modified soils become stable in terms of structure (i.e. $A>2/3$). However, the present study indicates that the lime modification is not necessarily helpful in the change of the stability category.

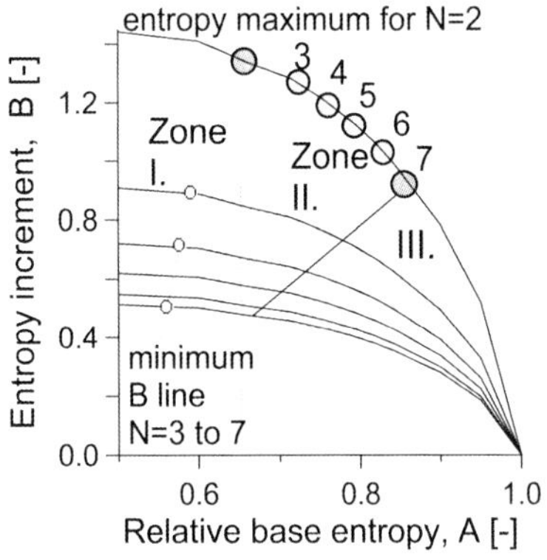

Figure 7 *Particle migration zones in the simplified normalized diagram (I-piping. II-stable. III-stable with suffusion) and maximum entropy points.*[1]

5 CONCLUSIONS

This study shows that the effect of lime modification on the grading curve evolution in the non-normalised grading entropy diagram is basically opposite to the effect of the usual particle breakage. Both can be related to the "entropy principle."

In the normalised grading entropy diagram the grading curve evolution path is the same except that discontinuity occurs upon the change in the fraction number. The entropy coordinate A can be used as a stability measure.

The discontinuous change in A results in a more stable structure for breakage and less stable structure if lime is added to a soil subjected to a small degree of plastic deformation. For soils with a very high clay content, the discontinuity with lime modification was stabilized.

Agglomeration may take place in the presence of a cementing agent like lime. If the failure of an agglomerated body is caused by the failure of bonds then the processes are symmetric. This study shows that the effect of lime modification is basically opposite to the normal particle breakage in terms of the grading entropy coordinates.

Acknowledgement

This project is supported from the National Research Fund Jedlik Ányos NKFP B1 2006 08 ('Biodegradation landfill technology').

References

1 J. Lőrincz, *Grading entropy of soils*. PhD Thesis, TU Budapest. 1986.
2 E. Imre, J. Lőrincz; Q.P. Trang, S. Fityus, J. Pusztai, G. Telekes and T. Schanz, *KSCE Journal of Civil Engineering,* 2009, **13,** 257.
3 E. Imre and J. Szendefy, *Hidrológiai Közlöny,* 2004, **4,** 61.
4 J. Lőrincz, E. Imre, M. Gálos, Q.P. Trang, K. Rajkai, S. Fityus and G. Telekes, *Int. J of Geomechanics* 2005, **4,** 311.
5 M.R. Coop, K.K. Sorensen, K.K. Bodas Freitas and G. Georgoutsos, *Geotechnique,* 2004, **54,** 157.
6 J. Lőrincz, E. Imre, L. Kárpáti, Q.P. Trang and S Fityus, *Proc of the 15th European Conference on Soil Mechanics and Geotechnical Engineering,* Athen, Greece, 2011, 215.